Going to Green

Going to Green

A Standards-Based Environmental Education Curriculum for Schools, Colleges, and Communities

Based on the PBS Series *Edens Lost & Found*

by Harry Wiland and Dale Bell

The Media & Policy Center Foundation

CHELSEA GREEN PUBLISHING
WHITE RIVER JUNCTION, VERMONT

Project Manager and Copy Editor: Margaret Pinette
Developmental Editor: Laura Strom
Proofreader: William F. Heckman
Text: Denise Kiernan and Joseph D'Agnese
Design: Sterling Hill Productions

Educational Consultants: Deborah Perryman
Elgin High School, Elgin, IL

Brett Ivers
Illinois Sustainable Education Project (ISTEP)
Illinois Department of Commerce and Economic Opportunity

Peggy Chamness
Illinois Sustainable Education Project (ISTEP)
Illinois Department of Commerce and Economic Opportunity

Illustrations by Ric and Wanda Scott of Ric Scott Design, Inc.

Printed in the United States of America
First printing, July 2009
10 9 8 7 6 5 4 3 2 1 09 10 11 12 13 14

Our Commitment to Green Publishing
Chelsea Green sees publishing as a tool for cultural change and ecological stewardship. We strive to align our book manufacturing practices with our editorial mission and to reduce the impact of our business enterprise on the environment. We print our books and catalogs on chlorine-free recycled paper, using soy-based inks whenever possible. This book may cost slightly more because we use recycled paper, and we hope you'll agree that it's worth it. Chelsea Green is a member of the Green Press Initiative (www.greenpressinitiative.org), a nonprofit coalition of publishers, manufacturers, and authors working to protect the world's endangered forests and conserve natural resources.

Going to Green was printed on 30-percent post-consumer-waste recycled paper supplied by BookSurge. This paper is acid- and chlorine-free and is certified by the Forest Stewardship Council (FSC). The ink used in printing is biodegradable and recycled and contains no lead or toxic chemicals.

Library of Congress Cataloging-in-Publication Data
Wiland, Harry.
The going to green curriculum : a standards-based, environmental education component to the PBS documentary series, Edens lost & found, for grades 9 - 12 / by Harry Wiland & Dale Bell.
p. cm.
"Based on the Edens Lost and Found PBS series."
ISBN 978-1-933392-88-2
1. Environmental education. 2. Sustainable development--Study and teaching. I. Bell, Dale, 1938- II. Edens lost & found (Television program) III. Title.

GE70.W53 2008
307.1'4160712--dc22

2008013261

Chelsea Green Publishing Company
Post Office Box 428
White River Junction, VT 05001
(802) 295-6300
www.chelseagreen.com

Based on the *Edens Lost & Found* PBS series

Also available:
The Going to Green Video Resource Library,
a 20-DVD set that supplements each of the
units in the academic curriculum

*Edens Lost & Found: How Ordinary Citizens Are
Restoring Our Great American Cities* Companion Book, Hardcover

Edens Lost & Found PBS TV Miniseries,
also available on DVD
Chicago: The City of the Big Shoulders
Seattle: The Future Is Now
Los Angeles: Dream a Different City
Philadelphia: The Holy Experiment

Visit www.edenslostandfound.org
or www.chelseagreen.com
for more information.

CONTENTS

Welcome to the *Going to Green* Curriculum. xi

Introduction: "From Hope to Practice, or Educating for Eden"
Tim Beatley, Teresa Heinz Professor of Sustainable Communities, University of Virginia xiii

Foreword: The Evolution of a Teacher—and a Curriculum
Deb Perryman, 2005 Illinois State Teacher of the Year. xv

Preface: *Going to Green:* The First of its Kind
Jennifer Wolch, Director, University of Southern California Center for Sustainable Cities. xvii

About *Edens Lost & Found* xix

About the Media and Policy Center Foundation xxi

How to Use This Curriculum: Strategies for Integrating *Going to Green* into Your Classroom xxiii

Service Learning in *Going to Green* **. . . . . .xxvii**

Handout: Service Learning Evaluation xxxiii

Unit 1: Understanding Sustainability 1

Topic Background 1
Group Icebreakers. 1
Activity 1.1: The Sustainability Game. 2
Activity 1.2: Making Connections in the Urban Ecosystem 3
Activity 1.3: Sustainable or Not Sustainable? 3
Activity 1.4: Global Perspectives 4
Activity 1.5: At the Podium. 4
Extensions 6
Resources 6
Reading Handout Unit 1: The Path to Sustainability. 7

Unit 2: Building Community 15

Topic Background 15
Group Icebreakers. 15
Activity 2.1: Thinking about Communities 16
Activity 2.2: Now and Venn 16
Activity 2.3: The Power of Intention. 17
Activity 2.4: Making Sense of the Census 17
Activity 2.5: Front Page 18
Activity 2.6: At the Podium. 19
Extensions 19
Resources 19
Reading Handout Unit 2: Real People, Real Lessons 21

Unit 3: Waste Management and Recycling 28

Topic Background 28
Group Icebreakers. 28
Activity 3.1: Don't "Waste" This Quiz 29
Activity 3.2: "Waste-ing" Away 29
Activity 3.3: Talking Trash. 30
Activity 3.4: Trash Audit .31
Activity 3.5: Trash Tops the Charts 33
Activity 3.6: Landfill Dispute 33
Activity 3.7: At the Podium. 34
Extensions 34
Resources 35
Reading Handout Unit 3: Young People in Urban Nature 36

Unit 4: Green Building. 47

Topic Background 47
Group Icebreakers. 47
Activity 4.1: Anatomy of a Green Home 48
Activity 4.2: Under the Green Roof. 49
Activity 4.3: A Green Light 50
Activity 4.4: Front Page 50
Activity 4.5: At the Podium. 51
Extensions 52
Resources 52
Reading Handout Unit 4: A City of Green Builders 53

Unit 5: Energy 62

Topic Background 62
Group Icebreakers. 62
Activity 5.1: Sun, Sky, and Earth: Where We Get Our Energy. 63
Activity 5.2: I Love My Car—I Just Hate Paying for Gas!. 64
Activity 5.3: Charting Energy. 65
Activity 5.4: Home Energy Survey 65
Extensions 65
Resources 66
Reading Handout Unit 5: French Fries in the Tank?. 67

Unit 6: Air Quality 82

Topic Background 82
Group Icebreakers. 83
Activity 6.1: The Air Up There 83
Activity 6.2: Smog, Ozone, and the Greenhouse Effect 84
Activity 6.3: How Much CO_2 Do You Contribute?. 86
Activity 6.4: Understanding the Greenhouse Effect 86

Extensions 87
Resources 88
Reading Handout Unit 6: Tree Boy 89

Unit 7: Water Quality 96
Topic Background 96
Group Icebreakers 97
Activity 7.1: Understanding Earth's Water Supply 97
Activity 7.2: Town Meeting: Watershed at Risk . . . 99
Activity 7.3: The Rundown on Runoff 100
Activity 7.4: Understanding Water Quality 101
Activity 7.5: Swales and Cisterns 101
Extensions 102
Resources 103
Reading Handout Unit 7: Restoring the Chicago River 104

Unit 8: Soil Quality 118
Topic Background 118
Group Icebreakers 119
Activity 8.1: Hydroponics to the Rescue 119
Activity 8.2: Alfalfa and the Importance of Cover Crops 121
Activity 8.3: Soil, Air, and Water 121
Extensions 122
Resources 123
Reading Handout Unit 8: Restoring Chicago's Calumet Region 124

Unit 9: Parks and Open Spaces 132
Topic Background 132
Group Icebreakers 132
Activity 9.1: Thinking about Parks 133
Activity 9.2: Open Space Around the World . . . 133
Activity 9.3: Mapping the Millennium 135
Activity 9.4: Design a Park or Piazza 135
Activity 9.5: One Space, Two Plans 136
Extensions 137
Resources 138
Reading Handout Unit 9: The Great Public Space 139

Unit 10: Transportation 146
Topic Background 146
Group Icebreakers 146
Activity 10.1: Passenger Costs 147
Activity 10.2: Pros, Cons, and in Between 148
Activity 10.3: One-Less-Car Challenge 149
Activity 10.4: Writing about It and Making Connections 150
Extensions 151
Resources 151
Reading Handout Unit 10: Living With One Less Car 152

Unit 11: Biodiversity 163
Topic Background 163
Group Icebreakers 163
Activity 11.1: Classmate Diversity Scavenger Hunt 164
Activity 11.2: Getting to Know Your Own Backyard: An Urban Survey 164
Activity 11.3: Overfished and Contaminated: Crisis in Our Seas 166
Activity 11.4: Ecosystem Swap Meet 166
Extensions 167
Resources 168
Reading Handout Unit 11: Restoring the Prairie 169

Unit 12: Urban Agriculture and Community Gardens 179
Topic Background 179
Group Icebreakers 180
Activity 12.1: Growing All Around 180
Activity 12.2: Zoning in on Hardiness 181
Activity 12.3: Food on the Move 182
Activity 12.4: Your Own Garden 182
Extensions 184
Resources 184
Reading Handout Unit 12: Philadelphia Eats What Philadelphia Grows 185

Unit 13: Urban Forestry 194
Topic Background 194
Group Icebreakers 194
Activity 13.1: Did You Know . . . ? Fun Facts about Urban Forestry 195
Activity 13.2: Getting to Know Your Trees 196
Activity 13.3: Trees and You: A Numbers Game 196
Activity 13.4: At the Podium 197
Extensions 198
Resources 198
Reading Handout Unit 13: The Magic of Trees 199

Unit 14: Urban Planning 207
Topic Background 207
Group Icebreakers 208
Activity 14.1: Our Town: A Closer Look 208
Activity 14.2: Penn's Plan: A Look at Old Philly 209
Activity 14.3: You're the Planner 210
Activity 14.4: Frederick Law Olmsted's Vision 210
Extensions 211
Resources 211
Reading Handout Unit 14: William Penn's "Green Countrie Towne" 213

Unit 15: Population Growth and Integrated Resource Management 221
Topic Background 221
Group Icebreakers 222
Activity 15.1: Pop Goes the Population 222
Activity 15.2: Growing, Growing . . . Gone? 223
Activity 15.3: Greening Your School 224
Activity 15.4: The Population Debate 225
Extensions 226
Resources 226
Reading Handout Unit 15: Water In, Water Out 227

Unit 16: Environmental Justice 248
Topic Background 248
Group Icebreakers 248
Activity 16.1: What Is Environmental Justice? 249
Activity 16.2: Environmental Justice on the Map 249
Activity 16.3: Environmental Justice Close to Home 250
Extensions 251
Resources 251
Reading Handout Unit 16: Toward a Wireless Chicago 252

Unit 17: Public Policy and Community Action 258
Topic Background 258
Group Icebreakers 258
Activity 17.1: Understanding Public Policy 259
Activity 17.2: Getting Organized: Making the Most of the Media 259
Activity 17.3: Putting It All Together 260
Extensions 261
Resources 261
Reading Handout Unit 17: A Lifetime of Community Action 262

Unit 18: Sustainable Commerce 271
Topic Background 271
Group Icebreakers 271
Activity 18.1: Interview a Green Business Owner 272
Activity 18.2: The Green Business Plan 273
Extensions 273
Resources 273
Reading Handout Unit 18: Meet a Green Banker 275

Unit 19: Green-Collar Careers 281
Topic Background 281
Group Icebreakers 281
Activity 19.1: The Green Workplace 282
Activity 19.2: Shadow a Green Mentor 282
Activity 19.3: Sharing What You've Learned 283
Extensions 284
Resources 284
Reading Handout Unit 19: All about Green-Collar Jobs 285

Appendix A: Answer Key 291

Appendix B: Scope and Sequence 311
The National Science Standards for Grades 9–12 (NSTA) 312
The National Social Studies Standards for Grades 9–12 (NCSS) 312
Guide to Science and Social Studies Standards for All Activities 316

Welcome to the *Going to Green* Curriculum

Nothing is more of an inspiration than the passions, commitment, and actions of people striving to make the world a better place. *Edens Lost & Found,* a four-part miniseries on PBS, celebrates the remarkable actions of people from coast to coast who have decided to take any action they can to rescue their communities from what they consider to be the brink of environmental and social disaster. Their stories and the resulting lessons they learned are the basis of this curriculum for high school students. The title of our documentary reminds us all that no matter where we live, we are entrusted with that precious plot of land and must strive to protect its beauty and natural resources.

About halfway through the Chicago film in our miniseries, one of the individuals we interviewed states, "You know, children are our future. If we can't get them interested in and excited about the environment, we're lost."

What makes this comment especially poignant is that the speaker is himself a teenager. He was just one of the many people we interviewed as part of our three-year project. While we encountered numerous innovative ideas set into action by adults, we feel that the future of America—and, indeed, our planet—lies in the education of our young people about the issues facing our environment and our communities today. In this spirit, we have developed this curriculum in the hopes of challenging minds and hearts to take up the cause. We have spent more than three years developing this teaching resource with the help of educators across the nation. Teachers who have tested these strategies in the classroom report great success, not only in high schools but also in university extension classes, community colleges, and even community organizations. The information and lessons provided highlight the connection between our actions and their effects on not only the planet's health but also on the well-being of our neighbors. This environmental curriculum seeks to empower students and teachers alike, as this journey toward a more sustainable future is one that we *must* all make together.

Harry Wiland and Dale Bell
for *Edens Lost & Found*

— INTRODUCTION —
From Hope to Practice, or Educating for Eden

Tim Beatley, Teresa Heinz Professor of
Sustainable Communities, University of Virginia

The PBS film series *Edens Lost & Found* was for me a watershed event, a beautiful rendering of the collective stories of four incredible cities, conveyed in visually effective and emotionally compelling fashion. These cities are not presented as abstractions but rather are seen through the lens of real people and real neighborhoods grappling with contemporary urban problems and in the end finding their way down a hopeful path. The films taken together are shiny pieces in a global mosaic of people, nature and place, and of course cities, and show us especially the *promise* of cities. There are many urban role models here, many citizen-stewards, if you will, and while there is apathy and inertia, certainly, the films together provide abundant evidence of the vitality, innovation, and hopefulness of urban living in 21st-century America.

For the full impact of this message to take hold, though, still more is required: a more developed practical sense of what to do, and what individuals and group *can* and *should* do, to improve and uplift their neighborhoods and cities and to recover the potential Eden that lies within reach everywhere. This is where a widespread effort at education and awareness raising is essential, and the academic curriculum presented here picks up where the emotional urgency and inspiration of the films leave off.

It is indeed an auspicious time for an academic curriculum such as this. Cities around the world are embracing a green agenda, as awareness of the essential role that cities play in addressing global climate change, loss of biodiversity, water scarcity, and other global and local challenges is on the rise. Here in the United States more than 910 cities have adopted the Mayor's Climate Action Agreement, committing them to meet or exceed the Kyoto targets. Cities are stepping in where the federal government won't. And our largest cities have recognized the essential need to address the greening of urban environments. Mayor Daley of Chicago has declared his intention to make that city the greenest in the country, and more recently Mayor Bloomberg has announced an ambitious green plan for New York City that will aspire to be the first ecologically sustainable city of the 21st century. Other cities are following suit, and while political leadership on green cities can be seen around the country and the world, this political agenda is still rather nascent and fledgling, and its success ultimately would require a larger cultural shift in the importance we attach to green urbanism. The curriculum herein should help immensely in this mission. We desperately need these kinds of educational tools if we are to educate the next generation of leaders, citizens, and professionals who will see this vision of green and hopeful urban life to its fruition.

And really what follows is, at the core, about the critical matter of educating a new kind of citizen. The curriculum presented here, if embraced and widely disseminated, could lay the broad foundations of a new *expanded* form of citizenship: a citizenship borne of awareness of the threats posed, global and local, and the challenges faced but more importantly of the ethical need to see oneself as more connected to others and the environment. It is a citizenship that understands that one's duties extend beyond a narrow notion of civics—that we actually have duties to (more) fully understand and account for how our choices about such things as food, energy, and travel affect the global and local environment and that we have affirmative obligations to reach out to and care for and actively steward over our cities and neighborhoods. Ultimately the impressive package of topics and curricular knowledge areas to follow is, to my way of thinking, about equipping, empowering, and facilitating this deeper, broader form of citizenship that we require more than ever today. The exercises and learning modules that follow reinforce this sense of the new ethical sensibilities needed and the inherent interconnectedness of contemporary life. They encourage us all to ask the right questions and to understand what we individually and collectively might do in our communities to address the problems faced and to at once create a better life and a healthier environment.

There are some terrific ideas here for getting started, for igniting curiosity, and for beginning dialogue and conversations on things that are often beyond the normal subjects off the day: What is the size of my ecological footprint? Where does the energy come from to light my home? How far does the food I eat travel? The knowledge and stories and case studies and exercises to follow have in turn the potential to convert these questions into tangible change: It is indeed possible to reduce my ecological footprint, for instance and, perhaps counterintuitively to some, improve my community and enhance my own health and quality of life at the same time. Icebreakers, handouts, games and activities, and profiles of good practice, all combine to make the following a potent set of tools for personal and community change.

The curriculum to follow challenges students (indeed all of us) to overcome the passivity of our times and it gives them (us) the knowledge and tools to become the kinds of *ecological citizens* we need more than ever today. I'm looking forward to seeing in my lifetime the changing awareness and the sense of personal responsibility taking hold that this creative curriculum can help to foster.

— FOREWORD —

The Evolution of a Teacher—and a Curriculum

Deb Perryman, 2005 Illinois State Teacher of the Year

On Valentine's Day, 1992, I landed my first teaching position at the Beaufort-Jasper Career Education Center in Beaufort, South Carolina. The CEC, as it was known, was for many students a last chance. The majority of these young people had gotten into too much trouble at their feeder schools and thus were reassigned to us.

As a first-year teacher, I, of course, thought I knew what I was doing. I had no idea that a first-year teacher should not be teaching six preps (biology; environmental science; physics; chemistry; forest, field, and stream; and math). To make my first teaching job even tougher, my classroom was a double-wide trailer with no lab, no equipment or supplies, and no textbooks. But like the true sucker I am, I took the job, ignoring the little red light blinking widely in the back of my mind.

I proudly tell you that my major accomplishment in that first year was realizing I was over my head. Having very few of the traditional resources, I had to develop my own. It turns out that my shaky start prepared me to be the teacher I am today. Like every other teacher in America, I learned to become resourceful.

I eventually left South Carolina and landed at Elgin High School in Elgin, Illinois. Elgin High is a traditional high school, and compared to my first teaching experience I knew I had hit the bonanza in terms of supplies, equipment, and colleagues. Truth be told, it was during my employment at Elgin High that I realized just how unprepared I was for the CEC.

Elgin High School had offered a class called "The Great Outdoors," and it was to be based in experiential education. I was informed by my supervisor that I was being hired to overhaul the class into a lab-based course. I was handed the textbook and sent on my way. Within the first week, I discovered that the text I was using was a college-level book and that the majority of my students, 11th and 12th graders, were not reading on grade level. To make matters worse, the course outline I had developed and my lecture topics highlighted only "doom and gloom." Fortunately for me, a brave student asked me, "If we can't do anything about it, why should we care?" That simple question revolutionized my teaching philosophy. From that moment on, my goal was to give my students hope and move them to action.

I started looking for resources and agencies that could help me develop an environmental science curriculum that empowered my students. Along the way, I had yet another revelation. I couldn't find teacher resource material that provided a wide variety of activities, for a wide variety of topics, in one book. I could find resource books on air, water, resource conservation, and solid waste management but not a single resource that delivered them all.

For years I thought to myself, "Someone should create a single resource that could provide a teacher with background information and activities on a variety of environmental topics," not realizing that years later I would in fact help develop such a curriculum resource. But here we are!

I am honored to be one of the content advisors that helped to develop the 20-lesson interdisciplinary program. Let me assure you that the *Going to Green* Curriculum Team had three goals in mind when creating this curriculum:

1. *Information:* We wanted to provide you with accurate background information on a variety of environmental issues in one resource.
2. *Inspiration:* We wanted to provide you and your students with the real-life stories of ordinary people who in fact made a difference for the environment.
3. *Action:* We wanted to provide you a tool to move your students to action.

Each chapter will provide you with a steady progression of activities that will take your students from rote memory to taking action in your community. This process is known as service learning, and I feel it is critical to reestablishing purpose in our educational process.

I have had many opportunities to ask people across the nation one very important question: "What do our students need from our schools?" A variety of answers have been given, some of which include understanding, support, an education, and the ability to act as a citizen with real-world problem-solving skills. The one response I have yet to be given: "Our students need to be able to pass a state standardized test!" (Oddly, I have asked several legislatures this same question, and they never think of standardized tests either.) We understand that the aforementioned are of great importance but are difficult to be measured. Further, if we were to research the mission statements of schools across the nation, we would find some mention of "creating citizenship." But when do our schools actually provide students with the opportunity to act as citizens? No Child Left Behind (NCLB) dictates that educators must help students meet or exceed state standards—a worthy goal, as citizens must be educated and motivated to participate in society. Why not use the time we have in schools to actually show our young people how each state's learning standards will actually apply to their lives? Why not allow our students to solve community problems and act as citizens? With the help of my students, I have found a teaching strategy—service learning—that allows me to fold learning standards into community action. Through these pages, supported by the Going to Green Video Resource Library, I share this teaching strategy with you.

The activities contained in the *Going to Green* Curriculum, to be used in conjunction with the soon-to-be released Going to Green Video Resource Library, are tried and true and meet national standards. The students of Elgin High, the people of *Edens Lost & Found,* and, I hope, *you,* will enjoy years of innovative and inspirational activities designed to help your students make your community more sustainable.

— PREFACE —

Going to Green: The First of Its Kind

Jennifer Wolch, Director, University of Southern California
Center for Sustainable Cities

It was the San Francisco Bay area, in the late 1960s, and Rachel Carson's *Silent Spring* had just been published, launching the environmental movement. Dissatisfied with voter drives and letter-writing campaigns around pollution and waste, my parents rolled up their sleeves and started the first all-volunteer community recycling program. For the next few years, the family spent weekends crushing cans and separating glass, chatting to neighbors about recycling, listening to talk about the economics, politics, and culture of materials recycling and reuse. Without really knowing it, I got an environmental education—by taking action, being involved, seeing how ideas got tested in the real world. In so doing, I learned firsthand that citizenship was not only about exercising rights like voting but also about engaging the community of which I would one day become a full-fledged member.

How will today's kids become tomorrow's environmental stewards? How can we help young people embrace what might be called "sustainable citizenship"—a set of values and everyday practices that collectively lead us away from a consumer society based on environmental exploitation and toward a more sustainable world?

As the realities of climate change begin to hit home and the scope of action needed to address its impacts becomes clear, the challenge is now that much more pressing. It is also more complicated. Throughout the 20th century, scientists, policymakers, and activists saw various types of social, economic, and environmental problems as largely discrete. Today we grasp the profound connections among poverty and social justice, environmental degradation and resource depletion, and economic fortunes or failures—connections that make the design of interventions more complex.

Despite the growing challenge of sustainability and its complexity, however, today's kids will most likely become sustainable citizens in much the same way I did—by some combination of formal and informal education. Increasingly, educators now understand that children learn best when they are provided with a rich array of resources—classroom presentations, discussions, and reading, to be sure, but also visual images such as photographs and videos, case studies from real-world settings, and direct engagement with the problem at hand. This engagement may be grounded either in science or society; for example, working on real experiments in the lab, learning how to restore a riverbank, or working with poor communities to build a new park.

Going to Green provides key tools—fundamental concepts, data and examples, group projects—for teachers charged with the education of our future sustainable citizenry. Designed for high school students, it is also perfectly suited to introductory courses for community-based organizations, public agency staff, and anyone else interested in understanding sustainability and how to achieve meaningful improvements in quality of life and the urban environment.

The *Going to Green* curriculum is based on an extraordinary partnership between documentary filmmakers, community-based organizations, local governments, and scientific experts that resulted in the PBS series *Edens Lost & Found*. The series of four one-hour specials explored how rather than waiting for grand plans or policies, everyday people and organizations in Chicago, Philadelphia, Seattle, and Los Angeles were recovering their city's environment and experimenting with new, more sustainable ways of urban living. Using an innovative model of communicating with the public, the filmmakers not only produced *Edens Lost & Found* as a documentary series but also convened expert conferences and created a companion volume replete with resources accessible to people wanting to make a difference.

Taking advantage of extensive research on pressing sustainability challenges, hundreds of hours of interviews, and on-camera investigations into some of the nation's most exciting urban projects, *Going to Green* mobilizes this wealth of visual and written materials and provides guidelines for creating service learning projects. It is teacher friendly, engaging for students, and premised on a holistic model of learning that emphasizes both knowledge acquisition and knowledge creation through active learning.

The result is an innovative, structured curriculum for high school and entering college students as well as community organizations. *Going to Green* is the first of its kind: connecting a motivational video series with a service learning–based academic curriculum. The project is unparalleled in its narrative inventiveness, compelling visual imagery, and intellectual rigor. Optimistic despite the odds, its sheer enthusiasm is infectious—hopefully mobilizing the next generation to go out and tackle the hard work of creating greener, fairer, and more prosperous urban worlds.

About *Edens Lost & Found*

The PBS Series

Edens Lost & Found is a four-part PBS documentary series highlighting urban sustainability—making cities more livable and less taxing on natural resources. Produced by the filmmaking team of Harry Wiland and Dale Bell, the four one-hour films investigate a fascinating phenomenon of urban renewal that is sprouting in four American cities: Chicago, Philadelphia, Los Angeles, and Seattle.

For the first time in American history, more people (80 percent) are living in cities than in rural areas. People move to urban areas for better job prospects and the pursuit of a better life, yet this geographic shift inevitably places an enormous strain on natural resources, such as air, water, and energy reserves.

For this human migration to work, we all must become wiser about the way we use our resources. This not only means we should conserve energy, air, and water; it also means we must intelligently redesign our infrastructure, including our mass transit system, and more intelligently manage and preserve our biological resources, such as parks and city forests. Did you know that cities would be markedly healthier places if more trees were planted in them? And did you know that programs for urban forestry, supported by the U.S. Forest Service, are already improving the property values of ordinary citizens? In a recent study, the Wharton School of business found that property values along city blocks increased by 15 percent when trees were planted near the curb.

This film, book, and video resource library project took three years to bring to fruition. Film crews crisscrossed the nation several times, chasing stories that tell the story of a greener, healthier America. For your students, this project offers an easy, engaging way to learn about an environmental movement, with opportunities for multidisciplinary learning.

The Companion Book

The print version of the miniseries is a slightly different entity than the film version, available on DVD. The companion book, *Edens Lost & Found: How Ordinary Citizens Are Restoring Our Great American Cities* (Chelsea Green Publishing, 2005) embodies the flavor of our miniseries but does not precisely adhere to the scripts of the shows. The book offers in-depth interviews with community leaders and greater depth and analysis. To make the most of the series, we encourage you to watch the films together and also assign students readings from the companion book. Of course all material is PBS approved and family friendly.

The Video Resource Library

The video resource library is a collection of 20 DVDs developed especially to supplement and bring to life the concepts discussed in each of the units in this curriculum. If you are studying the unit on "Parks and Open Spaces," for example, you can, as a prelude, simply have your class watch the contents of the "Parks and Open Spaces" DVD, and your students will instantly have powerful visual images to associate with the material they will be discovering as they explore the new unit. Again, as with the original four films and the companion book, it is not necessary to use the video resource library as part of your teaching, but it can greatly enhance the experience.

Edens Lost & Found in the Community

Edens Lost & Found is all about getting involved at home, at work, at school, and in neighborhoods. Getting involved may be as simple as planting flowers in public places or working with developers to incorporate state-of-the-art storm water treatment. Getting involved also means getting educated about issues that concern you and then lending your energy to organizations and coalitions that advocate for sustainable solutions. Perhaps you will be inspired to get the ball rolling and start a working group yourself!

The challenges that face your community probably have a lot in common with those "Edens" discovered in Chicago, Los Angeles, Philadelphia, and Seattle. Each of these cities has developed an action guide to blaze the trail for other communities—of all sizes—

that are seeking to build sustainable societies and healthy ecosystems. These four action guides can be downloaded free of charge from the *Edens Lost & Found* website and are intended to help you jump start community action in your town.

Edens Lost & Found also sponsors town hall meetings, where citizens and civic leaders come together to share ideas and develop strategies about how to make their communities a better place for all. For more information about how you, your students, and your community can become more involved with the Edens community nationwide and for suggestions about how to put simple strategies to work in your town, visit our website at www.edenslostandfound.org.

About the Media and Policy Center Foundation

The Media and Policy Center Foundation is a nonprofit media design and production studio and think tank that explores the issues of social welfare, public policy, education, the environment, and health care. The center creates, organizes, and produces innovative film, television, print, multimedia, and software programming dealing with these issues. The Media Policy Center sponsors and hosts national and international seminars, conferences, and forums devoted to the impact and importance of media as they relate to these significant issues. It fosters community engagement by focusing on how to increase awareness of, and investment in, our social capital.

Through these activities, and via its engagement with the community, the center helps determine what legislation is necessary to guarantee social justice for all citizens and convenes an annual conference of public and private sector decision makers and decision influencers who have the ability to see that essential measures are proposed, enacted, and enforced.

The Media and Policy Center's mission is to:

- Explore issues of social welfare, public policy, education, the environment, and health care
- Inform, challenge, and ultimately engage a responsive citizenry
- Encourage full and meaningful debate and participation across the political, social, and economic spectrum
- Use televised and coordinated multimedia projects to enlighten, educate, and empower those members of our society who seek to change their lives and their communities
- Press for enactment of legislation on the local, state, and federal level that ensures social justice for all citizens

— HOW TO USE THIS CURRICULUM —
Strategies for Easily Integrating *Going to Green* into Your Classroom

The *Edens Lost & Found* four-part PBS miniseries examines the role that ordinary citizens play in the revitalizations of their neighborhoods, towns, and cities. Throughout these stories runs a thread of sustainability, community, and environmental awareness that provides the perfect basis for educational materials. This curriculum project has grown out of a desire to bring these ideas and the tremendous examples set by the documentary's subjects to the classroom, where they can inspire new generations of thinkers and doers.

Whom Is This Curriculum For?

The *Going to Green* curriculum is designed to be used as part of a sustainability education or environmental education program. However, any science or social studies teacher can easily dip in and out of this resource to augment his or her own curriculum. Every teacher wishing to expand her or his repertoire and connect environmental education into the curriculum will find this resource valuable and doable. Teachers do not need to be environmental experts to use the curriculum and can choose to what extent they wish to incorporate these lessons into their yearly plan. All units include background information for the teacher. This is so teachers do not have to depend on outside material to use the curriculum. They can read the background and then decide how to present the material to their class. Some teachers may wish to copy this material and distribute it to students.

Do I Need the PBS Videos or the Video Resource Library to Use This Curriculum?

No, though they are recommended. This curriculum works in conjunction with the *Edens Lost & Found* films, the video resource library, and companion book, though none of these three is necessary to put this curriculum into practice in your classroom. Sustainable education crosses many disciplines and is an excellent way to get your students thinking like educated, responsible young adults. They will call on math and science skills, English, composition, arts and literature understanding, and current events, local history, and civics. Students also will hone their library and Internet research skills and learn how to make use of community resources.

Will This Curriculum Help Me Meet Curriculum Standards?

Yes. The units presented in this curriculum correspond to national standards in learning for science and social studies, grades 9–12, as outlined by the NSTA and NCSS. In addition to the science and social studies standards addressed throughout the curriculum, there is also a wide variety of cross-curricular activities, with a focus on literature, math, and art tie-ins. Please see the Scope and Sequence, Appendix B, for a list of the standards as well as a correlation to what is covered in this curriculum. Teachers can also refer to this chart to make their lesson plans. The chart conveniently shows which standards are satisfied in each activity. If they wish, teachers can seek out the standards first, to find the corresponding activities that best satisfy their needs.

How Is the Material Presented?

The curriculum is thematic and multidisciplinary. Throughout the curriculum, there is a focus on connecting the themes presented. This is to say that none of these topics exists in a vacuum, and none can be tackled without affecting or seeking to deal with any of the others. This multidisciplinary approach to sustainability education will enable teachers to truly instruct their students about the web of life in which humans exist and how that web must be repaired. The key to sustainability is not only understanding the importance of the pieces but fully comprehending the effects that each topic has on the others. Though information is presented in discrete themes, the connections among these issues are constantly being reinforced and explored.

How Are the Units Presented?

The *Going to Green* curriculum is divided into 20 units, each focusing on a different environmental theme. These themes are also discussed in the *Edens Lost & Found* companion book and DVD series. Extension activities and reproducibles are provided and can be used by a teacher year after year.

Each unit is broken down into the following sections:

- **Topic Background,** to give the teacher an overview of the topic, with **Group Icebreaker** class discussion questions.
- Reproducible **Reading** adapted from the *Edens Lost & Found* companion book, with follow-up **Questions** to ask students.
- **Activities,** with materials list and multidisciplinary connections clearly noted. Most activities include **Reproducibles** for making overheads or photocopiable **Handouts.** Some overhead Reproducibles are best used in color; color versions will be available at the *Edens Lost & Found* website.
- Each unit ends with **Extension** activities and **Resources.**

How Are the Activities Paced?

The activities are presented in a way that is easy for any teacher to implement. Activity lengths and content are varied and flexible. This allows teachers to pick and choose activities that not only correspond to the topic being covered and standards needing to be met but also work within the time they have available. Some activities can be done in as little as 20 minutes, while other, lengthier extension activities may carry on over several weeks or even an entire school year. Teachers with limited time can do as little or as much of these activities as they wish and have the option of either setting aside class time for research or assigning it as an out-of-class requirement.

Some activities take students and teachers beyond their classroom and into their communities where they can see the change that their learning can effect!

What Types of Activities Are Offered?

The activities in this curriculum were designed to be rich and layered. Some activities are simple readings and handouts to be completed in class or used for homework. Other activities consist of multistep projects requiring outside research, group or pair work, art projects, and classroom presentations.

Activities range from reading comprehension and graph interpretation to in-class experiments and topical debates. Students will be exposed to a variety of text features and hands-on learning opportunities that make use of multiple modalities. Choices are provided to appreciate the varied learning styles within a classroom.

Each activity lists the materials needed and the main subject areas it covers, as well as the science and social studies standards it satisfies. At least a year's worth of educational activities is provided. Teachers can simply pick and choose what works for them.

Most of the activities offer optional extension activities, ranging from journal exercises to full-scale presentations to your local government. Extensions that conclude each chapter tend to be long-term projects, offering suggestions for everything from class speakers to community service learning projects.

Can Students Actually Improve Their Communities while Using This Curriculum?

Yes! One of the key aspects of this curriculum and the *Edens Lost & Found* approach is its emphasis on service learning. Just as the individuals featured in the films and companion book are taking action to improve their communities, so too are students encouraged to take specific actions in their communities.

Service learning is a teaching approach that stresses education through doing and giving back. It differs from work-study in that the work being performed is not being done for pay but instead to help educate others in the community. In the process of educating others, students reinforce and augment what they have already learned. The projects are designed to improve the way your classroom functions and strengthen your students' information retention. You will find examples of service learning in virtually every unit in this book. Any time an extension or activity suggests that you and your students take what you have learned outside the classroom and share it with others, you are pursuing service learning.

Where Are the Answers?

All answers are at point of use after group icebreakers and in the appendix for all questions posed in reproducible materials for the student. Because of the do-it-yourself approach of many of the activities and extensions, many answers are of the "answers-will-vary" variety, but we have given you guidelines for assessing even these types of responses from students.

My School Purchased the Companion Book and DVDs for My Class. So How Do I Best Make Use of Them?

There are several ways to use the curriculum in conjunction with the book and film series, should you choose to do so:

- You could choose to watch an entire disc from the series, such as "Chicago: The City of the Big Shoulders," and go to the activity about Chicago's Millennium Park, titled "Mapping the Millennium," in Unit 9: Parks and Open Spaces.
- You could scan the curriculum first, searching for topics that interest you or that supplement a teaching unit you may already be exploring with your class, such as urban forestry. You could then watch selected portions of the films, read selected chapters, and follow up with specific curriculum activities or extensions that pertain only to your established interests and current syllabus.
- If you need homework ideas, summer vacation assignments, field trip suggestions, career day ideas, or fresh projects to engage your students for a single period or a whole semester, you can simply flip through the curriculum to pick one of the many featured extended activity ideas that appeal to you.

While we recommend you use the book and DVDs to make these activities more meaningful, the level to which you use these additional products is obviously up to you.

Is There a Website to Get More Information or Let You Know How Our Class Is Doing with the Curriculum?

Yes, and we suggest that you familiarize yourself with our website, www.edenslostandfound.org. We will make available additional handouts and other materials from time to time. Please check in with us frequently. Also, be sure to let us know how your experience goes with the *Going to Green* curriculum. We'll post success stories on our website, and maybe you and your students will be the subjects of a future documentary or book!

As far as other websites recommended in this resource, please be aware of the supervisory role necessary when using the World Wide Web. We have investigated all recommended sites but cannot be responsible for changes or disappearances that have occurred after publication of this resource. We cannot be held responsible for the content of any site other than our own. Please be sure to check all sites mentioned for appropriateness before assigning material to students. We also cannot be held responsible for all material that students may come across in their research.

Service Learning in *Going to Green*

Throughout this curriculum, you will notice that a number of the activities and extensions incorporate an element of action on the part of students. These may be projects outside the classroom, focused on specific topics affecting your community. This teaching strategy is called *service learning*.

In this book, all of the activities suitable for service learning satisfy a number of state and national learning standards in science and social studies. You will find a list of standards satisfied in the Scope and Sequence section, Appendix B. Throughout the curriculum you will also find reproducibles for overheads and handouts to help you implement your own service learning projects in your classroom.

The Purpose of Service Learning

Service learning is a growing movement in the United States and abroad and is not limited to the study of the environment and sustainability. The purpose of service learning is to bring together education, service, and community, giving students a chance to enhance their learning and retention through service and action in their own communities.

As described by the National and Community Service Trust Act of 1993, service learning

- Helps students learn and develop by participating in thoughtfully organized service that is conducted in and meets the needs of communities.
- Is coordinated with an elementary school, secondary school, institution of higher education, or community service program, and with the community.
- Helps foster civic responsibility.
- Is integrated into and enhances students' academic curriculum or the education components of the community service program in which the participants are enrolled.
- Provides structured time for students or other participants to reflect on the service experience.

And according to the National Commission on Service Learning, service learning

- Links to academic content and standards.
- Involves young people in helping to determine and meet real, defined community needs.
- Benefits both the community and the service providers by combining a service experience with a learning experience.
- Is usable in any subject area so long as it is appropriate to the learning goal.
- Works well for all age groups, even with young children.

To be clear, service learning is a teaching strategy, not a course in and of itself. The teacher has the freedom to tie the topic or issue at hand into a community-based project that relates to the curriculum. Although, as previously mentioned, this strategy is not limited to the arena of environmental studies, the strategy works very well here. For example: If a class works on a monthly basis throughout the year to pick up litter near a creek, this service can be tied to study of the Clean Water Act and fits nicely into an environmental curriculum. Students can then work to seek the source of that litter and perhaps even outline a permanent solution to the problem. The students explore a particular community issue (litter and how it affects a community's creek), while meeting state and national learning standards.

Service learning differs from community service—another valuable experience for students—in that community service does not tie in directly to the curriculum.

The relationship to curriculum is what makes service learning such a powerful teaching strategy, especially when a component of that service involves teaching others. The Learning Pyramid for Average Learning Retention, as described by the National Training Laboratories in Bethel, Maine, clearly demonstrates the relationship between doing, teaching, and student retention. As you can see, while students retain a mere 5 percent of what they are given in a lecture, the retention rate soars to 75 percent when they practice by

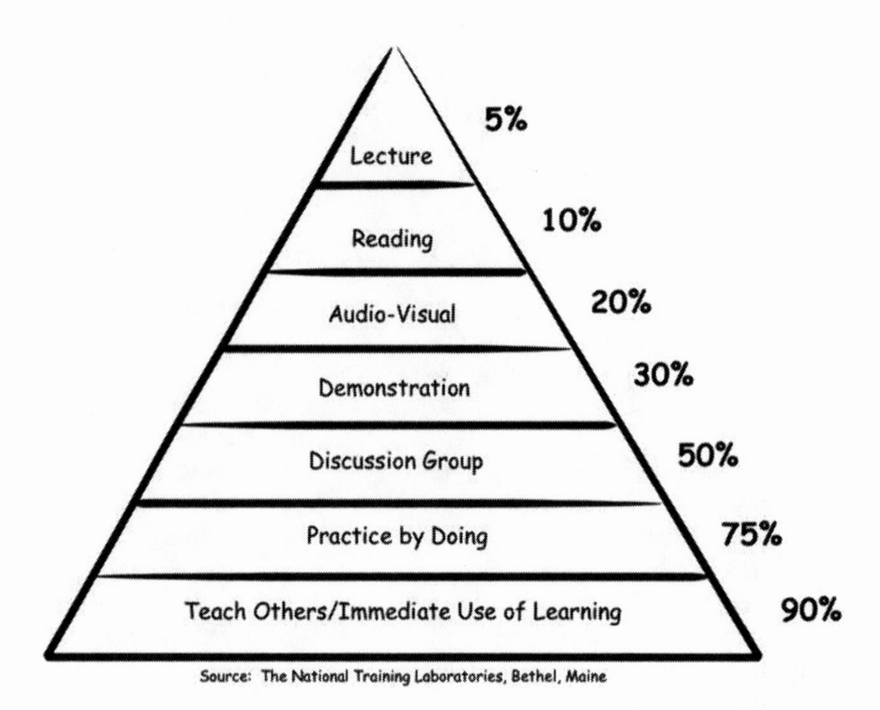

Source: The National Training Laboratories, Bethel, Maine

doing and then to a remarkable 90 percent when they immediately use that learning and/or teach others what they've learned.

Other key components to a successful service learning project are student reflection and community impact. Reflection allows the students to internalize "what they did" with "what was learned," making the learning more real and tangible, more defined and clear. Also, students participating in service learning are making real contributions to their school and community. The students develop self-worth and a sense of belonging to a greater whole. They also will vastly improve their school's relationship between the community.

Getting Started with Your Own Service Learning Project

Planning and implementing your own service learning project involves answering some simple questions and then putting those answers into action with your students:

- *Who* should be involved?
- *What* issue do my students want to work on?
- *Where* and *when* will this project fit into my lesson plans and also work best for the community it serves?
- *How* will the project be evaluated and by whom?

For the best possible results, adhere to the following standards outlined in *STANDARDS of QUALITY for School-Based and Community-Based Service-Learning*, published by the Alliance for Service-Learning in Education Reform:

The Standards: School-Based and Community-Based

I. Effective service learning efforts strengthen service and academic learning.
II. Model service learning provides concrete opportunities for youth to learn new skills, to think critically, and to test new roles in an environment that encourages risk taking and rewards competence.
III. Preparation and reflection are essential elements in service learning.
IV. Youths' efforts are recognized by those served, including their peers, the school, and the community.
V. Youths are involved in the planning.
VI. The service students perform makes a meaningful contribution to the community.
VII. Effective service learning integrates systematic formative and summative evaluation.
VIII. Service learning connects the school or sponsoring organization and its community in new and positive ways.
IX. Service learning is understood and supported as an integral element in the life of a school or sponsoring organization and its community.

X. Skilled adult guidance and supervision are essential to the success of service learning.
XI. Preservice training, orientation, and staff development that include the philosophy and methodology of service learning best ensure that program quality and continuity are maintained.

Ideas for service learning projects can come from almost anywhere, and involving students in this planning is a key element. Look at your local newspaper for important issues and attend community meetings or city council meetings not only for ideas but also to make contacts. Community members may want to partner with your school in your service learning projects. Speaking with municipal agencies, such as the Department of Public Works, can be particularly helpful because many are mandated by federal law to create and implement outreach programs specific to storm water. As you continue to develop service learning projects, you will continue to gain partners for your classroom.

Voices from the Classroom

By Deb Perryman,
2005 Illinois State Teacher of the Year

I have found service learning to be the great equalizer of my classroom. It doesn't matter if a student is average or gifted, below poverty level or wealthy. This is a teaching strategy that truly engages a student despite race, religion, or ability level. Many teachers say to me, "Sounds great, but I have LD, BD, or ESL kids in my classroom." I usually want to say, "So?" But that wouldn't be very nice of me, as this can be a legitimate concern. Children who have learning disabilities and behavioral disorders can participate in service learning projects with great success. Even children showing little motivation respond positively to service learning projects. I have noticed a marked improvement in attendance and attitudes when students are participating in service learning projects.

I know what you are thinking: "That works for her school and district, but it will never work in mine." School District U46 is the second largest in Illinois, behind Chicago Public Schools. Elgin High School is considered an urban school, despite being tucked away in the northwest suburbs of Chicago. We serve approximately 2,200 students who speak 28 languages fluently. Our 2006 state report card states that 38 percent of EHS students are receiving free and reduced lunch, and we are on that dreaded "Academic Warning List." Our community didn't always receive our students well. Of all of the schools in District U46, Elgin had the poorest reputation. Many in our community considered our young people troublesome. Our staff has been working very hard to build a bridge linking the community and our school. Service learning has been an important support beam in that bridge.

During the 2003–2004 school year, my environmental science students logged over 1,200 hours, teaching nearly 6,500 younger learners about Illinois natural history. This does not include our environmental literature volunteer program or our community outreach programs. That's 1,200 hours outside of class time, engaging themselves in science!

My students became so used to identifying and addressing issues in the community that they began bringing projects to me. (This is actually my goal, by the way: The truly authentic service learning projects come from the students.) I mentioned earlier that EHS is on the Academic Warning List. Two years ago, our local newspapers listed the schools on this list. One of my students brought in the article and showed our class that of the eight schools on the list, five fed into EHS. The students became very upset and started to talk openly about what this list meant for our school, community, and themselves.

They seemed most upset about the reading scores and asked if they could "do something" about it. They formed a research team and began looking into the problem by interviewing local reading experts and researching educational journals. When they reported their findings, they found that the number one determinate of whether or not a child would be a good reader was whether that child had been read to aloud. The answer didn't lie in fancy books or tons of money invested in a reading curriculum. Did the parents, and grandparents, read to their child aloud?

New questions arose from their findings. Did the family have the resources to read to the child aloud? Could the parents and grandparents read? We brainstormed several possible reasons and attempted to come up with solutions for each. That is where I came in as the teacher. I had them select one aspect of the problem that was solvable. My students decided that they could develop a volunteer reading program. They spent a lot of time planning, finding partners, and getting funding for the project. My job was to find a way to make this project appropriate for science class.

My stipulation was that they had to read a children's literature book that was linked to science, and they had to create an activity that would reinforce the scientific principle found in the book.

My students decided that they would need training in order to read aloud to an audience. They enlisted the assistance of three partners: a preschool director, the community readership coordinator, and a local bookstore owner. All three of these ladies came to EHS and held training sessions for students who wanted to volunteer as readers. The high school students learned how to hold the book, what to do if a child gets "crabby," and how to make the book interactive. A second team of partners came in and helped the EHS environmental students develop activities that went along with the book they selected. The end result is a small army of EHS students who read aloud in elementary classes, libraries, nature centers, preschools, and day care centers. They are volunteering after school, on weekends, and even on days off from school. If you are in a unit district, you know that sometimes the high school students are off while the middle and elementary students are in session. My students were volunteering even during this time.

Your next question is likely to be, "How did you grade that?" I require my environmental science students to create a portfolio each semester. The portfolio is worth 150 points. I have created a menu of items they can select from to earn their points. Community projects are worth 25 to 30 points. To get the points, students must turn in a lesson plan complete with goals, objectives, and state standards met. They also have to fill out a reflection sheet that details how the lesson went, how it will be improved, and how participating in the project made them feel. I assign points according to their lesson plan and reflection sheet. This year we are hoping to add an evaluation portion for the classroom teacher or organization to fill out about the high school students' performance. We aren't sure if we will assign points based on the comments or just use it to coach the student in lesson preparation.

Ms. Perryman's Service Learning Grading System

Portfolio	**150 points**
Community Project	**25–30 points**

(May add additional points for teacher evaluation and the community organization's evaluation.)

Not only are the students performing a meaningful service, they are having fun building their college application as well as learning science! I also think they will discover the importance of reading aloud to *their* children when that time comes. Perhaps that will be the impetus for the creation of a new reading paradigm in Elgin!

"Move Youth to Action"

Deb Perryman has used service learning strategies to enliven her classroom, motivate students, and forge strong bonds between her school and the community. Below is a reprint of a flyer describing some of their service learning projects:

Move Youth to Action!

Service learning is a teaching method by which students improve academic learning and develop personal skills through structured service projects that meet community needs. Service learning builds on students' service activities by providing them with opportunities to prepare, lead, and reflect on their experiences.
—Corporation for National Service Learning

How Effective Is Service Learning?

Studies prove that the average retention rate for lecturing is 5 percent, reading 10 percent, audiovisual 20 percent, demonstration 30 percent, discussion group 50 percent, practice by doing 75 percent, and teaching others is 90 percent (National Training Laboratories). Service learning incorporates all of these teaching strategies and emphasizes the last three.

What Does Service Learning Look Like?

If it looks as though kids are actively participating in a project that is addressing a need in your community, it may very well be service learning. Consider the following projects and decide if this is the type of learning in which you would like to see the youth of your community engaged.

Kid's Earth Day: The students of Elgin High School prepared activities that would highlight the beauty of their school's 35-acre nature center and provide an opportunity for preschool children to learn about Illinois natural history. The high school students taught concepts they learned in environmental science class.

River Watch: Elgin High School participated in the Illinois Department of Natural Resources Ecowatch program. Students were trained as Illinois Citizen Scientists and took water quality data on Poplar Creek. These data were then submitted to local, state, and federal data clearinghouses. Once students were trained as citizen scientists, they were then eligible to teach local school children about water monitoring and the importance of clean water.

Mighty Acorns: This curriculum for fourth, fifth, and sixth graders includes student field trips to a natural area three times each school year. This Chicago Wilderness Program moved the elementary children to action by conducting habitat restoration activities. Typically, a professional nature center or forest preserve would run this program, but the high school students were given the leadership role and had to find funding and provide training for the classroom teachers.

Storm Drain Stenciling Program: During the 1999–2000 school year, the environmental science classes were learning about water quality. The students were shocked to learn that storm drains led directly to Poplar Creek and the Fox River. They began sharing stories about witnessing substances being dumped into the drains and the effects of nonpoint pollutants. It didn't take long for these students to connect that the city of Elgin gets its water supply from the Fox River. Out of this discussion a community campaign was created, called "2,000 in 2000." The students pledged to paint "Do Not Dump" warning signs on 2,000 storm drains by the end of the school year. To date, more than 4,000 drains in the Fox River and Poplar Creek Watersheds have been painted using easy-to-use, mail-order stencils. The program was so effective that EHS now serves as the Conservation Foundation's clearinghouse for storm drain stenciling supplies and information.

Environmental Reading Program: Stunned by the news that some of Elgin High School's feeder schools had failed to meet "No Child Left Behind" reading standards, students began asking what they could do to help. After procuring a BP America Leader Award in the amount of $30,000, the students planned and implemented a volunteer reading program that is now a permanent part of the Elgin High School "School Improvement Plan."

During the 2002–2003 school year, these young people logged over 1,200 volunteer hours (that's an average of 120 hours per student) teaching more than 6,300 younger learners. They volunteered in 115 classrooms and provided environmental education and outreach to countless community organizations. This story is not over. The students of Elgin High have increased their services to the community and their test scores are rising. Elgin High School is now on the "No Child Left Behind School Choice List." Elgin High School students and staff are proud to share that they met the prescribed benchmarks for the 2003–2004 school year.

Resources

To Find Out More about Service Learning

Consult these websites and books to learn more about how you can bring service learning into your classroom, and network with other education professionals throughout the country:

The National Service-Learning Clearinghouse
www.servicelearning.org/

National Service-Learning Partnership
www.service-learningpartnership.org/site/PageServer

Learn and Serve America
www.learnandserve.org/

International Partnership for Service-Learning and Leadership
www.ipsl.org/

Peace Corps Service Learning page, a part of the Coverdell World Wise Schools program
www.peacecorps.gov/wws/service/

The Complete Guide to Service Learning, by Cathryn Berger Kaye, M.A. (Free Spirit Publishing)
An excellent, comprehensive book that examines service learning across a number of disciplines.

How to Evaluate a Service Learning Activity

If your students participate in a service learning activity, have them fill in the Service Learning Evaluation on the following page.

SERVICE LEARNING EVALUATION

Name ______________________________ Date ______________

Project Name ______________________________

Project Start Time ______________________ Project End Time ______________________

Describe the goals and purpose of the project: ______________________________

____

____

What steps were taken to meet the goals of the project? ______________________________

____

____

What part of the project did you like the most? ______________________________

____

____

What part of the project did you like the least? ______________________________

____

____

What could be done to improve the project for next time? ______________________________

____

____

How does this project relate to environmental studies? ______________________________

____

____

Was your project successful? How do you know? ______________________________

____

____

Your signature ______________________________

Name of person organizing the event ______________________________

Job title ______________________ Phone number ______________________

Signature of person organizing the event ______________________ Date ______________

Going to Green

UNIT 1

UNDERSTANDING SUSTAINABILITY

TOPIC BACKGROUND

Sustainability is a challenging concept for many students—and indeed adults—to understand. Chances are, your students could easily define *recycling* and tell you why it's important for cities and towns to *reduce, reuse,* and *recycle*. But recycling cans, bottles, and newspapers is only a small part of sustainability.

Sustainability means that an action, system, or process can be maintained indefinitely or even forever. In our reading, we ask students to imagine a walk in the woods, where trees "feed" themselves by fertilizing their roots with layers of rotting leaves called humus. As long as a tree gets sunlight and water, it can produce food forever or until it dies. Nothing is wasted in a forest, and nothing a tree produces pollutes the forest. Because leaves, nuts, berries, flowers, and other tree-made products are organic (made of natural materials), everything that falls to the ground eventually breaks down and returns to Earth with no harmful effects. A forest environment is sustainable: individual trees may die off, but the forest could and should live forever.

We can learn a lot from this process if we study human behavior or a product and ask ourselves, "Is this sustainable? Can we keep this up forever?"

A perfect example is the gasoline we pump into our cars daily. Petroleum oil is a *finite* resource; only a fixed amount of it exists on the planet. It was formed in the planet Earth over millions of years, and it has its origins in the time of the dinosaurs. (That's why petroleum-based fuels are nicknamed "fossil" fuels.) When the world's oil reserves are depleted, we will have to find other fuels to power our cars, heat our homes, and run the world's industries. On top of that, each time we burn fossil fuels, we pollute the atmosphere with carbon dioxide, soot, and other toxins. To answer our own question, we can't keep this up forever. Unlike the trees in the forest, we pollute our environment every time we use fossil fuels, and sooner or later we *will* run out of them.

Go easy on your students, and yourself, in this unit. It's easy to look around your community and conclude that very little we do is sustainable. It's also easy to become frustrated because sustainability encompasses so much that is seemingly out of our hands. Try to keep in mind that, like so many things humans strive for, sustainability is an ideal. It's a worthwhile goal that many towns and cities in the world are now shooting for. Your class may well be the first to spread that message to your community. Eventually, they will realize that ordinary citizens can change their societies, if they resolve to do so.

Purpose

Students will begin to develop an understanding of the term *sustainability* and see how it applies to their lives. In ever-widening circles, they'll begin to gauge the sustainability of their school, neighborhood, town, and ultimately, their nation. They'll see that sustainability moves beyond recycling into areas such as water resources, energy, mass transportation, parks and forests, building construction, and waste disposal. After years of absorbing the lessons behind recycling, your students are mature enough to grab hold of the "Big Story": sustainability.

GROUP ICEBREAKERS

1. Ask students to define *recycling*. (The process by which collected materials are reused as the "raw" materials for new products.)
2. Have students brainstorm things people do to help the planet. (Shutting off the faucet while brushing teeth, switching off lights when not using them, cleaning up trash, and the like.)
3. Ask, "What's the logic behind doing these things? What is our goal? What are we shooting for? Is it conservation? Is it reducing waste?" (Allow the class to respond with their understanding of the reasoning behind the green movement.)

4. Write the word *sustainability* on the board and then ask students to listen as you read this definition aloud: "The practice of providing for the needs of the present population in a way that doesn't jeopardize the ability of future generations to provide for themselves. This involves the use of resources that can be replaced, reused, and/or renewed so they are not depleted. It affects every aspect of modern-day life, from economic and technical development to food production." You might need to reread each part of this definition a couple of times to let it sink in.
5. Ask, "What do you think it means to *sustain* something?" Ask one of the students to look it up and read off some of the various definitions. Which definition or definitions seem to fit sustainability?
6. Ask if they can think of some examples from nature of a sustainable action or process. (Use our example of trees in a forest to start the ball rolling.)
7. Sustainability is a worthwhile goal, but it requires hard work. Ask them, "Do you think it would be easy or hard to make the world—as lived by humans—sustainable?" (Accept that some students will say "easy." Others will say "hard.")
8. Propose an imaginary scenario that helps students figure out how to practice sustainability in their own lives. For example, ask them to imagine they wanted to practice sustainability the next time they got a new pair of sneakers. Their goal would be to dispose of the old sneakers so that they "returned" to Earth as naturally as possible and buy a pair of sneakers that had as little impact on Earth as possible in its manufacture. Ask them to think about all the research they'd have to do: They'd need to know how the new sneakers were made, what materials were used, where those materials came from, how much energy was used to make them, and so on. It would be a pretty hard endeavor!

ACTIVITY 1.1
The Sustainability Game

Materials

10 chips or counters per player
paper plates
a portable stereo
recording of some fun, upbeat music
healthy treats such as bags of popcorn

Primary Subject Areas and Skills

science, social studies, math, critical thinking

Purpose

By playing a strategy game, students learn how Earth's resources must be shared wisely by all humans.

How to Play

1. Divide students into groups of three to six players, seated around a common surface—a table or on the floor.
2. Place a paper plate on the table in the middle of each group. Then place 10 chips or counters per person on each group's plate. (For example: A group with three students will have 30 chips or counters on its plate.)
3. When the music starts, players should grab for as many chips as they can. When the music stops, players must stop reaching for chips.
4. One treat or bag of treats will be awarded for every 10 chips that a player has.
5. No talking is allowed during the game! But body language and note taking are permitted.

Playing the Game

1. Start the music. Let it play for a minute or two and then stop it.
2. Take note of the situation at the tables. In most groups, the first round of play is usually mayhem, with students scrambling for as many chips as they can get their hands on as quickly as they possibly can.
3. Go around the room, give a treat to every person who has collected 10 chips, and collect those 10 chips. Extra chips go back into the plate.
4. Tell the students that you will double the number of chips left in their plate (up to their max number of chips). Groups with no chips left get nothing.
5. Play another round. Some groups will not get to play, because they have no chips left. Talk

about *why* they have no chips left. Why is it that some groups have chips left over and others don't?

6. Once the students understand the lesson—that cooperation ensures that everyone eventually gets a treat—you may wish to start the game over and see if the outcome is different.

Discussion Questions and Follow-Up

1. Is it possible that everyone gets a treat? How? (If players take only one chip.)
2. What happens if there is just one stubborn person in each group who does not want to cooperate with the limited resources (chips) that are available? (Then that person ends up with all the treats, and the group cannot play future rounds.)
3. How does this game reflect the environmental problems facing the real world? (Even if just one person or group decides to deplete a resource, everyone is affected.)

ACTIVITY 1.2
Making Connections in the Urban Ecosystem

Materials

pen and paper; Reproducible 1.2A

Primary Subject Areas and Skills

science, social studies, math, reading comprehension, critical thinking

Purpose

Students learn how a problem in one area of an ecosystem usually causes problems in other areas. They discover that most environmental issues are interconnected.

1. Show Reproducible 1.2A on an overhead projector.
2. Have your students study the images alone or in small groups.
3. Discuss the images with the class and pose the following questions:
   a. Which of these images resemble situations facing your town or city? (Students may not be aware that their city is facing energy shortages, for example, so you may need to prompt them. If you are not aware of situations like this in your town or city, adopt the position that situations like these go on in many towns and cities.)
   b. What steps could your town or city take to become more sustainable? (Add more trees to absorb rainwater and excess heat, invest in renewable energy, and the like.)
   c. Do you think it's possible for cities and towns to change, or is that too much to ask? (It's completely possible! Cities everywhere are becoming greener in order to save money and lives.)
   d. What are the challenges to "retrofitting" a city? (The cost of adding parks and green spaces, as well as retrofitting rainwater drainage systems, is high. Politicians are sometimes reluctant to do these things because expensive city projects are not popular with voters.)
   e. Which of these challenges do you think would be the easiest to fix? (Add more trees and parks, encourage energy conservation, and the like.) If changes are not already being done, what do you think is preventing the changes? (Costs and political will, as mentioned above. Also, many citizens are not educated about the interconnected nature of environmental problems.)
4. Have students research which departments in your community or local government oversee such projects. You may wish to invite one of these individuals to be a guest speaker in your classroom.

ACTIVITY 1.3
Sustainable or Not Sustainable?

Materials

pen and paper; copies of Handout 1.3

Primary Subject Areas and Skills

science, social studies, math, writing, critical thinking

Purpose

Students decide if a long list of everyday behaviors is sustainable or unsustainable in the context of Earth's future environmental well-being.

1. Copy Handout 1.3 on page 12 and distribute to students.
2. Instruct students to read each of the statements

and then write an answer that explains what they feel is sustainable or not sustainable about each scenario.

3. Have a class discussion in which students share their impressions and ideas about each of the statements.
4. You may also wish to first assign this as homework in preparation for your in-class discussion.

ACTIVITY 1.4

Global Perspectives

Materials

pen and paper

Primary Subject Areas and Skills

social studies, math, reading comprehension, critical thinking

Purpose

Students use a bar graph to determine which regions of the world have the largest ecological footprint.

Sustainability and sustainable development are often described as the practice of providing for the needs of the present population in a way that doesn't jeopardize the ability of future generations to provide for themselves. This involves the use of resources that can be replaced, reused, and/or renewed so that they are not depleted. It affects every aspect of modern-day life, from economic and technological development to food production.

A phrase often used to describe the effect that societies have on their environment is "ecological footprint." Analyzing the ecological footprint of societies, cultures, and the global community helps determine how we can become more sustainable and put less strain on the world's resources. Many organizations have developed programs and questionnaires that help individuals, organizations, and governments calculate their ecological footprint, or personal impact on the environment. Simply put, there is a limited amount of ecologically viable and productive land on the planet, and it has to support present and future generations.

How much land would be needed to support just one individual's demand for resources and still be able to absorb the waste that person will create in a lifetime? Estimates put the number between four and five acres per person. But is that the amount that we're using?

Show Reproducible 1.4 on an overhead projector. Have students look at the chart and then discuss the answers with your students.

Extension

How big is *your* ecological footprint? Discuss the chart and the idea of the ecological footprint with students. You may wish to visit www.ecofoot.org/ and have students take the ecological footprint quiz there. (See also the Resources section at the end of this unit.)

Instruct students to describe, in approximately 400 words (or a length of your choosing), what a day in their life would look like if they strived to live sustainably and leave a smaller "footprint."

You may wish to use the following writing prompts to get things started. The prompts may encourage students to think about the impacts that *all* aspects of their daily lives and actions have, rather than just the more obvious ones (though those are important, too).

- Where would your food come from? (The most sustainable thing is to eat what is grown and raised locally.)
- How would you travel to school? (Walking and bike riding are the most sustainable, healthy modes of travel for students.)
- What would you wear? Where would you get your clothes? (Sustainable garments are those that come from local sources and are made from fibers that have not been raised in a way that damages the environment. For example, fibers that have been sprayed with copious amounts of pesticides may not be considered sustainable.)
- How would your normal routine be affected? (Before making any choice about purchasing, travel, eating, and so on, we would all have to think carefully about our decisions. It's worth conceding to students that this is extremely difficult to do.)

ACTIVITY 1.5

At the Podium

Materials

library and Internet research materials

Primary Subject Areas and Skills

reading comprehension, writing, research, critical thinking

Objective

Students will begin to refine their overall thoughts and feelings regarding sustainability through participating in two debates.

Possible point and counterpoint arguments are provided. The first statement for debate is a general one:

> *We all, each of us, have a responsibility to future generations and other communities to live and act sustainably.*

Point

Yes, I agree. It is irresponsible and even criminal to squander the resources of future generations simply because we refuse to acknowledge responsibility toward the welfare of our fellow citizens by confusing mindless consumerism with personal freedom.

Counterpoint

No, I disagree. Individuals must do their best to ensure the best possible life that they can for themselves and their families. If everyone just focused on doing a good job of taking care of her- or himself, everything else would fall into place now and in the future. There's no need to tell law-abiding citizens how they can and cannot live.

1. Divide students into two groups of at least four students each. Depending on class size, be sure to leave a group of at least five (preferably an odd number) of students to act as judges. Assign each group to Point or Counterpoint as described above.
2. Instruct students to research and report on their point of view. They should devise statements in support of their position and be prepared for rebuttal from the opposing team.
3. For the debate itself:
   a. Judges should be concerned with organization, evidentiary support, and overall presentation, as well as politeness and poise. As judges will not be researching, their responsibility prior to the debate is to meet together with the teacher to decide on two or three criteria (in line with those listed here) that they will use to judge the speakers. Judges can use the rubric on Handout 1.5 to apply points. *Note:* You can use this rubric throughout the curriculum for all "At the Podium" activities.
   b. Debaters should clarify who will speak for the group. (Debaters may take turns, if that is decided beforehand.)
   c. The affirmative or Point team will speak first for three to five minutes to present their argument. The Counterpoint team will then have three to five minutes to present their argument.
   d. Each team will be given three to five minutes to respond to the other team's argument.
4. Have judges rate the arguments of each group and present their findings to the classroom.
5. Discuss among yourselves the emotions and ideas that came up during the debate. Did anyone change his or her mind about the topic at hand?

The second issue for debate is a more specific one, designed to elicit a more targeted response from students as they begin to think about their role in their communities and the role of "rules" and mandates as a part of change:

> *The department of public works will no longer remove leaves, clippings, or other organic materials from residents' properties. The material(s) should be kept and used to compost plants growing on the property or brought by the resident to a city composting station.*

Point

Yes, I agree. The amount of gasoline alone that is spent carrying around leaves and grass is wasteful and not cost effective. The organic material would be better used to naturally fertilize and feed lawns and plants. This would cut down on not only gasoline consumption but also on the use of harmful fertilizers.

Counterpoint

No, I disagree. The families in our community work too hard and pay too much in taxes to not have their trash picked up, even if that trash is organic. They don't have the time to learn how to compost, and the city shouldn't stop doing something it always has done for its residents. Pursuing this will result in unkempt lawns and a decline in property values.

Follow the same instructions for the first debate.

EXTENSIONS

Sustainability Critique

Consider mounting a yearlong project to gauge the sustainability of your school, town, or city hall. Using the activities in this and following units, your students should be able to accurately critique how well the entities closest to them are doing with respect to the health of the planet. Students will need to decide what criteria they will study and how they will measure it

Ecological Footprint Survey

Students could print one copy of their favorite online survey, make copies, and use them to interview their parents, other adult relatives, and local officials. Back in the classroom, students can input the data from their paper surveys into the online program and calculate a footprint for every person they interviewed. This information could be used to create tables, charts, or graphs. Students could also offer "footprint counseling" by advising their interviewees on how they can shrink their footprint to a more sustainable size.

Sustainability Collage

Create an easy-to-read bulletin board display in a public place in the school. This "Sustainability Collage" could contain simple reminders to passersby of the small things they can do to leave a smaller ecological footprint. Later, challenge the students to condense their information and publish it in the form of a brochure that can be distributed at a local farmer's market or other appropriate venues on Earth Day or Arbor Day.

Art of Sustainability

Consider hosting a sustainability art exhibit at your school that might include photos, drawings, and art made with recycled, reused, or organic materials. Auction the artwork to raise money for a cause of your choosing.

RESOURCES

Students can measure their ecological footprints using a calculator tool found on the following web pages:

- www.bestfootforward.com/footprintlife.htm
- www.mec.ca/Apps/ecoCalc/ecoCalc.jsp
- www.ecofoot.org/

This website offers tips on how individuals, schools, business, and community organizations can reduce their ecological footprint:

- www.rprogress.org/newprojects/ecolFoot/reducing.html

Reading Handout Unit 1:
THE PATH TO SUSTAINABILITY

In this reading the authors of *Edens Lost & Found* share their views about the concept of sustainability. After you've finished the reading, answer the questions.

If you wander in the woods in the springtime, you will catch the forest at its most magical time. The long period of winter dormancy is over, and the green shoots of leaves are sprouting everywhere. The smallest plants leaf first, taking advantage of the abundant sunshine. As the weeks pass, taller trees and plants yawn to life. Finally, the tall giants awaken and shelter the forest with a dense, green canopy. All summer long, the sun shines, and temperatures can soar. It makes no difference to the woods. Shielded by vast, leafy umbrellas, the forest floor remains cool and shady.

When the leaves fall in autumn, they drop to the ground and break down into soft, black earth. Dead plants and trees also become soil. This rich, new mulch fertilizes everything it falls on, and not a molecule of the old organism goes unused. When it rains, this soft layer of earth acts like a giant sponge, sucking up rainwater and storing it for later use. Deep underground, a complex network of tree roots also soaks up rain, so much so that a mature oak tree can capture 57,000 gallons of water.

Perhaps more than anything else, the forest demonstrates sustainability. The forest wastes nothing, and the needs of the present generation of trees, animals, and other organisms are met without depleting the resources needed by future generations. Everything that goes into the system remains in the system, in some form or another. Each dead tree feeds another. Every raindrop is harvested, and as long as average weather conditions persist, this cycle can go on forever without ever running out.

While most humans enjoy the woods, let's face it: We don't live there. In fact, the majority of Americans, 80 percent of us, live in "urban ecosystems," that is, large cities and nearby suburbs. Designed for commerce and human habitation, these ecosystems have often grown quickly without sustainability in mind. When it rains, all that precious rain—drinkable fresh water that cannot be absorbed by metal, concrete, and pavement—is funneled out to rivers and seas instead of being captured for human use. Apartment buildings, offices, and skyscrapers are giant heat magnets, soaking up sunlight and relying on expensive fuels to keep them habitable. (We cool buildings down in the summer and heat them up in the winter.) To get around, most people must drive, thus using even more fuels that also pollute the air we breathe and threaten our health.

We've strayed a long way from the perfection of the woods. Many facets of city life are unsustainable. If a city has squandered its water, it must buy water from another city. Because oil is a limited resource, every drop used literally means one drop less for future generations. Because most vehicles don't burn cleanly, excess residue fills the atmosphere in the form of planet-warming carbon dioxide (CO_2), harmful smog, and soot. In continuing these wasteful practices we harm ourselves, and, increasingly, we must spend more and more money to maintain our lifestyles.

But all is not bleak. Today, ordinary citizens, government officials, architects, urban planners, and others are banding together to reverse the trend toward destructive city life. They are trying to make cities sustainable.

It's easy to see how the word *sustainable* applies to the woods. But what does the word mean when applied to a city? In 1987, the World Commission on Environment and Development, in its report to the United Nations, developed a definition of sustainability that has been widely adopted:

> *Sustainable development is development that meets the needs of the present without compromising the ability of future generations to meet their own needs.*

Early in the filming of the *Edens Lost & Found* project, we had an interesting talk with a top executive at a major engineering firm. His company builds roads, bridges, and massive storm drain and solid waste treatment projects for city governments all over the world, including all four cities in this miniseries. We asked him if it was hard to get city officials, residents, and politicians behind the notion of sustainability. After all, the principles of the environmental movement have been contested, resented, and dismissed by people with opposing agendas or viewpoints, and the two sides have often been at odds. His answer surprised us.

"There's far less conflict in my world than there would have been 20 years ago," he said cheerfully. "People are more interested in learning how they can plan a city more efficiently, without wasting resources like water and money. In this country, the Environmental Protection Agency is getting tougher on what can and cannot be dumped into waterways. Years ago, you could get away with a lot more. Now, it costs you a lot, and it's recognized as wasteful. For years we designed cities and suburbs in this country to get rid of water. Now we're thinking, 'Well, maybe you don't want it to go away.' If you can use the water that rains on you, you don't have to go somewhere else to get it. You're more independent. You save money. Everyone I meet these days is thinking this way. So I'm very hopeful."

Each tree that grows in the forest is an independent entity. Each day it harvests food, sunlight, and water from its environment to make a living for itself. Yet planted alone in a middle of a barren field, the tree won't live long. No taller trees rise to shade it. No dead leaves fall to feed it. No grass or smaller plants exist to slow down and capture the water it needs to drink. The tree may be independent, but it thrives best when it lives with others of its kind to create a stable, nurturing, habitable community.

Each of us is that tree. This is the stirring metaphor behind the sustainability movement. Nature not only serves as a biological model for bringing balance to our lives but offers a very real personal guide as well.

Americans today are trying to live in *relationship* to Earth and each other. For some of us, that may mean spending our spare time planting trees and tending our compost bin. For others, it will mean turning our energies to community activism, alerting others to the benefits of creating a sustainable urban ecosystem, forming grassroots and neighborhood groups to support policies that are environmentally beneficial, and letting elected leaders know what the public wants.

Today our country is at a crossroads. Our populations are growing, our city budgets are constantly strained, and some segments of the populace are always in need. If large cities are going to be part of the solution, if they are going to live up to their promise as centers of culture and ideas, they must become sustainable. This is the next big step in the great American experiment. We have an obligation as the world's wealthiest nation and its biggest consumer to show the world that harmony with one's natural surroundings is not only attainable but also profitable and desirable. But how do we get to there from here?

Questions

1. How is a forest sustainable?

____

____

2. What is an urban environment or an urban ecosystem?

____

____

3. What is a forest able to do that a city environment cannot do? For example, can a city absorb rain? Why or why not?

____

____

4. Why are city buildings not ideally designed to cope with hot temperatures?

____

____

5. Do you think that there can be such a thing as a sustainable city?

____

____

6. Can you think of ways that cities can adapt themselves to mimic forests?

____

____

7. Write the United Nations definition of sustainability in your own words.

____

____

REPRODUCIBLE 1.2A: A City Suffers

Making a city sustainable is challenging because everything affects everything else.

Here are just a few of the common difficulties encountered in urban centers, and towns and cities as well:

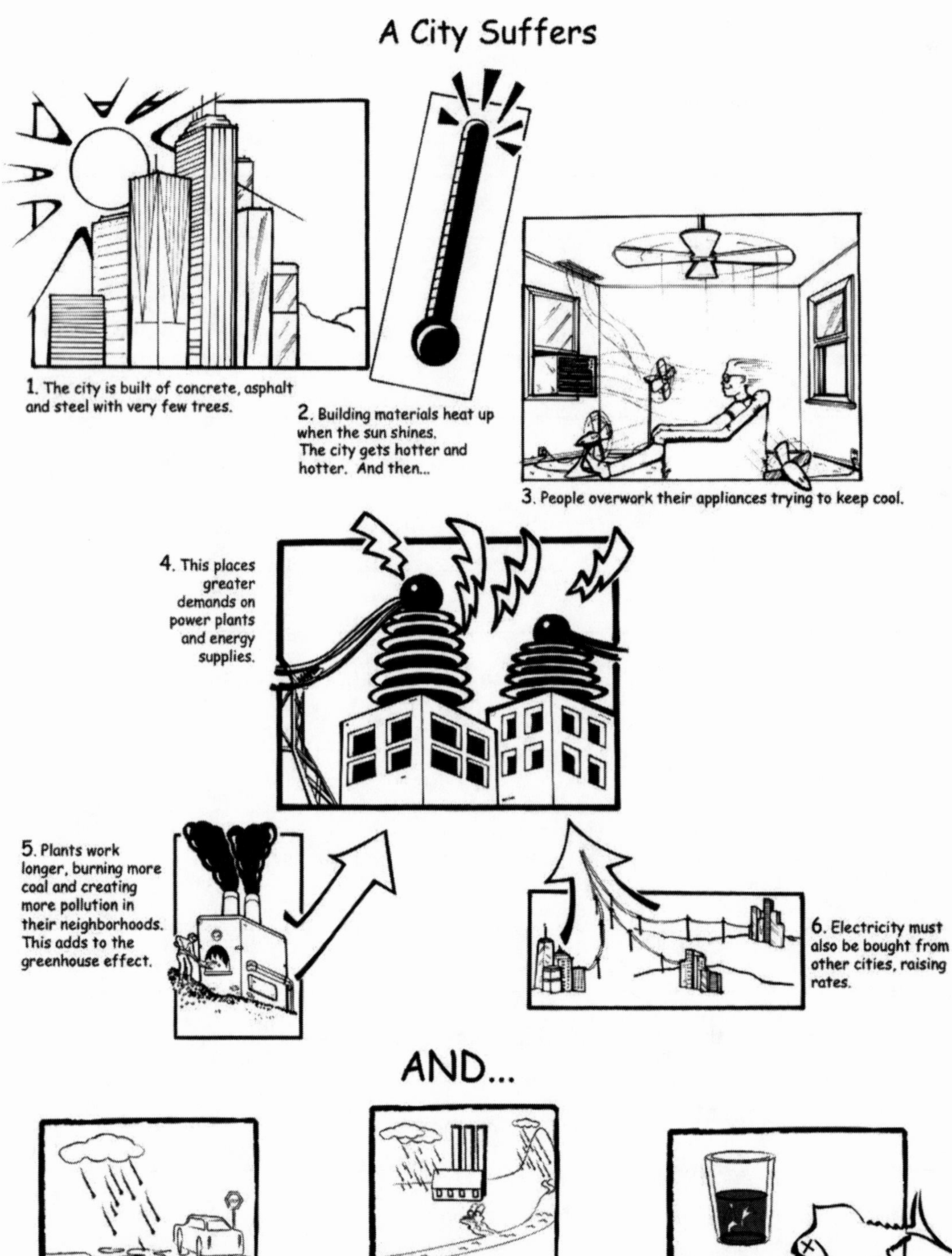

REPRODUCIBLE 1.2B: A City Comes Back

The citizens of Philadelphia, Chicago, Los Angeles, and Seattle, profiled in *Edens Lost & Found,* worked to make their urban ecosystems more livable and, in the process, more sustainable. Here are some keys to making urban ecosystems more sustainable:

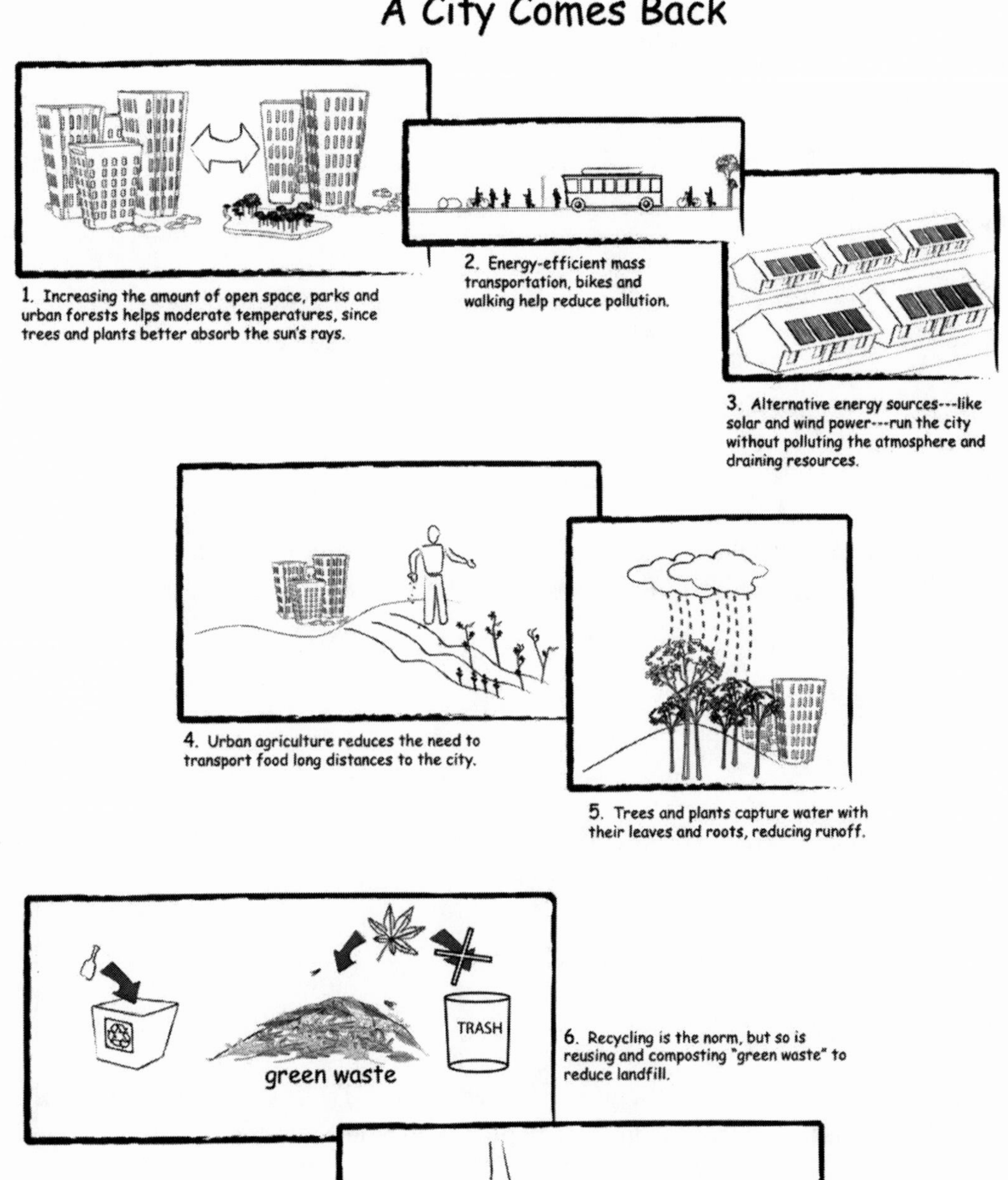

7. Green building techniques, such as installing compact fluorescent bulbs (CFLs) and better insulation, reduce energy costs and pollution.

Name ______________________________ Date ____________________

HANDOUT 1.3: Sustainability in the Gray Area

Read each of the statements below. How are the actions described below sustainable? How are they not sustainable? Write your answers in complete sentences.

1. You mow your lawn twice a week with a gas mower. You gather all the clippings and leaves together and use them to mulch and fertilize your garden.

   a. How is this sustainable? ______________________________

   b. How is this not sustainable?______________________________

2. You have installed low-flow faucets, toilets, and showerheads in your home. Each morning, you run the water while brushing your teeth and take a 30-minute shower.

   a. How is this sustainable? ______________________________

   b. How is this not sustainable?______________________________

3. You change your oil on a regular basis to make sure your car runs well and is fuel efficient. When you're done, you get a hose and wash the leftover oil down the driveway and into the nearest sewer grate.

   a. How is this sustainable? ______________________________

   b. How is this not sustainable?______________________________

4. You have found a grocery store with a wide variety of organic food shipped from all over the country. Your drive there at least three times a week to go shopping.

   a. How is this sustainable? ______________________________

   b. How is this not sustainable?______________________________

5. You've worked hard and saved up enough money for a new SUV. You decide to purchase an "environmental" license plate from your state, which costs you more but will donate money to protect natural resources in your area.

   a. How is this sustainable? ______________________________

   b. How is this not sustainable?______________________________

Extension

How do you feel about the ability to act more sustainably? Is it easier to do as an individual or as a group? Why and why not? Why do you think there are sometimes no "easy answers" when it comes to trying to do the right thing?

REPRODUCIBLE 1.4: Global Perspectives

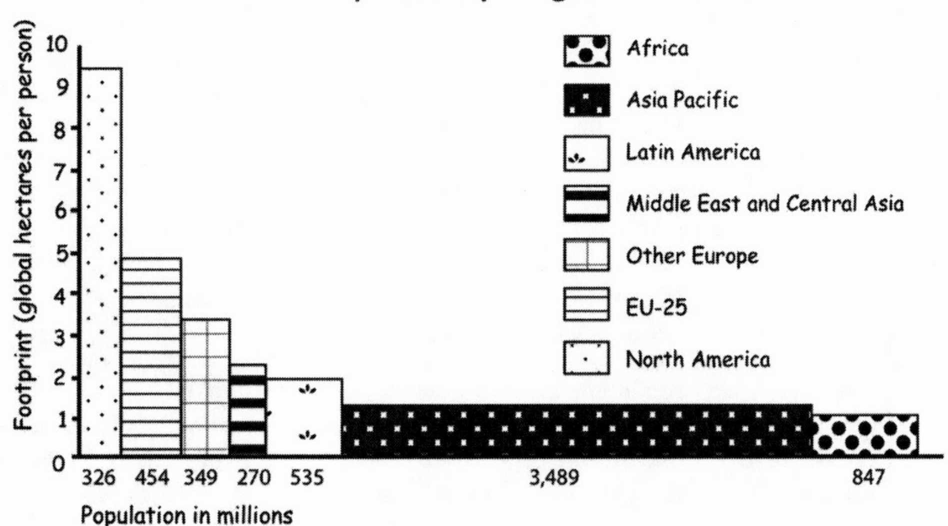

1. Which two regions have the largest ecological footprint?

___

2. Which region has the lowest?

___

3. Which region has the largest population? Does this region have a large or small footprint compared to the regions in questions 1 and 2?

___

4. The United States is located in the North American region. Which region is closest to this region in population? About how many times greater is North America's footprint compared to this other region's footprint?

___

5. What do you think might happen if the people who live in the Asia-Pacific region adopted a lifestyle that was closer to that of citizens in North America?

___

6. The United States and Europe are some of the world's most technologically advanced and seemingly prosperous societies. Are you surprised by the size of their footprints? Why or why not?

___

HANDOUT 1.5: Judges' Evaluation Sheet

Argument Component	0–4 points	5–9 points	10–14 points	15–19 points	20–24 points
Organization of argument Points ____	Argument was poorly conceived and difficult to follow. We had no idea what their main points were.	Argument was somewhat organized. Difficult to follow at times. (They lost us two or more times.)	Argument was fairly organized. We were unable to follow the argument once or twice.	Argument was well organized. We thought that they did a decent job but still could have improved.	Argument was very well organized.
Evidentiary support Points ____	Argument provided no reference to valid data or support at all.	Argument provided only one reference to valid data.	Argument provided two to three references to valid data.	Argument provided four to five references to valid data.	The argument had more than five references to valid data.
Overall Presentation Points ____	Presentation lacked interest and evidence.	Less than half of the presentation was interesting and well supported.	About half of the presentation was interesting and well supported.	Most of the presentation was interesting and well supported, but it dragged in spots.	Overall presentation was interesting and well supported throughout.
Poise and politeness Points ____	Team members constantly interrupted each other and spoke out of turn.	Team often spoke out of turn and interrupted the flow of the debate.	Team interrupted no more than two times.	The team interrupted the debate only once and was fairly well spoken.	The team was respectful throughout the entire debate and very well spoken, delivering arguments clearly.
Total Points ________					

UNIT 2

BUILDING COMMUNITY

TOPIC BACKGROUND

Human beings are social creatures. We thrive when we make connections with other people, build friendships, and cultivate a sense of belonging to a group. In all the cities we visited for the *Edens Lost & Found* project, we noticed an interesting pattern: By pitching in to clean up a dilapidated park, plant a garden, or clean up a local river, people made friends with others and felt rewarded by a greater sense of community. This equation never failed. It was present in every situation we observed. The word *community* has a couple of slightly different meanings. Consider the following statements.

1. A community is a group of people *living in the same area*. These people may not like each other very well, or they may not even interact, but by virtue of their geography, they constitute a community.
2. A community is a group of people *with a common background* (all people who descend from a country in Europe or Latin America, for example) or *with shared interests* ("the chess-playing community," or people who want to see more trees planted in a specific neighborhood).
3. A community includes nonhumans, too. Plants or animals that thrive best in a certain ecosystem, or specialized environment, also constitute a community.

As the stories in our project demonstrate, active participation goes a long way toward making a person feel happy, productive, and connected to others in his or her group. When people feel a "sense of community," they are more likely to act for the benefit of the group. They are more willing to get involved, take chances, volunteer, and work to make things better for all concerned. The word *community* pops up constantly throughout *Edens Lost & Found,* as we profile ordinary people who chose to act to better their city neighborhoods by doing simple things such as clean up parks, plant trees, and reach out to their neighbors in a new way.

Purpose

In this unit, students will learn and understand the definitions of various types of community. Community is an important concept for your students to understand and recognize. Every day, young people move freely between different social circles: the community of their families, close friends, school campus, and their cities, towns, or neighborhoods.

GROUP ICEBREAKERS

1. What is the difference between a community, such as a school or neighborhood, and having a "sense of community"? (A person can belong to a community without feeling invested in its survival or well-being. A "sense of community" means that a person feels that he or she has a stake in that community and will work hard to help it thrive.)
2. During the making of *Edens Lost & Found*, filmmakers Harry Wiland and Dale Bell noticed that when people pitch in to make their neighborhoods and surroundings a better place, they are rewarded with a greater sense of community. Why do you think this is so? (When people pitch in to make their neighborhood a better place, they meet their neighbors, enjoy more neighborhood events, and feel more connected to the place where they live.)
3. Do you think volunteering to help improve your neighborhood is a good thing to do or a waste of time? What are some easy things you could do right away to make a difference? (Answers will vary, but try to engage those students who think volunteering is a waste of time. Have them explain why they feel the way they do. Cleaning

a park or planting some trees or plants in their neighborhood is one easy thing they can do to make their neighborhood a better place to live. This curriculum will give you and them plenty of opportunities for more involvement.)

4. Classroom debate: Can communities be negative or harmful? What is the difference between a community and a clique? What about online or Internet communities, including chat rooms, blogs, and so on? (Unsupervised teenage groups can indeed be negative in nature, if they exclude other students on the basis of superficial characteristics, such as appearance or popularity. The Internet question is hotly debated in American society: Some students feel strongly connected to online groups, though many feel that without face-to-face contact, online groups are no substitute for a conventional community.)

ACTIVITY 2.1
Thinking about Communities

Materials
pen and paper; board or overhead projector

Primary Subject Area
social studies, writing, critical thinking

Objective
Students create a chart listing as many different types of communities as they can think of, and their attributes.

1. Instruct students to list five types of communities on a piece of paper.
2. On the board or overhead projector, create a chart like the following. An example entry is done for you.

COMMUNITY	TYPE	COMMON PURPOSE OR GOAL
Your School	Educational, Geographic	A place for people to learn, teach, develop

3. Ask students to share their lists. Write each community on the board. For each community that is listed, ask for a show of hands as to how many other students also listed that community.

For each community listed, discuss the type of community it is and what common purpose or goal members of the community share. For each type of community, encourage your students to stretch their ideas about what a community is. For instance: Using the example above, consider that certain schools may also be examples of a religious community as well.

Time Saver Option
You may also assign this activity for homework, asking students to come to class with a list of the different types of communities that exist in the world.

ACTIVITY 2.2
Now and Venn

Materials
pen and paper; Reproducible 2.2 optional

Primary Subject Area
social studies, math, writing, critical thinking

Purpose
Students use a Venn diagram to graphically represent various communities to which they belong.

PART I: A Closer Look

1. If necessary, review Activity 2.1 and the preceding Group Icebreakers with your students.
2. Have students list the communities of which they feel they are a part. Encourage them to think about some of the topics you have already discussed. For each community they list, have them think about the following questions:
   a. In what two ways are you involved in that community?
   b. Would you like to become more involved or less involved in that community? Explain your answer.
   c. Are there any communities with which you would like to become involved?

PART II: Graph It
After students have completed Part I above, instruct them to create a graphical representation of their discoveries using a Venn diagram. A Venn diagram

illustrates the relationships between different groups of items that share characteristics. The overlap and intersection between two or three types of communities, for example, can be represented by a Venn diagram. Have students create Venn diagrams illustrating the commonalities between two or three types of communities in which they participate. This activity can be done several times, comparing different communities. Reproducible 2.2 on page 25 provides an example.

Extension

Ask students to think of another way to graphically represent their own, personal involvement in different communities. For example, they could make a bar graph that compares the number of hours per week spent with each community with which they are involved.

ACTIVITY 2.3
The Power of Intention

Materials

Internet and library resources

Primary Subject Areas and Skills

social studies, writing, research, critical thinking

Purpose

Students research and learn about intentional communities, groups of people who have chosen to live together, usually to achieve some environmental purpose.

An *intentional community* is primarily used to describe a housing community in which residents usually share common values, responsibilities, and resources. Examples of intentional communities include cohousing, ecovillages, residential land trusts, urban housing, communes, ashrams, and housing cooperatives.

1. Ask students if they have ever heard the phrase *intentional community*. What does an "intentional community" sound like to them? What kind of connotations does the term bring up for them? You may wish to list responses on the board during your discussion.
2. Discuss and define the different types of intentional communities listed above. What do students think about them?
3. Place students into groups and assign each group to research one of the following intentional communities; or they may research their own. Be sure students pay particular attention to the missions, purposes, and membership requirements for each of these communities:
   a. Earthhaven Ecovillage, Asheville, NC www.earthaven.org/
   b. Ecovillage at Ithaca, NY www.ecovillage.ithaca.ny.us/
   c. Findhorn Foundation in Moray, Scotland www.findhorn.org/home_new.php and the Ecovillage Findhorn www.ecovillagefindhorn.com/
   d. Heathcote Community, in Freeland, MD www.heathcote.org
4. After students have done their research, have them take turns sharing their findings with classmates.
5. Discuss the following questions:
   a. What is appealing about intentional communities?
   b. What is unappealing about intentional communities?
   c. What do intentional communities have in common? How do they differ?
   d. What challenges might living in an intentional community present to you?
   e. Could you imagine living more than a year in an intentional community? Why or why not?

Extension

Instruct students to write an essay describing "A Day in the Life" in their ideal community. This community does not have to be an intentional community; however, if students wish to borrow some of the practices of intentional communities that they have researched, they may.

ACTIVITY 2.4
Making Sense of the Census

Materials

the website www.census.gov
encyclopedia or other library resources
Handout 2.4

Primary Subject Areas and Skills

social studies, math, critical thinking

Purpose

Students read a table of U.S. Census material to expand their understanding of community to include demographic communities.

PART I

1. Copy and distribute Handout 2.4 on pages 28–29
2. Have students answer the questions and then discuss as a whole group.

PART II

1. Assign students partners or put them into small groups.
2. Using the library or Internet, students should research census information about their geographic area and one other area of their choosing. The area could be as simple as the number of people who work at home (see chart example on page 26). Then have students answer the following questions:
   a. What kind of communities do you see represented in the census?
   b. What kinds of communities are not represented in the census?
   c. How do you feel about being classified according to your age, gender, or race?
   d. Can you make any assumptions based on the information you found? What are they?
   e. What are the major differences between a geographic community, such as your town or city, and a community based on shared interests or values?
3. Have each group prepare a graph to be shared with the class that illustrates data from their research.
4. After each group presents a graph, discuss each group's findings as a class. Brainstorm possible parameters for a new kind of census that would paint a clearer picture of your community. How would you gather the information? Could the information be used to improve the quality of life for people who live in your town or city? How or why not?

ACTIVITY 2.5
Front Page

Materials

phone books and/or web research

Primary Subject Areas and Skills

social studies, writing, research, critical thinking

Purpose

Students research and understand the role that various community groups play in their towns or cities.

1. Have students brainstorm the names of community organizations in your area. What do the groups do? Who are their members? (It may be helpful to prompt them to look up environmental, religious, or hobbyist organizations in their local phone book.)
2. As a class, discuss the idea of community centers and/or community coalitions. What are they, and what role do they play in communities? (Community centers allow people to come together who share similar interests or goals.)
3. Ask students to compare student groups at their school to community groups in your area. How are they similar? How are they different? What causes people to be involved in these groups? (School groups often include sports or hobbyist groups; and adult community groups meet to participate in the same things, too. It may be helpful to have them create a Venn diagram to graph the similarities.)
4. Have each student choose a member of a community group in your area to interview. Students should contact the group and individual and ask to interview them—over the phone or in person—about the role they play in their community and what it means to them.
5. Assign students to write a profile of that person. First, have students study profiles of individuals who have been written about in your local newspaper. What format do the profiles follow? What aspects of the individuals are shared? (In general, newspaper profiles start by describing the person's present-day activities, then trace those interests back to the past, and proceed to give a chronological report of the person's involvement to this day in the chosen field or activity.) Have students make a list of interview questions before calling their subject.

6. Once the profiles have been completed, they should be shared with the class. If practical, each profile can be copied and distributed to all students. If possible, several minutes could be allotted per student to present her or his subject and share her or his experience.

Extension

Consider publishing the community profiles in a school newsletter. Or you might be able to serialize the profiles and have one run each week in the school newspaper. This is also an ideal activity to incorporate into a classroom blog. Students may also wish to bring their interview subject to class for a career-day-type minipresentation.

ACTIVITY 2.6

At the Podium

Materials

library or Internet research materials;
copy of the "Judges' Evaluation Sheet" found on page 16.

Primary Subject Areas and Skills

social studies, reading comprehension, writing, research, critical thinking

Purpose

Students will debate the following statement about the responsibility of citizens:

> *Increased community involvement among all citizens is a necessary part of responsible citizenship.*

Point

Yes, this is true. A society cannot thrive unless *all* citizens find a way to pitch in and make sacrifices for the common good of all.

Counterpoint

No, this is false. The hallmark of American society is based on individual freedom of expression and action. Individuals should strive to be the best they can be, and as long as they do no harm to anyone else, they should be able to choose their level of involvement.

1. Divide students into two groups of at least four students each. Depending on class size, be sure to leave a group of at least five (preferably an odd number) of students to act as judges. Assign each group to Point or Counterpoint as described above.
2. Instruct students to research and report on their assigned point of view. They should devise statements in support of this position and be prepared for rebuttal from the opposing team.
3. Using the Judges' Evaluation Sheet as a tool, students should be concerned with organization, evidentiary support, and overall presentation, as well as politeness and poise.
4. Debaters should clarify who will speak for the group. (Debaters may take turns, if that is decided beforehand.)
5. The affirmative or Point team will speak first for three to five minutes and present their argument. The Counterpoint team will then have three to five minutes to present their argument.
6. Each team will be given three to five minutes to respond to the other team's argument.
7. Have judges rate the arguments of each group and present their findings to the classroom.
8. Discuss the emotions and ideas that came up during the debate. Did anyone change his or her mind about the topic at hand?

EXTENSIONS

Get Involved

Suggest that students volunteer briefly for the group of their choice. Ask them to keep a journal about how participating in this activity makes them feel. Recall that the words *community* and *common* have the same root. What do students have in common with the people they've met in a chosen community?

In Touch with Community

As a group outing, have the class locate and visit three different "community" centers in their area. What does each of them do? How would students describe the feeling generated at one of these centers?

Put Your Money where Your Community Is

Ask students to find and visit a "community development" bank. Have them interview a bank officer to find out how this institution differs from other banks. What do these banks invest in? How do their investments promote change and improve the local community?

RESOURCES

- **U.S. Census Bureau**
  www.census.gov

- **Scholastic's Kids USA Survey**
  http://teacher.scholastic.com/kidusasu/

Find out what kids have to say about timely topics—and check out charts, graphs, and math activities based on the results of the vote. Examples of topics include: all about families, censorship, favorite sports, school uniforms, top issues facing the United States, violence in the media, and saving the environment.

Reading Handout Unit 2:
REAL PEOPLE, REAL LESSONS

In this reading, you'll learn about the work of a philanthropic organization called Philadelphia Green (PG). PG is devoted to helping people, usually by donating money and professional gardening advice. After each reading, you'll be asked to answer some questions about how PG helped foster a sense of community in each of the neighborhoods you read about.

Philadelphia Green is an organization made up of people who didn't believe they had what it took to round up their neighbors and get them committed to a task. But little by little, these urban warriors found the strength to inspire others.

Doris Gwaltney is one of them.

Ms. Gwaltney grew up in a Philadelphia neighborhood called Carroll Park, named for a four-acre park built, owned, and cared for by Eugene Carroll, a wealthy realtor. When Ms. Gwaltney was a little girl, she played in the park with her brother Donald. She grew up hearing about how her parents used to meet for dates in the park back in the 1930s. It was beautiful place back then, with a massive flowerbed in the center and the edges of the park lined with proud Philadelphia row houses.

The neighborhood was made up mostly of Irish and Italian families, though some African American families like Ms. Gwaltney's began to move in. Over time, the neighborhood changed. Ms. Gwaltney's family stayed. They watched the decades pass in Carroll Park. Mr. Carroll turned the deed to his park over to the city of Philadelphia. Soon after, the park began to look less beautiful. In the 1960s, the city ripped out the flowerbed and installed a spray fountain, which families enjoyed in summers. But they could see that the trees were beginning to look overgrown, the remaining flowers hadn't been tended, and weeds were piling up in the beds. For a while a neighborhood association, comprised of local elderly people, tried to take care of things. But as each member died, there were fewer and fewer residents willing to take care of the place.

Drug dealers began hanging out in the park, and longtime neighbors learned to avoid the place. "It just wasn't a safe place to be," says Ms Gwaltney. "It was just something that was there, and you didn't really think about it anymore."

Occasionally neighbors would complain to the parks department, which would send out a crew to clean up, but as soon as they left, the criminals moved back in. "I tried to pick up trash in there from time to time," Ms. Gwaltney recalls. "But it was just too much. It was wall-to-wall glass from broken bottles in there."

In 1996, Philadelphia Green started a program to rescue nine parks around town. They called a meeting at a local church, which Ms. Gwaltney—who was well on her way to becoming a community leader—attended. There, for the first time, she met neighbors like herself who desperately wanted to see "the criminal element," as they called the drug pushers, gone for good.

"So we went to the church to see what everybody had to say about Carroll Park once again," recalls Ms. Gwaltney. "We went to the church, and there was a young woman making a presentation. Once again, they're going to save Carroll Park. So, we're sitting there skeptically, at least I was. 'Well, how are we going to save Carroll Park this time?' I wondered. And the woman said, 'We have

$30,000 grant for you through Philadelphia Green.' My ears perked up. 'Ah, they have the money,' I thought. This is what made the difference. Carroll Park's time had come."

In the summer of 1997, residents came to the park with brooms, rakes, and paint for a day of work. They painted the benches, the spray pool, and the storage-shed building. At the end of that day, Carroll Park was on the path to transformation.

"It was a hard day's work," recalls Ms. Gwaltney, "but the community knew: That's a new group taking back Carroll Park."

Money in hand, the neighborhood came together. Those days in the park, says Ms. Gwaltney, you could almost feel the sense of satisfaction that something was *finally* being done. Philadelphia Green provided the money and lent the expertise; Carroll Park residents provided the human power.

It's never easy reclaiming a park. The key, experts agree, is for law-abiding citizens to make their presence known, day after day. Joan Reilly, associate director at Philadelphia Green, calls it an uphill battle. "These folks like Ms. Gwaltney were like warriors. If Ms. Gwaltney and her group of women were to stop coming out here, in a matter of time, the park would revert back. When parks get neglected, they don't simply become eyesores. Neglected spaces become havens for negative activity. And that preys on a community."

Ms. Gwaltney and the others visited the park regularly, even when it wasn't comfortable to do so. "I spent the winter of 1999 in Carroll Park," she boasts. "I came out here every Monday at four o'clock to pick up trash. Even in the ice and snow I was out here. I was out here picking up trash because I wanted the neighbors to look out their windows and see that that crazy woman is still out here picking up trash every Monday at four o'clock. I wanted the neighbors to know that we were here, and we were here to stay."

That was seven years ago, and Carroll Park is now the hub of a thriving community. "We have a little group that gets together every Monday at 4 P.M. to pick up trash," says Ms. Gwaltney. "And then once a month we get together to do the bigger chores."

On October 10, 1998, residents built an entire playground in a single day. At 6:30 A.M., a huge flatbed truck showed up with a load of steel parts. "They looked like big Legos," says Ms. Gwaltney, giggling. A few neighbors who were ironworkers started welding, and by nightfall, says Ms. Gwaltney, "They had slapped that puppy together."

With time, local citizens cleaned up the spray pools, planted flowerbeds, and began the arduous task of tending or entirely replacing the old trees first planted in Mr. Carroll's heyday. Landscape architects from Philadelphia Green helped the community pick out the right species—maples, oaks, hawthorns, lindens—that were right for the region. One day, the news spread around the neighborhood that the local Citizens Bank would donate the last 22 trees the park needed to round out its planting. The director of the bank himself was on hand to plant those trees.

"A partnership is a three-legged stool," says Ms. Gwaltney. "If one of the legs on that stool is broken, that stool is useless. So everyone must do his and her part to keep the partnership alive and vital."

These days, even local children are invited to take part in that partnership. After school on Wednesday, the park sponsors a Park Patrol class for kids, who pick up trash and participate in environmental education classes. The park is home to Earth Day festivities every April and a concert series in the summer and is the site of meetings of the local garden club.

Carroll Park citizens are proud to report that their homes are worth more on the real estate market since the park's revitalization. "One woman was moving out to New Jersey," says Ms. Gwaltney. "She put a sign out in front of her house on a Monday and two days later it was gone. I thought she had changed her mind. 'Oh, she's staying after all.' But no. That's how fast she sold her house. Another house around the park was boarded up for years. But someone came along, bought it, rehabbed it, and there's a young family living there now."

Some of the changes in Carroll Park were hard to describe. Ms. Gwaltney wasn't a gardener before this experience, but now that she has taken courses given by Philadelphia Green, she is qualified to prune and care for any of the trees in the park. Today, at the age of 66, Ms. Gwaltney can say she has spent her whole life watching the seasons come and go in Carroll Park. In fact, she still lives in the same house with her mother, who is 91 and remembers well those days when she met her girlfriends or had dates around the big flowerbed.

"The young people are still dating in the park," Ms. Gwaltney is happy to report. "Oh yes, indeed! Just like old times."

Questions

1. Why do you think Ms. Gwaltney was so skeptical about the city of Philadelphia finally doing something about Carroll Park? What changed her attitude?

____

____

2. Ms. Reilly says that if people don't use the park or take care of it, criminals will take over. Why do you think that might be so?

____

____

3. How does a simple act such as picking up trash tell people that a park is off limits to criminals? What message does a neat, orderly park send out to people?

____

____

4. Why do you think there is a greater demand for homes in Carroll Park today?

5. What do you think Ms. Gwaltney and her neighbors learned about community? Were they a community before?

6. What does Philadelphia Green get out of helping people and giving them money to fix up their neighborhoods? What do philanthropic organizations do, and why do they do it?

7. Before Philadelphia Green donates money to a park, they insist that 85 percent of the people living in the neighborhood must promise to participate in the project. Why do you think they have this rule?

8. How does having a place to gather together build community?

REPRODUCIBLE 2.2: Venn Diagram Sample

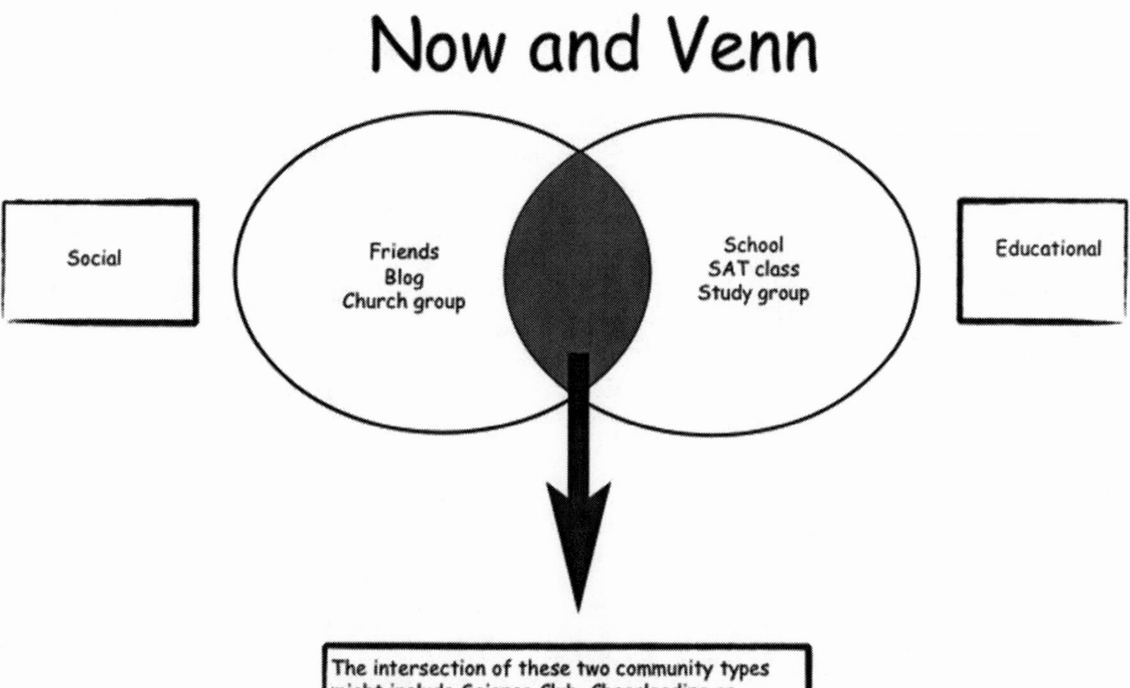

HANDOUT 2.4: U.S. Census (2004)

General Characteristics	Estimate	Percent
Total population	*285,691,501*	
Male	139,782,818	48.9
Female	145,908,683	51.1
Median age (years)	36.2	(X)
Under 5 years	20,008,152	7.0
18 years and over	212,767,197	74.5
65 years and over	34,205,301	12.0
One race	280,285,784	98.1
White	216,036,244	75.6
Black or African American	34,772,381	12.2
American Indian and Alaska Native	2,151,322	0.8
Asian	12,097,281	4.2
Native Hawaiian and other Pacific Islander	403,832	0.1
Some other race	14,824,724	5.2
Two or more races	5,405,717	1.9
Hispanic or Latino (of any race)	40,459,196	14.2
Household population	285,691,501	100.0
Average household size	2.60	(X)
Average family size	3.18	(X)
Total housing units	122,671,734	
Occupied housing units	109,902,090	89.6
Owner-occupied housing units	73,754,173	67.1
Renter-occupied housing units	36,147,917	32.9
Vacant housing units	12,769,644	10.4

Social Characteristics	Estimate	Percent
Population 25 years and over	*186,534,177*	
High school graduate or higher	(X)	83.9
Bachelor's degree or higher	(X)	27.0
Civilian veterans (civilian population 18 years and over)	23,756,268	11.2
Disability status (population 5 years and over)	37,858,580	14.3
Foreign born	34,279,756	12.0
Male, Now married, except separated (population 15 years and over)	61,315,221	56.4
Female, Now married, except separated (population 15 years and over)	59,727,165	51.4
Speak a language other than English at home (population 5 years and over)	49,632,925	18.7

Economic Characteristics	Estimate	Percent
In labor force (population 16 years and over)	*145,437,824*	*65.9*
Mean travel time to work in minutes (workers 16 years and over)	24.7	(X)
Median household income (in 2004 inflation-adjusted dollars)	44,684	(X)
Median family income (in 2004 inflation-adjusted dollars)	53,692	(X)
Per capita income (in 2004 inflation-adjusted dollars)	24,020	(X)
Families below poverty level	(X)	10.1
Individuals below poverty level	(X)	13.1

Housing Characteristics	Estimate	Percent
Owner-occupied homes	*73,754,173*	
Median value (dollars)	151,366	(X)
Median of selected monthly owner costs	(X)	(X)
With a mortgage (dollars)	1,212	(X)
Not mortgaged (dollars)	345	(X)

(X) Not applicable.

Source: U.S. Census Bureau, 2004 American Community Survey

Questions

1. Name three different communities—ethnic, financial, or other—that are represented on the census chart.

   ____

   ____

2. Name at least three communities that are not represented on this chart. Do you think they should be? Why or why not?

   ____

   ____

3. How accurate do you think this information is? What might affect the accuracy of a census?

   ____

   ____

4. How might the census information be used? How do you think it shoul be used?

   ____

   ____

UNIT 3

WASTE MANAGEMENT and RECYCLING

TOPIC BACKGROUND

Every city grapples with garbage and recycling issues. The challenge will never go away. As long as there are humans, we will produce trash that must be disposed of in some way—and that costs money. As fuel costs rise and landfills are becoming stricter about what they can and cannot accept, towns and cities are taking stronger measures to persuade citizens to change the way they handle their refuse. Thanks to education programs and strong laws, Americans have grown accustomed to separating glass, aluminum, plastic, paper, and other items because they know these things can be remanufactured and reused.

But recyclables are just the beginning. For decades, Americans didn't think twice about tossing out "green waste"—grass clippings, brush, tree trimmings, and the like—and hauling it to a landfill. But now cities are convincing homeowners to chop up this material and use it to mulch their flowerbeds. (Mulch, a thick layer of organic woody or grass material, protects plants by absorbing rainfall and shielding delicate roots from the harsh sun.) Home center stores now sell "mulching" mowers, which chop up grass so finely that you can simply scatter the mulch on your lawn. This saves having to bag the cuttings and take them to the dump. Many city halls are also distributing compost bins for free, in the hopes that residents will happily discard their organic kitchen waste—vegetable and fruit peels, coffee grounds, and the like—in a bin and use the resulting compost in their gardens.

These are all fine first steps, but there is much work to be done. New challenges arise every day. For example, at this writing Americans own 200 million computers, 150 million mobile phones, and 200 million televisions. Each year, millions of these items are tossed out. Off they go to the landfills, along with our printers, microwaves, PDAs, fax machines, and other "stuff." Many of the materials (lead, cadmium, mercury) found in these devices are toxic and should be recycled. But little legislation is in place to make this happen.

In the *Edens Lost & Found* series, you'll see how Americans in four different cities are trying to deal with waste. In Chicago's early days, for example, businesses used to dump garbage and sewage into the river. The water whisked it out of sight and mind. At that time, the river flowed north and emptied into Lake Michigan, from which Chicago drew its drinking water. As the city grew, officials realized that this practice was unsanitary. In 1900, a massive engineering project reversed the river's flow, now sending water down the Mississippi River. This was fine for Chicago but bad for its neighbors downriver who complained about the garbage flowing past them on the way to the Gulf of Mexico. Chicago had to clean up its river for good, and it did.

In Los Angeles, where vegetation grows abundantly even in winter, people are grappling with the rising costs of hauling green waste to dumps. At the same time, city leaders know that the region's salvation lies in planting still more trees to shade the city and capture precious rainwater. Is there some way to reach a happy medium? Mulching green waste for gardens instead of hauling it to the dump would alone save L.A. County $30 million annually! These are some of the challenges you can see presented in the films.

Purpose

In this unit, students will learn and understand how American cities are dealing with their waste. The topics of recycling, landfills, composting of yard waste, and toxic waste will all be discussed.

GROUP ICEBREAKERS

1. Ask students how much trash or garbage they think they and their families throw out every day. (Keep it simple. They may say one or two trash bags full, more if there is a weekly collection for yard waste or recyclables.)
2. How much trash do they think the school

disposes of in a day? In a week? (It's fine if there is no definitive answer or consensus answer here. In the course of the unit, they will research better numbers from the school.)

3. Without doing any research, what do students think happens to all the stuff they throw out? Where does it go? (They may guess that it is hauled to a landfill in their town or in another state or even incinerated. The point is, they probably don't know for sure.)
4. In answer to the previous question, they might say that all their city or town's trash goes to a *dump* or *landfill*. If so, challenge them to define these terms. For example: Is a dump different from a landfill? If so, how? (Most people use these terms interchangeably. But a *landfill* is scientifically managed in such a way to minimize future harm to the community from that waste. *Dumps* tend not to be so rigorously managed.)
5. Do they know, off the top of their heads, what procedures are followed at their closest dump or landfill? For instance, is the trash simply left there forever? Is it buried? Is it allowed to accumulate for a short period of time and then trucked to another site? (All of the choices mentioned above are possible, but students won't know for sure unless they speak to someone at their local public works department. Some cities merely collect recyclables, then have them removed by a private contractor. Some cities do not manage their own landfill but hire a private company to do it for them.)
6. Ask how many people in the class have ever recycled an item such as a glass or plastic bottle. Next, ask for a show of hands of how many times people have intentionally purchased something made from recycled materials, such as 100 percent recycled paper. What was the product? How did they know the product was recycled? (You will find that most students have never thought to purchase a product made from recycled items. Manufacturers of such items usually label the product clearly, but not always. Many fleece garments are made from recycled plastic products.)

ACTIVITY 3.1

Don't "Waste" This Quiz

Materials

pen and paper; Handout 3.1

Primary Subject Areas and Skills

science, social studies, critical thinking

Purpose

Students take a short quiz to assess how much they know about recycling.

1. Distribute Handout 3.1 to students to begin a discussion about the uses of recycled products. Alternately, you may wish to copy this onto an overhead projector for a class activity. The quiz may also be used as a take-home assignment in preparation for a class discussion.
2. Before beginning, ask students how much they think they know about recycling. Where did they learn it? School? TV? Newspaper? At home?
3. After students take the quiz, call out the answers listed in the answer key and have them self-correct their papers. Then have them discuss what they learned. Were there any surprises?

ACTIVITY 3.2

"Waste-ing" Away

Materials

pen and paper

Primary Subject Areas and Skills

social studies, science, writing, critical thinking

Purpose

Students take a much closer look at items they see every day and assess the likelihood of their being reusable or recyclable. Much of what we consider to be trash may, in fact, have uses beyond those that we normally consider.

1. Together with your class, brainstorm as many different "reuse" possibilities as you can for the following items. Encourage students to be as creative as possible. Write the possible uses on the board or overhead.
   a. Plastic soda bottle
   b. Old CDs

c. Aluminum cans
d. Old newspaper

2. The amount of waste that human beings generate can be an astounding thing. Instruct students to keep a "trash journal." In this journal, challenge them to list every single item that they throw away, such as the wrapper from a piece of candy, the peel of a banana, an empty can of soda, or a piece of paper.

At the end of a distinct period of time—a week, month, or semester—have students write a report and, perhaps for extra credit, devise a way to graph the amounts of various types of waste they produced.

Write the following table on the board and use it to help students get started. Students can track not only what they throw away but whether or not they could have recycled the item. If they could have recycled but didn't, why not? You may also wish to post an ongoing table, such as the sample below, on the wall of your classroom.

Here are some examples:

Item	Type of Waste	Is it recyclable?	Did you recycle?	Why?
Apple core	green compost	yes	no	no compost bin at school
Soda can	aluminim	yes	yes	

ACTIVITY 3.3
Talking Trash

Materials
pen and paper; Reproducibles 3.3A, B, and C

Primary Subject Areas and Skills
science, social studies, reading comprehension, critical thinking

Purpose
Through a variety of overhead presentations or handouts, students will learn how a landfill works, what types of trash are suitable for disposal in one, and what humans routinely do with their trash.

The ins and outs of waste are fairly intricate, and the diagrams and charts that follow are designed to serve as an introduction for you and your students.

A Crash Course for Teachers on Trash

Dumps are often confused with landfills, but the two are quite different. The first dumps existed thousands of years ago in the time of the ancient Greeks, when certain areas of town were designated to accept waste from the cities. Dumps are often open spaces where trash is deposited and open to the elements, resulting in the easy contamination of land, air, and water. Illegal dumpsites remain a serious issue in the United States.

Early landfills were somewhat better, as waste was covered over to help prevent the spread of odors. It wasn't until the early 1990s that landfill construction and operation were regulated. Today's modern landfills are designed to be safer and more sanitary than dumps, but they still present a danger, and precautions need to be taken.

As you will see in Reproducible 3.3A, a landfill contains special linings and mechanisms to collect dangerous liquids and gases. It is covered over every day with soil, and it is designed to prevent as much leakage and contamination as possible.

The fastest-growing problem today is the disposal of electronic waste, such as old computers, cell phones, printers, fax machines, and peripherals. While much of technological trash can be recycled for parts or the metals and minerals they contain, many of them still end up in landfills, where their plastic parts and dangerous chemicals—lead in a computer screen and mercury in the circuit board, in addition to many others—can pollute the soil and water for generations.

Many counties have special hazardous waste collection sites where these items can be brought, but it is often up to homeowners and businesses to seek them out. When hazardous materials—such as the alkaline, carbon zinc, lithium, and/or mercuric-oxide found in household batteries—are discarded, it creates two of the main contaminants produced by dumps and landfills. These contaminates are *leachate* and *landfill gases*. Both of these can damage air, soil, and water in and around a dump or landfill.

Leachate is a liquid. It occurs when water in the form of rain, snow, or other moisture seeps into a dump or landfill and becomes contaminated. The resulting leachate becomes very concentrated, much more so than even raw sewage. It can pollute the surrounding soil and groundwater.

Landfill gases (LFG) vary, with the most commonly occurring gas being methane (CH_4). Gases produced by decomposing waste are the result of either *aerobic* (with oxygen) or *anaerobic* (without oxygen) processes. First, there is a lot of carbon dioxide (CO_2)

created, as *aerobic* decomposition occurs. But when there is no more oxygen available—especially further down in the refuse pile—the *anaerobic* decomposition begins, which releases methane. Although it may take one to two years before methane begins to be emitted, it may be continuously released for over 50 years! Other LFGs emitted from landfills, such as benzene, are known carcinogens (cancer-causing agents).

1. Show Reproducibles 3.3A, B, and C on an overhead projector.
2. Have your students study the images and charts alone or in small groups.
3. Discuss the images and charts with students and pose the following questions.

Discussion

1. What might affect the rate at which something breaks down? (In general, objects made of organic materials—such as paper or food—are more likely to break down and return to the earth.)
2. In many cases, scientists investigating the contents of landfills find that biodegradable substances remain basically intact many years more than expected. What role do sunlight, moisture, and oxygen play in this process? (Items are more likely to break down when they are exposed to the elements. But there is little sunlight and oxygen in a landfill because items are buried. Consequently, many objects never "rot" in a landfill.)
3. What kinds of hazardous wastes probably end up in your household trash? How do you dispose of things like old AA batteries? (Students may wish to consult their trash journal. Household chemicals such as cleansers, bleach, pesticides, herbicides, and paint products are very common; electronic trash is increasingly common. Hazardous trash should be disposed of only at your city's special drop-off centers or on scheduled drop-off days. Many electronics manufacturers now take back old equipment as well.)
4. What are five things you can do today to help reduce the amount of waste you generate and the amount of harmful pollutants you may be releasing into the environment? (Organic material like vegetable scraps and yard waste can be composted. People can pledge to purchase only toxin-free products that can be safely tossed out in the trash. And they can reduce the amount of all products they use in their home.)

ACTIVITY 3.4
Trash Audit

Materials

recycling bins
gloves for handling recyclables
a scale or balance
markers and poster board
pen and paper

Primary Subject Areas and Skills

science, math, writing, research, critical thinking

Purpose

Students conduct an audit of the recyclable materials disposed of by their class in a specific period of time and keep track of other trash they discard. They use the data to brainstorm ways of changing their future behavior.

One of the most powerful ways to teach students about waste production is to have them measure it for themselves. The trash audit is a popular tool used by many educators, including Deb Perryman, the Illinois teacher featured in *Edens Lost & Found*. It involves gathering and classifying the waste accumulated in the classroom and then assessing ways that that waste can be reduced, reused, or recycled.

Trash will be collected for one to two weeks—longer if you choose—and then classified and weighed. Students will determine how much waste they are producing and, perhaps more importantly, the *kind* of waste they are producing.

1. This project can be done with one specific class or with all of your classes for a larger group project.
2. If you decide to do this project with more than one class—or if, for example, the students in your class have different lunch periods—keep a designated "trash audit can" in your classroom or another chosen spot.
3. Set up recycling bins in your classroom for paper, plastic, metal cans, and any other objects you wish to track, such as electronic equipment or batteries. You will not collect food or personal trash such as tissues in this activity. That trash should continue to be discarded

properly. Hang a sheet of paper over the regular wastebin and have students write what they have discarded each time they toss something out. For privacy concerns, accept general descriptions, such as "tissue," "sheet of paper," and the like.

4. If you haven't already, ask students to keep a "trash journal," listing all the different types of trash they toss out in a certain period, such as a week. Journals need not be fancy; a small notebook that can fit in a pocket will do nicely. Each day after lunch, before students toss out their lunch trash, ask them to jot down whatever they are discarding. If lunch is eaten in your classroom, have them discard food items and wrappings in the proper trash but rinse out and collect all recyclables.
5. After one week's time, have your students analyze their data. Have them create a chart that divides what they've discarded into three possible categories:
    - To be recycled
    - To the landfill or incinerator
    - Could be composted
6. Have them weigh each of the bins containing recyclables and record the weight. This will allow them to associate certain weights with the items being recycled.
7. Have groups of students devise a graphical representation of the amount of trash generated by the class. Some groups may choose a pie chart, others a bar graph, others a table. The types and amounts of each type of trash collected should be represented.
8. Once students have done this, ask them to evaluate their data.
    - Are there materials that *could* be composted or recycled that are not? If not, why not? (Most likely, items that could be composted are not because compost bins are not available at school or would present sanitation problems. Recyclables are often discarded with regular trash because a recycling bin is not immediately handy.)
    - Are there things that students could do to reduce the amount of trash? (For example, they could bring real cutlery to school instead of using and discarding plastic forks and spoons or carry a lunch box or cloth bag to school instead of using a different paper bag every day. They could choose not to drink so many soft drinks, which are often unhealthy anyway.)
    - Ask students what they thought about the amount of food packaging that they used and discarded. Can they think of ways it could be reduced? (They could buy groceries in bulk, for example, using the same container over and over.)
9. Challenge students to create a list of alternate behaviors and practices that could be employed by them, their schoolmates, and their school administration. Perhaps one of the students could write an article about the experience for the school paper. For extra credit, they could prepare a report for the PTA or school board recommending specific environmentally friendly practices to be put into place in schools throughout your district.

Activity Variation

If you wish to expand this activity, you could dispense with the recyclable collection and log-in sheet above the trash bin, and simply and have students collect trash from a number of classrooms near the end of *one day,* then sort and analyze it. (You may need permission from school officials or parents to allow students to participate.) Ms. Perryman reports that she has done this with her students for several years without incident. She does, however, follow five rules:

1. Collect no trash from any bathrooms/nurse's offices (for health reasons).
2. Collect no trash from the main office or other school officials (for confidentiality reasons).
3. Collect no trash from the cafeteria (to avoid food waste).
4. Have students immediately discard tissues and paper towels that they may find as they sort.
5. Have all students wear gloves, as mentioned above.

"Beyond that," she writes, "it really is not a big deal. Most of the trash you are going to get is 'fresh' and recyclable, inert stuff."

Extension

You may wish to use the information you gather to devise a program at your school or even in your district that would provide for the proper recycling bins and even composting bins on the premises.

ACTIVITY 3.5

Trash Tops the Charts

Materials

pen and paper; Handout 3.5

Primary Subject Areas and Skills

math, social studies, science

Purpose

Students study bar and pie graphs to learn how much waste is produced by Americans and what percentage of it is being recycled.

How much waste do we produce, and what percentage of it is being recycled? The charts that follow will give you and your students a clear picture of the waste production in the United States. This information will also be fruit for discussions about waste, recycling, and related social issues.

1. Distribute Handout 3.5.
2. Instruct students to answer the questions alone or in pairs. You may also wish to assign the handout for homework.
3. Discuss their answers in class. You may wish to ask students about the following:
   a. Were they surprised by the information presented in the trash chart? If so, what were they expecting? (People are often amazed that paper makes up the largest percentage of all trash. It is surprising because we have all heard that we are moving toward a paperless society. It certainly isn't true.)
   b. Whose responsibility is waste management? Should there be more governmental control? More laws? Higher penalties? (Waste management is the responsibility of all citizens, but most of us hand over that responsibility to our local cities and towns. Regardless of laws and penalties, all Americans need to work harder to recycle many of the items they use in everyday life.)
   c. Did students assume Americans were better at recycling? Why? (Recycling has been such a fact of life for many decades that it is hard to imagine someone tossing out plastic and glass containers, for example. Yet these very items are the ones with the lowest recycling rates.)

ACTIVITY 3.6

Landfill Dispute

Materials

library or Internet research materials
the Judge's Evaluation Sheet found on page 14
pen and paper
additional presentation materials of the students' choice
Reproducible 3.6

Primary Subject Areas and Skills

science, social studies, writing, research, critical thinking, public speaking

Purpose

Students research how and where a fictional landfill should be sited to learn about the various issues surrounding placement of a real municipal waste landfill. Students must evaluate the objections that arise when a landfill needs a home and none of the choices available for location are ideal. How does one decide? How does a community make the best possible choice when there is no perfect solution?

Dealing with trash at the municipal level is a complicated and difficult task, one that affects more communities, families, and businesses than you might imagine. This project provides the opportunity to introduce the concept of "NIMBY," or "not in my backyard." This is quite a prevalent sentiment throughout the nation. Many people pay little attention to environmental issues until a specific situation brings issues directly into their community, or backyard, and moves them to action.

In this activity, students must evaluate the objections that arise when a landfill needs a home and none of the choices available for location are ideal. How does one decide? How does a community make the best possible choice when there is no perfect solution?

1. Copy Reproducible 3.6 to use as an overhead.
2. Tell students that a new landfill is needed for this community. The maps they are looking at show the only two possible spots within town (A and B) for that landfill. Option C involves paying to ship the garbage out of state to another landfill.
3. Instruct students to study carefully and consider the three options. You may wish to discuss this briefly as a class. What are the students' initial reactions?
4. Divide the class into three teams. Assign each of the teams one of the landfill choices. Tell them

they need to research and present an argument for their location in preparation for a debate. Why is their site the best and least damaging location? What proof and support of their position do they have? Can they quantify their proof or give support that the environmental damage done to the community will be minimized by their option?
5. Give each group a limited amount of time, during which one elected individual will present their argument to the class.
6. In this debate, there does not have to be a "winner." In fact, the lesson here is that there are no easy answers. The objective is that students think about and investigate all the possible ramifications of the placement of a landfill and express those ideas fully and clearly. You may wish to invite members of the community to come in and evaluate the arguments presented by each group. This is a good idea if you feel students will have a hard time being impartial. It may help to use the Judges' Evaluation Sheet found on page 14 to help sort out the different issues that judges need to assess.

ACTIVITY 3.7
At the Podium

Materials

library or Internet research materials; Judges' Evaluation Sheet, page 14.

Primary Subject Areas and Skills

science, social studies, reading comprehension, writing, research, critical thinking

Purpose

In this activity, students will debate this statement about their responsibilities as an individual toward reducing waste.

The weight, volume, and content of municipal waste an individual is permitted to throw away needs to be strictly limited. Only a limited amount of nonrecyclable and noncompostable items should be accepted by trash collectors.

Those individuals who exceed their weekly trash limits should be fined.

Point

Yes, I agree with a fine. Encouraging people to reduce, reuse, and recycle is not working fast enough. People will only act responsibly when forced to do so. It's human nature.

Counterpoint

No, I disagree with a fine. It isn't my fault that I discard as much waste as I do. I don't package the products that I buy. If the city wants to fine someone, they should go after the big guys—the product packagers and major companies—that make this stuff in the first place, not the consumers who buy it from them.

1. Divide students into two groups of at least four students each. Depending on class size, be sure to leave a group of at least five (preferably an odd number) of students to act as judges. Assign each group to Point or Counterpoint as described above.
2. Instruct students to research and report on their point of view. They should devise statements in support of their position and be prepared for rebuttal from the opposing team.
3. Judges should be concerned with organization, evidentiary support, and overall presentation, as well as politeness and poise.
4. Debaters should clarify who will speak for the group. (Debaters may take turns, if that is decided beforehand).
5. The affirmative or Point team will speak first for three to five minutes to present their argument. The Counterpoint team will then have three to five minutes to present their argument.
6. Each team will be given three to five minutes to respond to the other team's argument.
7. Have judges rate the arguments of each group and present their findings to the classroom.
8. Discuss among yourselves the emotions and ideas that came up during the debate. Did anyone change her or his mind about the topic at hand?

EXTENSIONS

Get into Techie Trash

Challenge students to institute and operate their own electronics recycling program as a fund-raising program for their class. Ideally, local citizens and businesses would drop off their used electronics at your school, along with a check for a specified fee. The fee would cover your class's costs for packing and mailing items to a recycling center or for hand-delivering items to your county's local waste recycling center. Obviously, to make this work, students will need to seriously research all their options, the time commitment, and costs involved. If they do not have a mode of transportation to a local post office or waste recycling center, they'll need to be more creative. For example, the post office will pick up properly addressed and postage-paid parcels at your school office.

Trash Tour

Track your trash: Have student research the location of the nearest landfill and schedule a class visit. (Many landfills are set up to provide tours of their facilities.) After their visit, challenge students to use what they've learned to create their own diagram, sketch, collage, or other detailed visual representation of a landfill site.

Landfill Study

Have students research the locations and operations of all the landfills in their state. Because all are under different jurisdictions, there are likely to be different operating procedures. Have students compile and report on the similarities and differences of each. They'll need to use their skills as reporters and researchers.

School Composting

Encourage students to begin a schoolwide composting program to turn their school's leftover vegetable and fruit matter into nutrient-rich garden soil. They should check with their local city hall to see if any compost bins are available free through their town or city's solid waste programs. If not, they can consider building their own using cast-off lumber, chicken-wire fencing, or shipping pallets. (Consider asking for donations; pallets will be especially abundant at local businesses.) Once the composting bins are built on school property, students will need to take turns turning and monitoring them. They'll need to research ways to keep the pile "cooking" during winter months. And they'll need to seek out good uses for their "black gold." Some schools have sold the compost to gardeners in spring sales; others have used theirs to enrich school-run gardens.

RESOURCES

- The EPA's Official Recycling Site for Young People
  www.epa.gov/recyclecity
- The EPA's Official Waste Site for Young People (wonderful teaching materials)
  www.epa.gov/epaoswer/education/teens.htm

Composting
- www.mastercomposter.com

Composting with worms
- www.niehs.nih.gov/kids/worms.htm
- www.recyclemore.org/article.asp?key=49

Many computer and printer manufacturers sponsor recycling programs, which you can learn more about online. Some require a fee. See your manufacturer's website for details.

- Recycle your cell phone by donating it to a charity
  www.collectivegood.com
- National Recycling Coalition Electronics Recycling Initiative website
  www.nrc-recycle.org/resources/electronics
- Find a rechargeable battery recycling site near you (free teaching materials)
  www.rbrc.org/call2recycle/
- To locate your nearest recycling or Household Hazardous Waste Collection Center
  www.earth911.org
  www.collectivegood.com/other.asp
- Goodwill Industries
  www.goodwill.org/page/guest/about/howwe operate/recycling
- National Cristina Foundation
  www.cristina.org/
- eBay ReThink Program
  http://rethink.ebay.com

Check with your favorite retailer and see if they sponsor recycling efforts. If they don't, encourage them to participate.

Reading Handout Unit 3:
YOUNG PEOPLE IN URBAN NATURE

In this reading, you'll learn how the problem of illegally dumping was threatening a community. You'll meet the man who changed all that. Michael Howard helped his neighborhood fight crime, educate its young people, clean up toxic waste, and build an amazing new park. After the reading, you'll be asked to answer some questions about Mr. Howard's work.

On a June night some years ago, an explosive nicknamed a Molotov cocktail crashed through the window of Michael Howard's bedroom at 2 A.M., as he and his family lay sleeping. Seeing the flames, Mr. Howard jumped up to protect his wife Amelia and five children and ran for a fire extinguisher and a phone to call the police. Even as he was fighting the flames, Mr. Howard was shocked, but not surprised, that drug dealers in this Chicago neighborhood of Fuller Park had made him a target. Just the day before he'd found a poorly made bomb in his mailbox.

Mr. Howard, a big man with a big heart, has spent his life trying to help others help themselves. A successful building contractor, Mr. Howard persuaded his family to leave their affluent Beverly neighborhood in 1992 to return to their old neighborhood, Fuller Park, a poor African American neighborhood on Chicago's south side. They moved back, in fact, to the house where Mrs. Howard, a teacher, had grown up. Almost immediately, Mr. Howard was troubled by what had become of their once-proud neighborhood. Daily he griped about kids selling drugs on the corner. "I kept telling my wife, 'Somebody's got to do something about this!' And she challenged me a little bit. She said, 'So? What are *you* going to do about it?' "

That line, coming from a loving wife who knew her husband was incapable of doing nothing, spurred a man to action and transformed a neighborhood.

Mr. Howard was never one to stand still. He joined the Army as a young man and enjoyed those three years of discipline. Fresh from the Army, he thought he'd take some time off. He was bored before the week was out. With the help of a friend, he landed a job as a financial analyst, where he met his wife. Mr. Howard always enjoyed working with his hands, and so he eventually quit his office work to train as a carpenter.

One project led to another, and soon he was buying and fixing up buildings all over Chicago. It was good work—highly profitable work—but Mr. Howard wanted to help his community. In 1986, he closed his business and joined a local ministry program, dedicating himself to using his talents as a speaker, preacher, singer, and contractor to locate and create affordable housing for members of his church. Now, having moved back to his old neighborhood, he was being handed another huge challenge by his wife. Could he really stop drug dealing in Fuller Park?

He thought back to his days as a contractor. Not a day went by when someone didn't ask him for a job. Unfortunately, the person asking usually didn't have employable skills. All over Chicago, general contractors needed plumbers, electricians, carpenters, and the like. Mr. Howard called some of his friends in these trades and asked if they would consider teaching students, if Mr. Howard could find a place to train them. The result was the South Point Academy.

"It was a battle from the beginning," Mr.

Howard recalls. "Every chance I had, I was going out to the corner where people sold drugs to recruit students. I'd say, 'What are you doing this for? Why don't you come learn a trade and then you can get a job?' See, there are no other alternatives for these kids. Eighty percent of them can't read. Ninety-five percent of them were high-school dropouts."

Thanks to Mr. Howard's persistence, more than 300 students enrolled at his school and graduated with employable skills. "I have a lot of star graduates," he says. "We have some who have gone on to start their own businesses. We have some who have actually rehabbed the house right around the corner. One young lady rehabbed a house for a Chicago police officer who moved back into the community. That was really a good story to hear. Another young man has been working since he graduated, and he just bought a house a block away from me. We've got a lot of stars who are really doing well."

The firebombing happened early in Mr. Howard's tenure in Fuller Park. He could have given up and pulled up stakes. But he stuck it out. He worked closely with the police, who managed to conduct more and more effective raids on drug houses. Whenever a house was vacated, Mr. Howard and his students moved in to help rehabilitate the structure. His knowledge of finance and property law came in handy as he helped neighborhood residents buy some of these homes. The neighborhood steadily improved, and drug dealing appears to have disappeared.

Things were starting to come together, but many challenges remained. Mr. Howard read a report that said that Fuller Park—the city's smallest neighborhood, with little power in local government because of its small number of voters—had the highest lead levels in the city. A disproportionate number of residents were getting sick from various cancers.

Mr. Howard wondered: "My house is rehabbed. So where is the lead coming from? So we went out and had the water tested." They found that many of the water lines in the neighborhood dated to the days of the great Chicago fire in 1871! The pipes were either wood or lead. It was expensive to change those pipes, and without waiting for the city to get around to doing it, people in the neighborhood held a fundraiser and used the money to buy filtered water pitchers and faucet filters to strain lead from drinking water.

Next, they tested a nearby vacant lot, which neighbors suspected had been seriously contaminated by illegal dumping. Hauling construction waste to a landfill is expensive, so many builders and homeowners who are renovating their homes take a shortcut to disposing of their building refuse. They toss it in trucks or car trunks, wait until dark, and then dump the material in an abandoned lot. When this happens, lead paint, asbestos from old insulated pipes or window caulk, and other dangerous substances end up in neighborhoods where people are living, rather than encased in a professionally managed landfill. If the material is not buried, chips of paint and flecks of damaging materials blow around in the wind and can be breathed by people who pass by. Fuller Park residents tested the soil in this particular lot and found it contaminated with lead and asbestos. The Environmental Protection Agency confirmed it, but neither the city of Chicago nor the federal government did anything more. Cleaning up toxic waste is always expensive, and a small, powerless neighborhood can easily be ignored by politicians and city officials.

Mr. Howard organized the neighbors and phoned contractor friends, who donated earth movers, dump trucks, bulldozers, and backhoes. Together they picked up all the trash on the block, including 23 tons of concrete that had been dumped from old construction sites. They sent this refuse to a certified waste disposal site. Then, using about $3,000 earned in the fundraiser, Fuller Park residents bought several tons of topsoil and used it to cap off the existing soil. Slowly, they began to transform the block into Eden Place, a wildlife preserve and nature education center, complete with its own prairie, wetlands, nature pond, savanna, Indian village, farm yard with farm animals, and extensive vegetable garden.

At every step, Mr. Howard forced himself to learn more and more about nature to be able to teach neighborhood kids. He became a certified master gardener and learned to train birds of prey. He and his wife also became certified tree keepers. "A lot of us in the African American community think of the green movement as a white people's movement. We need to teach people that we *all* live in nature. It's part of all our lives," he says.

"When I grew up," he says, "I was an avid camper and avid fisherman, and I gained a true love for nature. And I used to take people on trips to the boundary waters where the United States meets Canada, to expose them to what I had found. When I can share nature, I can share all of the science, the beauty, the art, even the reading that you can find in nature with children, I really see a light go on in people's eyes and there's a connection. And for our community I think it's a great healer."

Today Eden Place is a magical site for students to learn about nature in their neighborhood, the state of Illinois, and the world beyond. Fuller Park has become a thriving community again. "In 2005, for the first time in 40 years, we finally got a new home built," says Mr. Howard. "We've got our first major development: a 100-unit senior retirement home, also the first in 40 years. We've got new businesses coming into the community. We finally got three gas stations. Now, that may not seem to be a big deal to you, but can you imagine living next to an expressway and not having a gas station in your neighborhood? For 30 years, we had no gas station, and now we've got three! We've got a little economy coming in. We hope to put ten families in business. Developers are moving in, and we want to be able to buy these new homes that are being developed."

All this hard work is tiring. "There are a lot of times when really I feel overwhelmed," he admits. "I get tired all the time. I'm tired right now, but I can't stop. I can't give up, you know, there's just too much to do. And I really think it's important to teach your children how not to give up just because you're tired. If you just stop for every road block that's going to come in your life, you won't ever succeed, and you won't make any progress."

Mrs. Howard thinks her husband will keep striving for the betterment of his neighborhood, because he knows how high the stakes are. "Michael doesn't see anything small. There are no small visions. And we're always telling him, 'Well, can't you do it this way?' And he'll say, 'No, no, no, I see it *this* way.' I think that's what I love most about him. Because in his heart he says, 'Why shouldn't my people have this?' Or, 'Why shouldn't my kids enjoy this? You know they deserve it.' "

"Michael is a dreamer," she adds. "He can look at an empty field and see a whole city. And I think because of his dream and because of his vision and because of his excitement and enthusiasm, he can get you to catch the dream too."

Questions

1. By your count, how many different problems does Mr. Howard try to tackle in his neighborhood? His daughter describes him this way: "He's a community leader; he's a nature man; he's a construction worker; he's a teacher; he's a preacher. Yes, he does everything. He never stops." Why do you think he commits himself to these projects? Why do you think "he never stops"?

   ____

   ____

2. Can you describe the process by which toxic trash ends up in neighborhoods like Mr. Howard's? Where does it come from? Who dumps it and why?

   ____

   ____

3. How did Mr. Howard and his neighbors combat the problem of lead in their drinking water?

   ____

   ____

4. How did the Fuller Park residents clean up the contaminated vacant lot in their neighborhood? What steps did they follow to get it done?

   ____

   ____

5. What did residents do with the vacant lot once it was cleaned up?

   ____

   ____

Name ______________________________ Date ____________________

HANDOUT 3.1: DON'T "WASTE" THIS QUIZ: How much do you know about recycling?

Answer the questions below to find out how much you know about recycling. Before you begin, think about the following: Do you think you know a lot, a little, or nothing about recycling? How about after the quiz? Do you know more about recycling after taking this quiz?

1. Which of the following is *not* considered one of the three "Rs"?

   a. Reuse

   b. Repair

   c. Reduce

   d. Recycle

2. Plastics are usually made from fossil fuels such as natural gas or oil.

   True or false? ___True ____False

3. Which of the following materials can be made from recycled plastic?

   a. Fleece pullover

   b. Carpet

   c. Film

   d. All of the above

4. Shredded and "crumb" rubber from automobile tires can be used in all of the following products *except*:

   a. Tennis courts

   b. Landscaping mulch

   c. Asphalt

   d. Glass tabletops

5. About how long does it take an empty soda can to biodegrade (break down)?

   a. 20 to 59 years

   b. 60 to 79 years

   c. 80 to 100 years

   d. 100 to 150 years

6. Millions of computers are destined for landfills as upgraded models become available. About what percent of computers can be reused or recycled into new products, if they are kept out of the landfills?

   a. Less than 25 percent

   b. 40 to 50 percent

   c. 75 to 80 percent

   d. Over 95 percent

7. Look at the products listed below. Place a checkmark next to the ones that can be recycled.

   __Wood

   __Acid batteries

   __Latex paint

   __Inkjet cartridges

   __Waxed paper

   __Steel cans

   __Motor oil

   __Plastic bags

   __Tires

8. Glass can be recycled over and over again.

   True or false? ____True ____False

REPRODUCIBLE 3.3A: Cross-Section of an Active Landfill

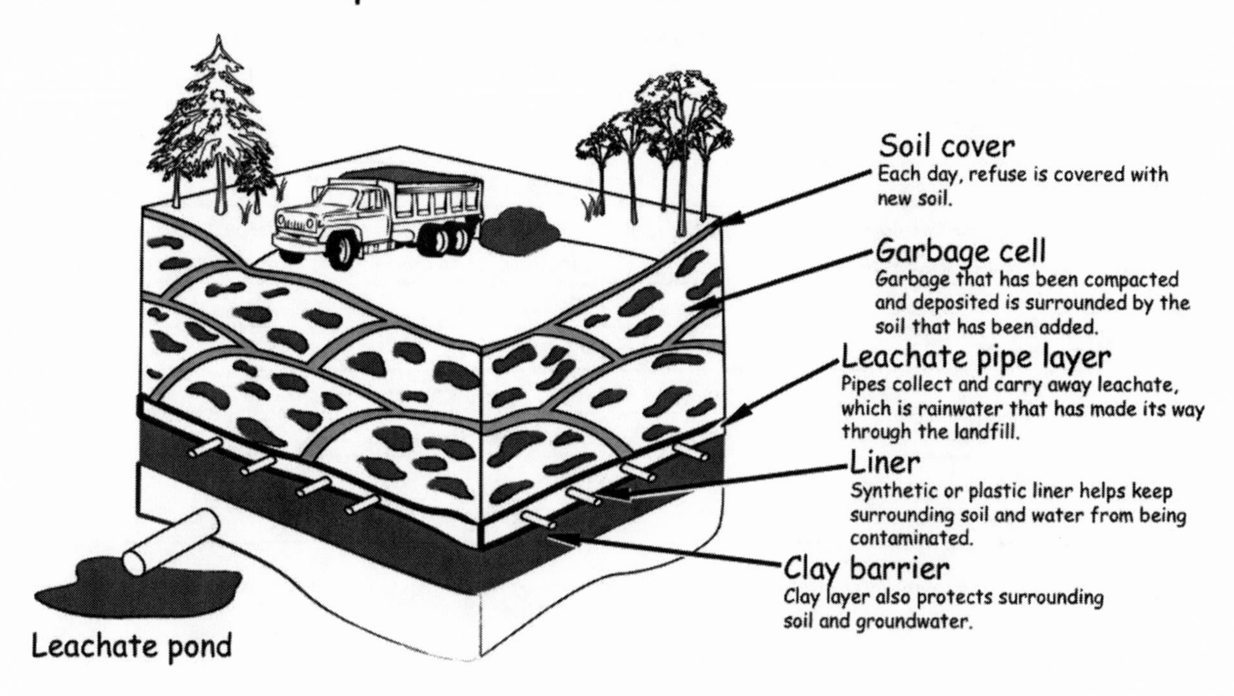

REPRODUCIBLE 3.3B: Is All Trash the Same?

TYPES OF TRASH

It is not always clear what is or isn't *biodegradable*. The definition of the word varies according to whom you ask. Biodegradability standards vary from country to country, as well. Many people consider a *biodegradable* item to be one that breaks down and disappears into the earth within six months. That is the standard used in this chart.

BIODEGRADABLE	NONBIODEGRADABLE
Can be decomposed or "broken down" by biological processes into raw, natural material	Cannot be decomposed or "broken down" by biological processes into raw, natural material
EXAMPLES	**EXAMPLES**
Fruits and vegetables	Plastic bags
Soap	Many chemicals and pesticides
Leaves	Synthetic clothing
Some types of paper	Paint
Wood	Cement, tar
Cotton	

THREE TYPES OF WASTE

Most cities recognize three different types of waste, shown below. Each is handled differently.

MUNICIPAL WASTE	INDUSTRIAL WASTE	HAZARDOUS WASTE
Defined as: Typical household trash and waste from small businesses and schools. **What happens to it:** Collected either by city employees or a company hired by your city. Disposed of in landfills; some materials recycled.	**Defined as:** Materials left over from a manufacturing process, a construction site, or large businesses. **What happens to it:** Strong laws mandate that it be disposed of properly, either in municipal or private landfill. Some dangerous waste goes to special landfills. Most companies hire private waste collectors to handle their trash.	**Defined as:** Household and industrial waste that contains highly dangerous chemicals and materials. **What happens to it:** Cities usually have special drop-off centers or drop-off events for hazardous waste. Waste is collected and sent to landfills specially designed to contain these items in such a way that they do not contaminate the environment for future generations.

REPRODUCIBLE 3.3C: The Breakdown Rates of Common Household Items

HOW LONG?

How long will it take for items in a dump or landfill to break down? This chart gives you an idea.

ITEM	TIME TO BREAK DOWN
Banana peel	3 to 4 weeks
Paper bag	1 month
Cotton sock	1 to 5 months
Uncoated paper	2 to 5 months
Lemon peels	6 months
Milk carton	5 years
Synthetic fabric	30 to 40 years
Tin can	80 to 100 years
Plastic bags	450 years
Aluminum cans	200 to 500 years
Disposable diapers	500 to 600 years
Glass	unknown; estimated 1,000,000 years
Styrofoam food packaging	unknown; possibly forever

WHAT CAN WE DO RIGHT NOW?

Sadly, a lot of waste that ends up in your local landfill doesn't need to be there. It could be reused, recycled, or composted. Everyone can begin to tackle this problem *right now*. Compost materials can be used in any garden to improve the soil and reduce the need for harmful fertilizers.

COMPOST MATERIALS	RECYCLABLES
Much food and plant matter, including things like banana peels, apple cores, grass clippings Coffee grounds, egg shells Leaves No meats, dairy products	Aluminum cans Newspaper Many other types of paper Glass

Name ______________________________ Date ____________________

HANDOUT 3.5: Trash Tops the Charts

Look at both charts below. Then answer the questions that follow.

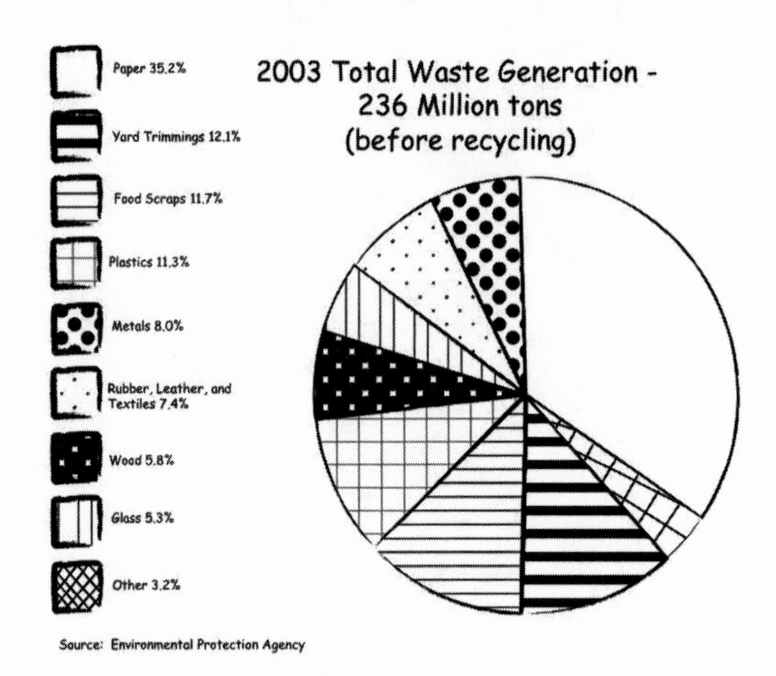

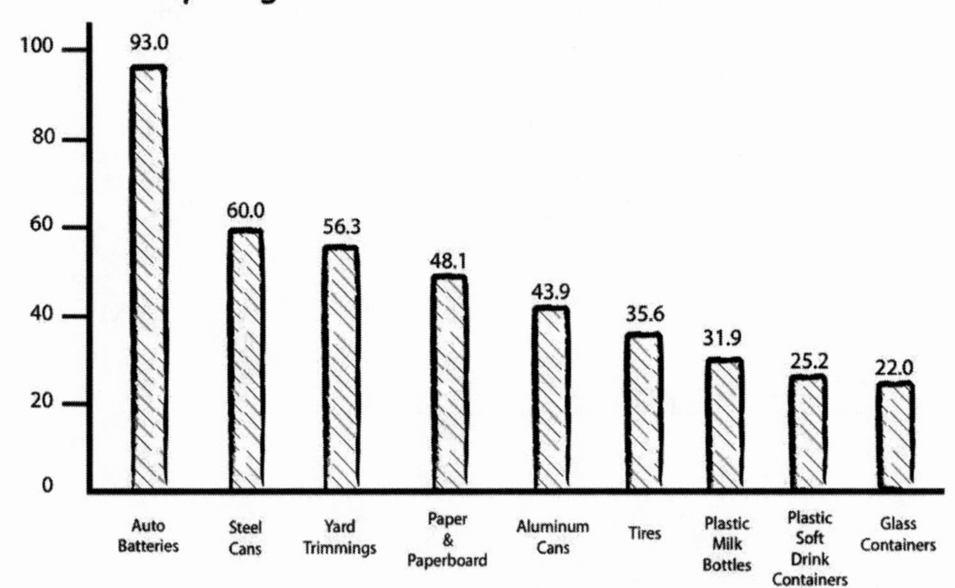

Source: The Environmental Protection Agency

1. How much greater is the percentage of paper thrown away than the percentage of glass?

2. What percentage of total waste generation is yard trimmings? How many tons is that?

3. Using your answer from question 2, how many tons of yard trimmings are recycled?

4. What percentage of total waste generation is metals? How many tons is that?

5. Which is greater, the percentage of total paper waste and yard trimmings or the percentage of total waste consisting of food scraps, plastics, and metals (aluminum and steel)? How much greater? How many tons greater is that?

6. What percentage of aluminum cans is not recycled?

7. What percentage of steel cans is not recycled?

REPRODUCIBLE 3.6: Landfill Map

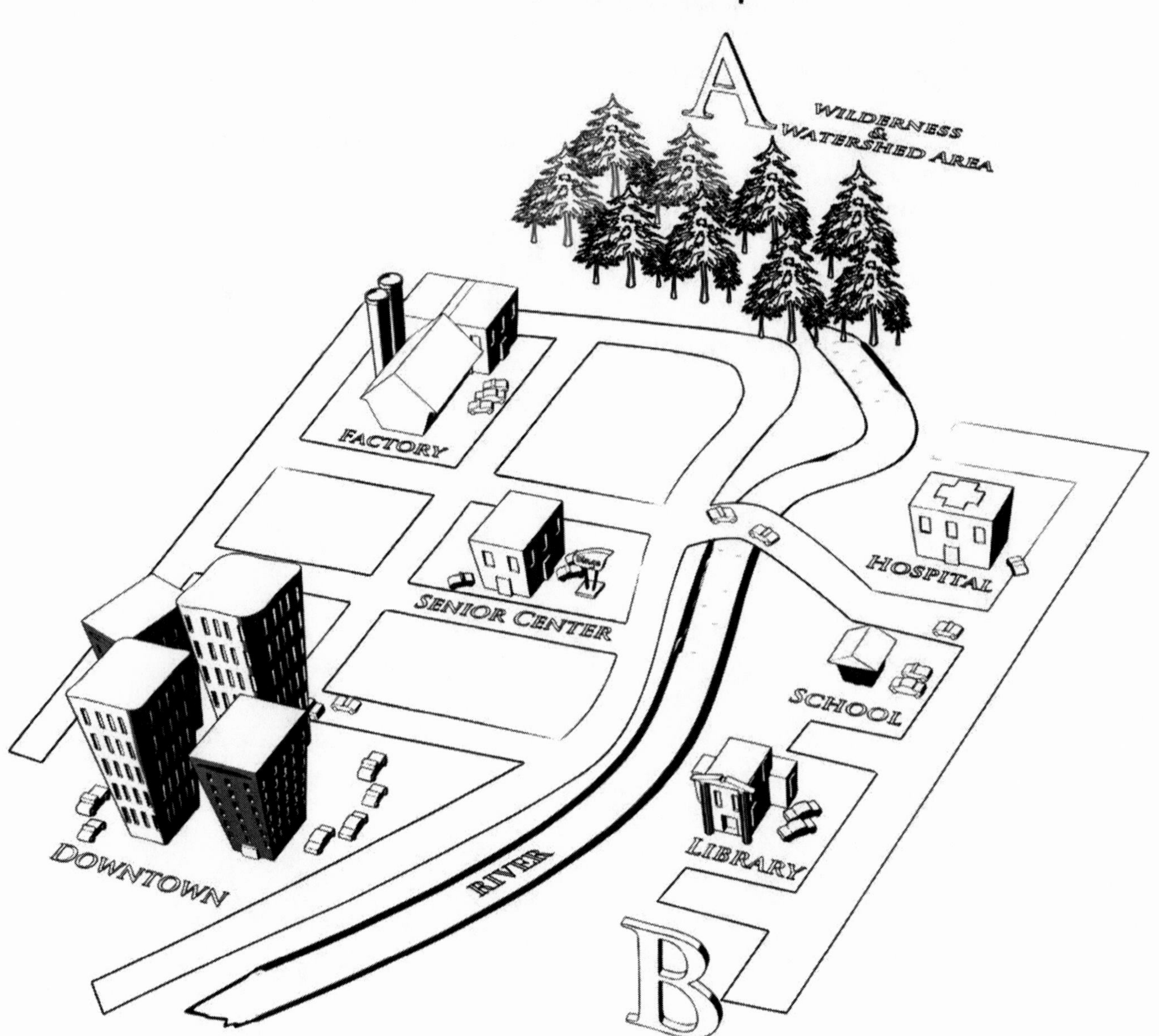

UNIT 4

GREEN BUILDING

TOPIC BACKGROUND

Home construction is one of the major industries in the United States. If you ask your students to scour the local business pages or watch business news programs on TV, they will notice that economists often link the health of the U.S. economy to the number of new housing "starts." As a nation we rely on home building to create jobs and start a chain reaction of purchasing associated with national prosperity. Each time a home is built, a builder must buy materials to erect the house. Then, once a family moves into the home, they usually need to buy furniture and other items to furnish it. But as we see in the Los Angeles film, rampant building can gobble up our landscape and cause unchecked urban sprawl. Moreover, the physical act of building has numerous environmental impacts: soil erosion, tree damage, and tons of debris shipped to landfills; and countless building materials—everything from paint to construction adhesives to pressure-treated lumber—are toxic and can find their way into waterways and the air we breathe.

Obviously, people need homes. But can we build homes that do not have as much impact on the environment? The "green building" movement seeks to minimize these impacts by choosing less harmful materials, protecting local trees and soil during construction, reusing and recycling as much existing material as possible, and using more energy-efficient windows, insulation, appliances, and so on. Green building is catching on as conventional builders try to court energy-conscious home buyers. Green building is currently a more expensive way to build shelter, but those higher up-front costs generally pay off down the line in energy savings.

We think this can be a very exciting topic for young people, many of whom aspire to be architects, builders, or interior designers or, at the very least, long to create interesting living spaces for themselves. Not all of them will be able to afford a solar home or be able to rip open their walls to beef up their insulation. Yet they can do simple actions such as replacing all the lightbulbs in their homes with compact fluorescent bulbs, checking for air leaks around doors and windows, fixing leaky faucets, or purchasing energy-efficient appliances that carry the U.S. Department of Energy's "Energy Star" seal.

You can also encourage your students to take what they are learning about green building to leaders in their communities. For example, by implementing green building practices, your school may be able to save money on heating and cooling costs that could be better spent on academic programs. Your local city hall, library, hospital, and senior center could save money as well. Unfortunately few people are aware of the green building movement, or else they are put off by the initial high costs to bring their structures into harmony with the environment. All these institutions could benefit from sincere presentations by young people who passionately believe in conservation.

Purpose

In this unit, students will learn about green building practices and discover ways that they can retroft their own homes or schools to conserve energy and reduce building material waste.

GROUP ICEBREAKERS

1. Ask students if they have ever heard the term *green building*. What do they think the phrase means, and why should we care about it? What do they think a home builder who is "building green" is really trying to do? (Green building is a fast-growing method of designing and building homes and commercial structures in such a way as to minimize waste, reduce damage to environmental resources, and result in an energy- and water-efficient structure.)
2. What is unsustainable about a drafty window or a leaky faucet? (A drafty window lets heated or

cooled air escape, wasting precious electricity, oil, or natural gas. Leaky faucets waste water.)

3. Ask students if they can think of some ways to save money in the home by being more careful about the way they use resources. (Thoughtful answers might include not turning up the heat or air conditioning, making sure doors and windows are closed properly when running those systems, maintaining appliances with annual checkups, and using every opportunity to improve a home's insulation.)
4. Green building tends to be more expensive than conventional building. Why do they think that is? (Extra work often requires more money. Green materials such as denim insulation tend to cost more than ordinary materials.) Can they think of some ways governments can help Americans who want to build a green home? (Tax incentives are already in place in a number of communities to help the home owners who buy green homes. If more people buy such homes, demand for them will grow, and their costs should subside to the level of ordinary homes.)

ACTIVITY 4.1
Anatomy of a Green Home

Materials

pen and paper
Reproducible 4.1
Handout 4.1

Primary Subject Areas and Skills

science, social studies, math, writing, critical thinking

Purpose

Students study a diagram to learn the attributes of a green-built home.

While some aspects of green building involve advanced techniques and the use of alternative forms of energy or materials, other aspects of the green building approach are just common sense. They're things that everyone should be doing anyway, not only to save energy and reduce the damage done to the environment but to save money, too.

For example: Almost *half* of household energy costs go to heating and cooling, and both of these expenses can be drastically reduced by simply using proper insulation in your home. Using compact fluorescent energy-saving lighting—explored further in the next activity—can drastically reduce the amount of energy used to light your home *and* save money. This is another readily available option for everyone.

And while some technologies have higher start-up costs (see Activity 4.6) before savings are realized, many green buildings cost no more at all than standard construction. Businesses are going green, too, not just to practice sustainability but because it makes good, solid financial sense as well. Some states will even cover the cost of installing items like solar panels, knowing that it will save the entire community money and resources in the long run.

So what goes into a "green home"?

1. Copy Reproducible 4.1 to show on an overhead.
2. Ask your students to describe the energy-saving ideas that come to them as they look at the diagram.
3. Copy Handout 4.1 and distribute to students.
4. Discuss the information it presents before leading them through the discussion questions that follow.

One diagram cannot possibly begin to describe all the available green building technologies. Links provided in the Resources section of this unit can lead you to numerous organizations that specialize in green building techniques.

Discussion

1. Which of the actions suggested do students think they could do right now? (The easiest things to do are to change all incandescent lightbulbs to compact fluorescent lightbulbs and to stop buying toxic products for home use.)
2. What happens to old building materials when a house is knocked down? (Normally they are deposited in a landfill.) What is a more eco-friendly alternative? (Reuse the material.) What are some other benefits to reusing and recycling old building materials? (If more building materials were recycled, reused, or disposed of properly, fewer tons of these materials would end up dumped illegally in poorer neighborhoods.)
3. Do students think the use of green building techniques and/or recycling of old building and housing materials should be mandatory? Why or why not? (Answers will vary. Arguments are good for both sides. Most builders favor a voluntary approach. You may also wish to assign this question as a prompt for an essay.)

Extension

Assign students to pick one green building product, such as bamboo flooring, and prepare a report on it. Ensuring that all students or groups pick a different green building technique will allow you to create a portfolio on green building.

ACTIVITY 4.2

Under the Green Roof

Materials

library or Internet research materials
Reproducible 4.2
Handout 4.2

Primary Subject Areas and Skills

science, social studies, math, writing, research, critical thinking

Purpose

Students study a diagram to understand how a green roof is constructed, then learn how such roofs can help cities combat soaring temperatures in summer and conserve energy costs.

When the mayor's office of the city of Chicago installed a green roof on top of city hall, more than a few Chicagoans thought it was money spent on beautification alone.

Nothing could be further from the truth.

A green roof is one of the most efficient techniques in green building. It reduces heating and cooling costs and extends the life of the roof structure itself, as well as absorbing rainwater that can be collected and used rather than contributing to runoff and flooding.

The leaves and canopy of a green roof absorb sunlight for food. In the summers, the trees and plants lend shade to the roof itself, lowering the temperature on the roof and reducing the need for air conditioning. And introducing a patch of green in the city helps to reduce the "heat island effect," a rise in temperature that occurs when masses of asphalt and cement absorb sunlight and radiate heat.

1. Copy Reproducible 4.2 to show as an overhead.
2. Look at the diagram together and discuss what you see.
3. Then copy and distribute Handout 4.2 to students. Divide the class into groups and assign them the task of assessing the pros and cons of installing a green roof on a local government building. You may wish to give them a class period during which to work on the project; or for a more intensive project, you can encourage them to work on it outside of class for a determined amount of time.
4. Discuss the questions with them or assign the handout as homework. Students may also work in pairs on the handout in class.

Extension

Tell students that they have been elected to a town committee. They are responsible for assessing the possibility of installing a green roof on one of the government buildings in town. Taxpayers don't appear to be willing to pay the small initial cost to install the green roof, despite the fact that they will save money and get tax breaks down the line. Instruct students as to ways they might make money from the green roof. Have them choose two of these ideas to research and investigate more fully. (Some possible ideas: growing flowers to be dried, arranged, and sold; growing fresh herbs for local restaurants; making decorative wreaths.)

As they come up with ideas, tell them to consider the following:

- How much money do they estimate can be made? Explain how this number was calculated.
- What are the itemized start-up costs for their money-making plans?
- What are the maintenance costs involved in their plans, such as gardening supplies, water costs, soil costs, etc.?
- If installing the roof costs $30,000, how long might it take for them to recoup the cost and begin to make a profit?
- What steps they would need to take to bring the project to fruition?
- How can they test how practical it would be to build the roof, then produce and distribute the product?

If you opt for a longer project with your students, encourage them to design a poster that shows how their different businesses will function and examples of costs versus income for their business schemes. (For example: If they will sell honey, like the city of Chicago, what are the costs involved in that operation? How much do they have to pay the beekeeper? How much will they charge per jar? Will they have to pay for advertising? What special supplies will they need?

Can they get any of these donated, or get experts to volunteer their time?) Students can then present their business plans to the class.

ACTIVITY 4.3

A Green Light

Materials

pen and paper; Handout 4.3

Primary Subject Areas and Skills

science, math, writing, critical thinking

Purpose

Students use math skills to complete a chart that explains the costs and benefits of using compact fluorescent lightbulbs.

The best way to light your home is with the greatest free source of all: the sun. Many green homes are designed to let in as much natural light as possible. But when the weather and time of day don't allow for it, having efficient, cost-effective, and low-energy lighting can make a big difference to the environment and your bank account.

Compact fluorescent lights, or CFLs, are a key element in both energy conservation and green building. These bulbs are one of the easiest ways to reduce energy use, energy costs, and pollution. Yet so many people still don't use CFLs in their homes and businesses. How are CFLs different, and what are their advantages? First, let's understand some electrical lighting basics.

Lighting Language

Lightbulbs are described by their number of *watts.* Watts, named after 18th-century British engineer and inventor James Watt, are a unit of power used to measure electrical and mechanical power. Another helpful and descriptive term is *lumen.* The unit lumen is used to measure the amount of light produced—a measure of its brightness, in a sense. It is based on the *candela,* which is a measure of "luminous intensity" of light emanating from a small source. This originally correlated to the amount of light emanating from a single candle. So, a watt measures the amount of electric power that a bulb needs to produce light, and the lumen refers to the amount or brightness of the light produced.

Basic Bulb Comparison

The most common household bulb is the incandescent lightbulb, which is the one with which you're probably most familiar. It was a brilliant invention by Edison, but it's less efficient than it could be. Electricity is used to heat up a filament inside the bulb, causing it to become white-hot and to glow, emitting both light and heat. In fact, an incandescent lightbulb emits more heat than it does light, and thus wastes a lot of energy.

A CFL is a fluorescent lightbulb—like the long, thin light tubes you're used to seeing in office buildings—but more compact. A fluorescent lightbulb does not use heat to emit light. Instead, the bulb features coated glass and contains mercury vapor in argon or neon gas that becomes excited by electricity. When this excitement of the vapor in gas takes place, an invisible ultraviolet light is produced, and this becomes visible when it hits the coating of the bulb.

In short, CFLs emit more light—or lumens—per watt. That means more light for less energy. For example: A 13-watt CFL can produce as much light and give off as many lumens as the very prevalent, standard 60-watt incandescent bulb. So, if a watt is the measure of energy needed, and lumen the measure of brightness, you're getting the same amount of brightness for less than one-quarter the amount of energy.

1. Copy Handout 4.3 and distribute to the class.
2. Use the preceding information to help them understand the terms *watts, lumen,* and *candela* and how CFLs can help us save energy.
3. Have them complete the worksheet and answer the questions.

ACTIVITY 4.4

Front Page

Materials

newspaper opinion pages; library or Internet research materials

Primary Subject Areas and Skills

writing, research, social studies, critical thinking

Purpose

Students read and analyze articles from the opinion ("op-ed") section of the local newspaper, then write their own opinion essay calling for more green building in their town or city.

1. As a class, discuss the idea of green building. What are the students' initial reactions to the

idea of following and applying the principles of green building in their daily lives?

2. Copy and distribute several articles from the opinion page (op-ed) of your local newspaper of your choosing—ideally those focusing on environmental issues. If there are none in your local paper, there are plenty to be found on the web or in the library.
3. Have students read a piece in class and then critique it as a group. Alternatively, you may wish to put students into groups and have them analyze the structure of the piece.
   a. How early in the piece do the authors make their strongest points? (Usually the first or second paragraph.)
   b. How do they support their arguments? (By citing examples from real life.)
   c. If students disagree with the opinion, do they at least appreciate the argument that has been made? (Answers will vary, but this should be the ideal outcome.)
   d. Have students discuss their findings with the class.
4. Instruct students to write an op-ed piece for your local paper about applying green building practices to the local school, library, and city government buildings. Students should do their own research. These pieces are usually no more than 1,000 words and enumerate the pro points supporting a particular position or argument. Explain that the more factual support to your opinion, the more convincing your argument. It's OK to dismiss an opposing point of view as well, but it should be done using logic and reasoning as opposed to reactionary and emotionally charged statements.

Extension

Consider publishing your profiles in a newsletter that can be sent around your school. Submit the op-ed pieces to your local newspaper and see if they would be willing to publish them.

ACTIVITY 4.5
At the Podium

Materials

Internet or library research materials; judges' evaluation sheet, page 14

Primary Subject Areas and Skills

reading comprehension, writing, research, critical thinking

Purpose

Students participate in a pro–con discussion, debating the wisdom of making compact fluorescent lighting mandatory in all U.S. buildings.

Have students debate the following statement.

The use of compact fluorescent lighting (CFLs) should be mandatory in all U.S. homes, businesses, schools, and city and federal government buildings.

Point

Yes, I agree. CFLs are a simple, cost-saving measure that, in the long run, will not cost individuals any more money and will reduce energy use and pollution.

Counterpoint

No. It's fine if you want to make city hall use CFLs, because my tax dollars are paying for that. But it's different when you come inside my home and force me to do something I may not want to do. What's next? Forcing me to install solar power, which is costly?

1. Divide students into two groups of at least four students each. Depending on class size, be sure to leave a group of at least five (preferably an odd number) to act as judges. Assign each group to Point or Counterpoint as described above.
2. Instruct students to research and report on their point of view. They should devise statements in support of their position and be prepared for rebuttal from the opposing team.
3. Judges should be concerned with organization, evidentiary support, and overall presentation, as well as politeness and poise.
4. Debaters should clarify who will speak for the group. (Debaters may take turns, if that is decided beforehand).
5. The affirmative or Point team will speak first for three to five minutes to present their argument. The Counterpoint team will then have three to five minutes to present their argument.
6. Each team will be given three to five minutes to respond to the other team's argument.
7. Have judges rate the arguments of each group and present their findings to the classroom.
8. Discuss among yourselves the emotions and

ideas that came up during the debate. Did anyone change his or her mind about the topic at hand?

EXTENSIONS

House Detective

Prepare a checklist and have students use it to investigate some of their own home's efficiency issues. Have them check such things as their toilets' gallons (or liters) per flush (this information is usually printed on the rim or inside the tank of the toilet by the manufacturer), how many incandescent lightbulbs are used in the home, how much their family might save in energy costs by switching to CFLs, or whether their faucets or showerheads are leaking.

Go Shopping

Have students visit a home center store on their own time. Ask them to gather some free consumer information on products such as energy-efficient windows or insulation. Back in the classroom, have them design some problems that would allow them to calculate the efficacy of these products. (Both window and insulation manufacturers support websites that demonstrate the mathematics and science behind these products.)

You're the Designer

After researching some components of a green-built home, have students design one. Encourage them to build a model or make a cross-sectional drawing of an energy-efficient home. If your students enjoyed designing their own home, consider having them display their work in a school art show. (Professional scientists often display academic studies in what are known as "poster sessions," where colleagues are invited to walk through a gallery and read large posters of the scientists' work.)

Talk to the Builders

Research green builders on the web and in your community. What kinds of services do they offer, and what do they charge? (Consider an e-mail exchange with the builder to answer some of these questions.) Invite a builder to come to your class to talk about reducing energy costs. Home energy auditors are now popular in many cities. For a fee, they'll come to a person's house and test the home's air, heat, and water efficiency. Because so much of their business depends on good consumer education, they may be delighted to visit your class.

Do the Math

Using the savings calculator on the Energy Star website, calculate how much money your school would save if it switched to compact fluorescent lightbulbs.

RESOURCES

- **U.S. Green Building Council**
  www.usgbc.org/
- **The LEED (Leadership in Energy and Environmental Design) Green Building Rating System**
  www.usgbc.org/DisplayPage.aspx?CategoryID=19
- **Virtual tours of "green" office spaces: World Resource Institute**
  www.igc.org/wri/office/index.html
- **The Center for Neighborhood Technology**
  http://building.cnt.org/tour/site.php
- **National Resource Defense Council (NRDC)**
  www.nrdc.org/cities/building/default.asp
- **Green Building Resource Guide**
  www.greenguide.com/
- **The Center for Energy Efficiency and Renewable Technologies (CEERT)**
  www.cleanpower.org/
- **Switch your school to a "green" power company**
  www.green-e.org/
- **Download a list of "green school" resources**
  www.nesea.org/buildings/greenschoolsresources.html
- **The Energy Star website on compact fluorescent lightbulbs**
  www.energystar.gov/index.cfm?c=cfls.pr_cfls

Reading Handout Unit 4:
A CITY OF GREEN BUILDERS

In this reading, you'll learn about the work of a carpenter and builder named Martha Rose, who is building energy-efficient homes in the city of Seattle. Her choices, materials, and approach to building are all part of a growing trend toward sustainable housing called "green building." After this reading, you'll be asked to answer some questions about how the work of a green builder helps the environment.

In recent decades, Seattle has become a more popular place to live. The great irony of this popularity is that so many people are flocking to the city that they literally are loving the place to death. Building homes for all these new people has an impact on the city. Any time you disturb the soil with large building projects, the land erodes and washes away, polluting waterways and stressing the wildlife. If a city is to grow in a sustainable manner, people must find new ways of building homes and businesses that do not affect or destroy so much of the natural world. The desire to do this has led to a movement in environmentally friendly construction and architecture called *green building*. Seattle has gained a reputation as a hotbed for this kind of work.

For Martha Rose, the philosophy of green building is the culmination of a lifetime of strong ideas about sustainable shelter. When she was only 19 years old, Chicago-born Ms. Rose took a job as a carpenter on a construction site in the Washington, D.C., area. "The work was hard," she remembers, "but I liked it."

The granddaughter of a Chicago builder, Ms. Rose had always had a strong reaction whenever she saw new houses going up on what was once farmland outside Chicago. "Every time I saw new development in rural areas, I didn't like it," she says. "And that was when I was young. Every place else that I went, I'd see the same thing, and my feeling was the same: I just didn't like it." There had to be a better way to make new homes for people than to keep gobbling up swaths of green, untouched land, she thought.

And there could be better ways to build, too. On her first job as a young carpenter, she was appalled when her boss had her install shoddy doorknobs on a door. "I think the only thing that was metal in those knobs were the screws holding them to the door," she remembers. "The rest of it was plastic. It was so cheap it made me sick. I would rather have a better-built, smaller house than a cheaply built big house."

Eventually, she would have her chance. In 1982 she took a job as a building inspector in Seattle, a job that allowed her to see countless vacant, near-vacant, or dilapidated parcels of land within the city limits that were crying out for intelligent renewal. If you were going to build homes, she thought, this was the place to do it. Inside the city, you don't have to affect farmland, or worse, raw nature. This type of construction has a name: *urban infill housing*. This refers to the filling-in of city lots. Such construction is usually speculative; the builder buys a piece of property, erects a townhouse, and then sells it at a profit. The practice is controversial because longtime residents often don't like the way the new structures alter the character of the neighborhood. But Seattle was a city on the move. Its population was growing, and newcomers especially liked the idea of moving into a vibrant city center.

Ms. Rose began building homes in downtown Seattle, looking for ways to build safer, more efficient, more ecologically sound structures. The start of any construction job is messy, she knew. A demolition crew sweeps in, levels everything to the ground, and hauls away the debris of the older structure to a landfill site. Ms. Rose and other green builders were beginning to see that much of the stuff they threw out could actually be reused or recycled. Big machines tore up the neighboring land, often ruining the topsoil and local vegetation and sending loose soil down storm drains when it rains.

Today Ms. Rose has adopted a new type of Earth-friendly approach to site management. Before the job starts, she hires a salvage company that picks through the existing structure for usable items, such as hardwood floors, cabinets, and mantelpieces. Each of these items is carefully pried out, donated to shelter organizations such as Habitat for Humanity, or sold to people restoring old homes. Next, the empty shell is demolished, and the resulting concrete, wood, and bricks are sent to a special recycling site that sorts "commingled debris." The wood is chipped for composting, the metal is recycled as scrap, and anything unusable is ground and used to cover landfills. Thanks to this method, only 5 percent of old buildings—instead of 100 percent of them—ever make it to a landfill!

Recently, Ms. Rose has started using demolition crews that use only biodiesel equipment to knock down buildings. "Believe it or not," Ms. Rose says, "the inventor of the diesel engine, Mr. Rudolf Diesel, actually designed his original engine to run on peanut oil." Biodiesel fuels, made from various vegetable products, are gentler on the environment and burn cleaner.

Next, the crew tries to save as many trees on the site as possible. They erect fences to halt further erosion. They place a layer of wood chips on the ground to give the machines a slightly firmer footing and more traction, so the wheels don't spin and tear up the earth. Because Seattle is a fairly mild climate, the crew builds walkways with a new type of porous concrete that allows water to seep into the sidewalks and eventually drip into the ground.

Green builders try to eliminate the use of pressure-treated lumber, which uses toxic chemicals to keep wood from rotting in damp or wet ground. Instead of installing roof shingles with a 20-year lifespan, they opt for 30-year products, which lessen the need for shingles later down the line.

In recent years, the goal among green builders is to reduce energy loss as much as possible. Most people assume that means using solar technology, but solar-powered heat and electricity are still quite expensive technologies. (Solar water heating is the most cost-effective technology at this time.) Instead, carpenters try to build larger spaces in the walls for thicker insulation. They even insulate the underside of the concrete slab on which the house sits. "You can pretty much build a house these days so efficiently that your heating costs go way down," says Ms. Rose. "I did the same thing in my office and home. I've been here a year, and I cannot believe how comfortable it is."

The Martha Rose Construction Company is still small by most city standards. Ms. Rose estimates that she builds three to eight houses a year in Seattle, but each of them has been certified as energy-efficient and environmentally friendly structures by Washington State's "Built Green" organization.

"What I do is always risky financially," says Ms. Rose. "I'll build a house really well, and I'm taking a chance that I will be able to make my money back, plus a profit, and that's tough. Homebuyers are not yet educated about what a green house is all about. You might spend more money upfront buying this house, but you'll save money every year you live in that house, compared to houses that are not built this way."

"The emphasis right now is on using more durable goods—things that last longer," she adds. "We are installing commercial-grade tile and carpets, or cedar siding, which will last forever. We install dual-flush toilets, which have two buttons on them, one for a light flush, and another for a heavy flush. Three-quarters of the time, you would only use the small button. That's a toilet that was developed in Australia. It helps homeowners save money in water and sewer costs, but it costs builders three times as much. You can use paints and adhesives that don't have harmful fumes, or VOCs [volatile organic compounds]. You can use fiberglass that is not made with formaldehyde. You can install an all-natural linoleum and Energy Star appliances and compact fluorescent light bulbs. All these things are available, if you look for them. But if you don't know why you should want eco-friendly features, then you're going to think this house is more expensive that others on the market."

It's estimated that only about 5 percent of the population is willing to pay extra for this kind of house, but nearly 20 percent of the homes going up in Seattle these days are Built Green, including several low-income housing complexes. Slowly, but surely, traditional builders are incorporating some of these energy- and resource-saving strategies in their structures. The cost to build a green house is about 5 percent more than conventional construction. That sounds small, but that tiny percentage translates to big bucks when you're building something as costly as a house. If it's not done carefully, the green house might price itself out of the neighborhood.

Still, Ms. Rose would not consider any other kind of work. "I think this is important," she says. "It's the difference between a lawyer who wants to be a public defender versus one who wants to work in a big law firm. Of the builders who do this work, some are doing it as a selling feature for those homes. They want to be able to say, 'Look, here's a green-built house for sale!' But most are doing it because they feel strongly about it. Hey, somebody's got to start doing it. Pretty soon it's going to be the norm. We're on the bandwagon before everyone else is on the bandwagon. I'm proud of what we're doing. I want to always be able to go back to a house I did in five years and have it be in good shape. I want to offer something that the masses aren't producing. A notch up in quality means a lot to me."

Questions

1. What is urban infill housing?

___

___

2. What's wrong with building outside a city in the countryside? Why do you think some green builders want to stay inside the city limits?

___

___

3. What are some steps Ms. Rose follows to prepare a site for construction? What happens to the debris of old structures that she and her crew remove from a building site?

___

___

4. What are "durable goods"? Can you cite some examples?

___

___

5. Say someone was shopping around for a $150,000 house. According to Ms. Rose, how much more would they be paying if that same house were "built green"?

___

___

REPRODUCIBLE 4.1: Elements of a Green Home

HANDOUT 4.1: Green Home Smarts

LOCATION, LOCATION, LOCATION

Let the sun shine! A home's location and position can determine how much solar energy it can harness.

Energy-efficient window placement allows for cooling breezes and cuts down on cooling costs in summer.

Materials

Local, recycled, reclaimed. From roof shingles to kitchen tiles, recycled materials are available for every part of the home. Homeowners these days can build their homes with reclaimed lumber and bricks, as well as salvaged flooring, banister spindles, and flooring.

Energy

Solar energy becomes more affordable every year, and using the sun's power to heat your hot water, for example, can involve installing just a few panels on your roof.

Installing compact fluorescent lightbulbs is one of the easiest ways to reduce energy use and costs.

GREEN FROM THE INSIDE OUT

Fabrics and Carpets, Floors and Furniture

Bamboo and cork are for more than just cutting boards and bulletin boards—they're for your floor, too. Fast growing, durable, and readily renewable, these are two of the most popular forms of responsible flooring.

Go with the Flow

Low flow, that is. Showerheads, toilets, and faucets all come with low-flow options that save resources and money.

Appliances

Everything from dryers to microwaves are available in low-energy and environmentally friendly versions. Look for the "Energy Star" rating wherever you buy large appliances.

House De-Tox

Paints, glues, carpets, shower curtains, and many everyday items contain chemicals that can release toxic gases that can damage indoor air quality.

GREEN FROM THE OUTSIDE IN

Planting native, drought-tolerant, deciduous trees and plants outside your home can make a temperature difference inside your home as well. In the summer, the shade from properly placed trees has a cooling effect on the home, reducing air conditioning costs. And in the winter, when the trees have lost their leaves, sunlight can easily penetrate strategically placed windows, helping to heat the home and reducing the need for heat.

The root structure of trees also helps retain rainwater and melting snow that seep into the ground. This helps prevent runoff and flooding, which in turn reduces the amount of toxins that run into sewers.

WATER, WATER EVERYWHERE

Cisterns underground or near houses can collect valuable rainwater to use in gardens. This conserves water resources and reduces the amount of water sent to treatment plants. This in turn lessens the stress on the facilities themselves, again saving more energy and creating less pollution.

REPRODUCIBLE 4.2: What's Under the Green Roof

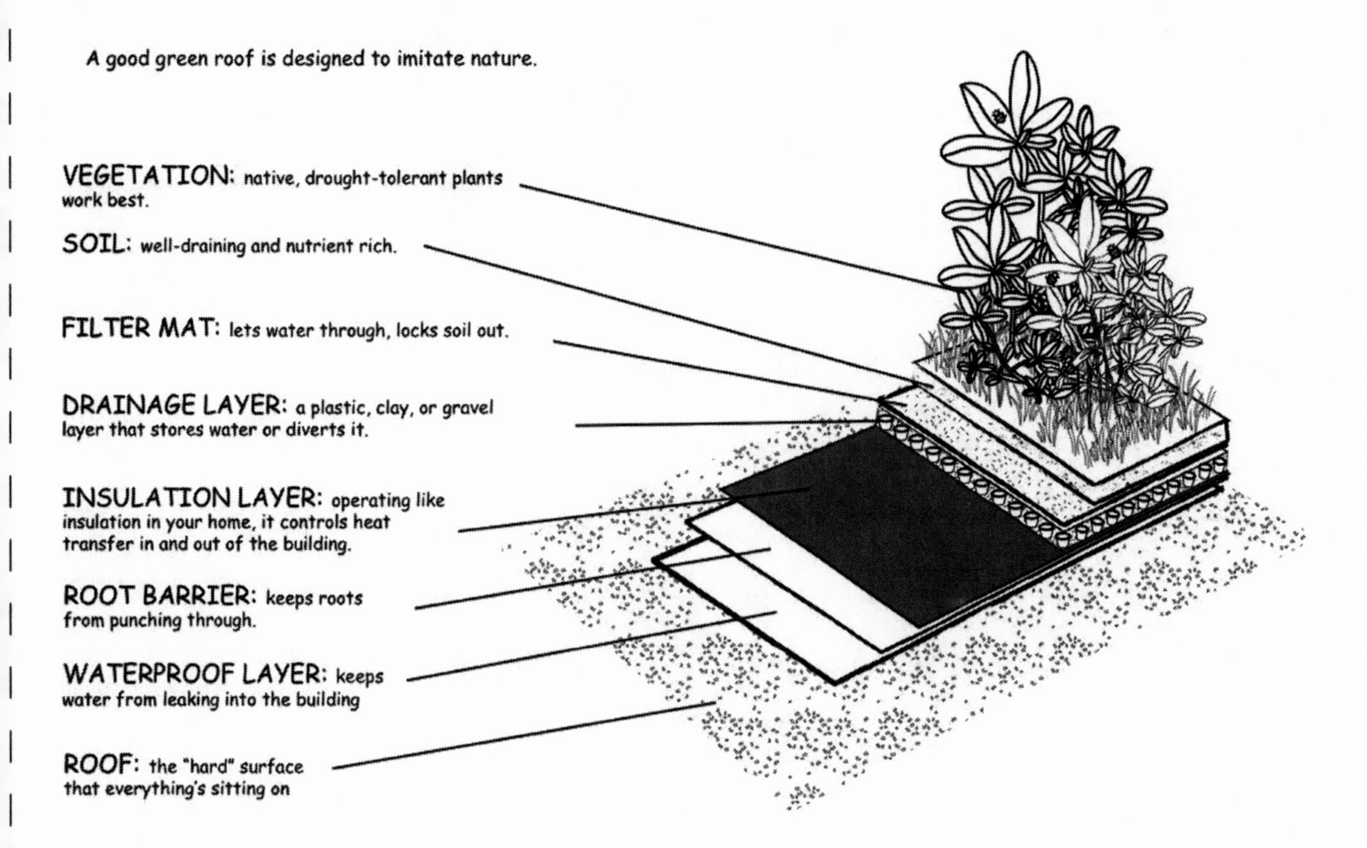

Name ______________________________ Date ________________

HANDOUT 4.2: Under the Green Roof

Answer the following questions. Then list the pros and cons of a green roof. Be creative!

1. Why do you think the soil needs to drain well?

____

2. How might weather conditions differ on a roof than on the ground? What effect might that have on your green roof?

____

3. Why should the plants chosen be drought tolerant?

____

4. Why should the plants be native plants?

____

5. Where could water from the drainage layer be collected? How might it be used?

____

6. What other purposes could the green roof serve?

____

Pros and Cons of a Green Roof

Pros	Cons

Name ______________________________ Date ____________________

HANDOUT 4.3: CFL Quick Facts

Look at the chart below. Fill in the missing information and answer the questions.

Type of Bulb	100-watt Incandescent	23-watt Compact Fluorescent
Cost Per Bulb (approximate)	$0.75	$11.00
Life of the Bulb	750 hours	10,000 hours
Number of Hours Burned per Day	4 hours	4 hours
Amount of time (in days, months, or years) 1 bulb will last		
Number of Bulbs Needed for three years (estimate)		
Total Cost of Bulbs	$	$
Lumens Produced	1,690	1,500
Total Cost of Electricity for three years (8 cents/kilowatt-hour)	$35.04	$8.06
Total Cost over three years		
Total Savings over three years with the compact fluorescent:		
Source: U.S. Department of Energy, Energy Information Administration		

1. How many lumens per watt are produced from the incandescent bulb? ______________

2. How many lumens per watt are produced from the CFL? ______________

3. How many times longer does a CFL last than the incandescent? ______________

4. What *percentage* of savings could be realized over the three-year period? ______________

5. Based on this information, will the percentage of savings decrease, increase, or stay the same over the next three years? Why? ______________

6. Why do you think people resist using CFLs? Do you agree? ______________

Extra Credit

How do you find the total cost of electricity used over three years? Show your work. (*Hint:* First get the total number of watts used, based on the total number of hours, then convert to *kilo*watt hours.)

MAKING CONNECTIONS

On a separate sheet of paper, answer this question: What is the relationship between energy production and pollution? How does reducing the amount of energy used to light your home reduce the amount of air pollution you contribute?

UNIT 5

ENERGY

TOPIC BACKGROUND

The dictionary definition of *energy* is "the ability to do work." Human beings expend energy 24 hours a day. Our bodies are constantly working and reliant on stored energy in the form of calories to keep us going. Every other part of our lives requires energy, too, though not in the form of calories. To run our homes, cars, schools, and businesses, we must consume some form of energy, whether it is natural gas, coal, oil, or numerous others. Unfortunately, many traditional forms of energy pollute the environment in which we live. Burning coal to make electricity is the leading cause of air pollution in the nation. Our cars and factories do not burn cleanly and spew excess carbon dioxide into the atmosphere—an action that many believe is causing global warming.

Americans have been told time and again that if they want their towns and cities to be sustainable, to keep on going forever without running out of energy, they must move beyond finite sources of energy such as oil. (Our nation burns through 21.1 million barrels of oil a day, much of it imported.) They are encouraged to explore renewable sources of energy such as solar and wind power or hybrid-electric vehicles. Admittedly, these are all promising technologies, but their costs have traditionally put them out of the reach of many Americans.

This is changing, however. The tragic events of September 11, 2001, the subsequent wars in Afghanistan and Iraq, and effects of Hurricane Katrina in 2005 have all inspired many Americans to come up with ways to reduce their dependence on oil. Many states are now offering tax credits and refunds to people who choose to install solar power equipment in their homes and businesses. In some states, requests to convert schools to solar power are pouring into government offices faster than they can be processed. Well-known American companies are jumping on the "green" bandwagon, pledging to overhaul their operations to become more energy efficient. Some have pledged to have a certain percentage of their stores run off wind power. In the auto world, car makers are releasing "hybrid" vehicles that run on a mixture of gas and electric battery power. Worldwide, *many* carmakers are exploring the possibility of running future cars on hydrogen—the most abundant element in the universe. Just think: Instead of spewing harmful carbon monoxide exhaust, cars will emit ordinary steam!

This progress is remarkable, but it can seem very overwhelming to the average teen who wants to help out but doesn't know how. Reassure your students that the quickest and smartest way to save the planet is through *conservation*. Every time you shut off a light switch, reduce your use of hot water, or close a drafty door, you are making a difference.

Purpose

In this unit, students will learn about how humans produce energy to run their homes, businesses, and vehicles. They'll learn how energy benefits humankind but is nevertheless threatening our precious Earth. We'll share some simple steps that students can take to make their home or apartment more efficient. We're hoping that after they take our home energy survey, they'll want to extend what they've learned to their school building, the local library, city hall, and possibly every home in town!

Remarkable career opportunities await young people in this realm. Home energy auditing is a growing field, as is energy-efficient car design and solar and wind power equipment, engineering, and installation. Your students stand on the threshold of a less oil-dependent new world. This unit may be the first glimpse they get of it.

GROUP ICEBREAKERS

1. Ask students to think of some ways to save money at home by conserving resources. (Thoughtful answers might include not turning

up the heat or air conditioning, making sure doors and windows are closed properly when running those systems, maintaining appliances with annual checkups, and using every opportunity increase a home's insulation.)

2. Make sure you allow students to brainstorm how they might be able to improve the use of resources in apartments. (Nearly everything mentioned in the previous answer applies to apartments as well, with the possible exception of maintaining appliances. If a student's family is renting an apartment, it is the landlord's responsibility to maintain heating-cooling appliances as well as all plumbing fixtures. It is the tenant's responsibility to use these things with care and efficiency and to let the landlord know immediately if an appliance or fixture is not working properly. A dysfunctional piece of equipment wastes resources.)
3. Ask students if they know what kind of energy heats their home or school. Does a different form of energy cool these buildings in the summer? (You might mention some of the possibilities: oil, electric, or gas.) What about food preparation and cooking? (Kitchens often run on electric or gas.) How many different types of energies are used to bring kids to school every day? (Most students travel by gasoline-powered vehicles, but some school districts are experimenting with electric, zero-emission buses. And some students simply walk.)
4. Can they think of any ways that their school wastes energy, directly or indirectly? (For example, leaving lights on in an empty room wastes energy directly. But wasting hot water is an indirect waste of energy because energy is used to heat that water—something not enough people are mindful of!)
5. Can they think of at least three different types of energy that are used or can be used to transport people around town? (Walking and biking both require human energy, and then there are gasoline-powered and electric vehicles. A fourth type, hybrid vehicles, is a combination of gas and electric power. A possible fifth type, not used everywhere, is biodiesel, the subject of this unit's reading.)

ACTIVITY 5.1
Sun, Sky, and Earth: Where We Get Our Energy

Materials

pen and paper
library or Internet research materials
Reproducibles 5.1A and 5.1B

Primary Subject Areas and Skills

science, math, reading comprehension

Purpose

Students will familiarize themselves with some traditional and alternative means of generating the energy that keeps our lives running every day.

The United States gets more than 90 percent of its energy from fossil fuels, a nonrenewable resource. Despite advances in a variety of alternative energy technologies, we continue to be dependent on a limited resource when other, renewable resources are available.

1. Copy Reproducibles 5.1A and 5.1B and place them in turn on the overhead. Read through the information with your students.
2. Reproducible 5.1A deals with fossil fuels. Ask your students if they know what fossil fuels are and how they're formed. Discuss the concept of a nonrenewable energy source. (Once that resource is used, it is gone forever. That's why fossil fuels are considered a *finite* resource.)
3. Reproducible 5.1B deals with renewable energy sources. Discuss the concept of renewable energy resources. (Such resources cannot ever be depleted, as long as the sun shines and the wind continues to blow. That's why renewables are considered *infinite* energy resources.)
4. Use the questions for discussion and extensions for further study.

Discussion and Follow-up

1. Coal is a source of energy. But getting coal from the ground to your local electric plant consumes energy, too. How? (First, we use energy digging the coal. Then we use energy processing it and shipping it to electric power plants.)
2. Everyone seems fascinated by solar and wind power, but those technologies still have some downsides. What are they? (They are still costly, and they rely on features of weather that are changeable. When the wind is not blowing, for example, you cannot use wind power.)

3. Have students trace the hypothetical route of coal, oil, or natural gas from a specific source, such as a particular mine in the United States or drill site in a foreign country, to its destination. The destination might be a local power plant or gas station, for example. Encourage the students to think about each step of the process and what is involved, especially from an energy-use standpoint, in getting energy to the light switch in your home or to the gas tank of your car.
4. As a longer activity, students could provide a step-by-step chart or collage that shows the journey that our fossil fuels make to our homes or cars.

Group Activity

Besides the forms of energy we discussed in the Reproducibles, other forms of alternative energy include:

- *Geothermal,* which uses reserves of hot water and steam from deep below Earth's surface.
- *Biomass,* which uses plant and animal wastes—cornstalks, sugarcane, wood chips, and much more—and converts them into fuel.
- *Hydro,* which captures the power of flowing water and converts it into electricity.
- *Nuclear,* which harnesses the power of the tiniest known particle of matter—the atom—for use in a variety of ways.

Divide students into groups and have them choose a form of energy out of a hat from all of the choices given. Let them research their selection and then present their findings to the class. Each report should include the following information:

- The type of energy
- How it works
- How the energy source is primarily used (heat, electricity, transportation, and so on)
- Whether or not it's renewable
- Whether or not it pollutes
- The advantages
- The disadvantages
- How cost effective it is
- Whether or not they would recommend this technology

Do not allow the students to use the reproducible material as research. Final presentations could be done in a "business-meeting" setting, in which students try to "sell" their technology to their classmates. Students can use diagrams, graphical representations, sculpture, or any other means to convey their findings.

ACTIVITY 5.2

I Love My Car—I Just Hate Paying for Gas!

Materials

pen and paper; Handout 5.2

Primary Subject Areas and Skills

science, social studies, math, research, critical thinking

Purpose

Through a series of engaging readings, students learn all the types of energies humans use to power their motor vehicles. Students are challenged to create their own humorous, yet thoroughly researched, readings.

You might think that gasoline is America's only automotive fuel option. But we actually have a number of choices. The following dialogues are intended as a fun jumping-off point to educate your students about the different fuels used in modern cars. You might consider having them act out the roles of the different characters. After they read the dialogues, have them answer the questions and tackle the suggested activities.

Extensions

1. Short dialogues such as these can hardly tell all there is to know about specific fuels. Have small groups pick one of the fuels and work with some classmates to find facts that these characters did not already mention. Let them draw their own educational cartoon using funny characters to get their point across.
2. Some fuels currently in use—such as petrodiesel or ethanol—are not mentioned in this handout. Ask students to research these and report back to the class on pros and cons of these fuels. They may wish to illustrate or act out their discoveries.

ACTIVITY 5.3

Charting Energy

Materials

pen and paper; Handouts 5.3A and 5.3B

Primary Subject Areas and Skills

science, social studies, math, critical thinking

Purpose

Students read a bar graph and double bar graph to understand which nations use the most energy, and they gain an understanding of the term *per capita*.

As the world's population continues to grow, one of our biggest concerns is that we do not have enough resources to support the number of humans on the planet. But it's more than mere numbers. The amount of energy each of those people is using, on average, is what needs to be examined.

This exercise provides an excellent opportunity to discuss the meaning of *per capita* with your students. Begin the lesson by discussing this topic and then continue with the graphing activity that follows.

1. Ask students if they are familiar with the term *per capita*. If some are, ask them where they heard it or in what context. It is very possible that they came across it in newspapers or on the news, perhaps in a financial context. Per capita is a unit of measure that simply means "per person." In Handout 5.3B they will study how much energy various nations use per capita, or per person.
2. Instruct students, using dictionaries if necessary, to research the Latin origin of this phrase. They will discover that *per capita* is Latin for "by heads." Each "head" stands for one person.
3. Copy and distribute the graphs and questions in Handout 5.3A and 5.3B. You may wish to do this during a class period or to send home as homework in preparation for a longer class discussion or debate.

Extension

Ask students if they feel that people who can afford to pay for their energy should be expected to use less of it. Also ask them if quotas based on energy use—independent of an individual's ability to pay for what he or she uses—should be imposed for the sake of conserving resources for future generations. Instruct students to choose either a pro or con position on one of these topics, to research that position, and then write a paper describing their thoughts and analysis.

ACTIVITY 5.4

Home Energy Survey

Materials

pen and paper; Handout 5.4

Primary Subject Areas and Skills

science, social studies, math, critical thinking

Purpose

Students conduct an energy survey to gauge their home's energy efficiency.

The first step to conserving energy is performing a thorough inspection of one's home. This activity will help students understand how their home may or may not be energy efficient.

1. Copy Handout 5.4 and distribute to students.
2. Tell them that they will conduct a home energy survey to see how or where their home may be wasting energy. Tell them to pick a day when they can do a walk-through of their home with their parents or an adult family member. They will use this sheet to guide them.
3. They will need to check off each of the points as they walk through their house. If they cannot correct an item then and there, let them leave it blank and make plans with their parents to address the situation later. When they are done with their walk-through, have them answer the questions, then share their findings with the class.

Extension

Ask students what would happen if everyone in their school did a home energy survey. Do they think their town or city would be on the way to saving a lot of money? How could they put such a project into place?

EXTENSIONS

Establish an Energy Task Force

Have students take turns monitoring the amount of energy wasted at your school. Let these "Watt Busters" take turns monitoring the school's halls, looking for such things as lights left on unnecessarily in classrooms and offices, thermostats set too high, leaky windows and open doors in winter, overworked or wasted air conditioning in summer. Students may even want to design and issue "citations" to classrooms where lights or other equipment were left on without reason.

Build a Solar Cooker

One of the best ways to illustrate the power of the sun to your students is to let them experience firsthand just how easy it is to use the sun's power to do perform a basic, everyday task such as cooking. Solar Cooking International and other organizations provide information about how to make very simple solar cookers using materials like cardboard boxes and aluminum foil: http://solarcooking.org/plans.htm.

School or Home Retrofit

For a long-term project that can provide long-term savings and learning for your students and your community, challenge students to figure out what it would take for your home or school to convert to solar and/or wind energy. Using information students gather from their parents and the school, they should be able to prepare a thorough breakdown of costs. Be sure to check with your local Department of Energy or local board of public utilities to find out what tax credits and rebates are available for those who adopt alternative energy technologies. Installation costs and other start-up costs can often be reduced significantly. Some states actually pay as much as 60 or 70 percent toward the purchase and installation of panels and turbines.

- Ask parents or school officials for copies of one year's worth of utility bills.
- Based on the information you gather, calculate the amount of energy you would need to generate with solar or wind technologies.
- How many years would it take to pay back the initial investment and start saving?
- Ask for bids from at least three solar contractors. How much would such a project cost?
- Using the activity in this unit, perform an energy audit. If the cost of solar is still out of reach, how much could be saved by simply insulating or replacing windows?

RESOURCES

- **National Energy Education Development (NEED)**
  www.need.org
- **The Department of Energy has information for both teachers and students.**
  www.energy.gov/foreducators.htm
  www.energy.gov/forstudentsandkids.htm
- **The American Solar Energy Society**
  www.ases.org/
- **American Wind Energy Association**
  www.awea.org
- **The Department of Energy** also has an Energy Efficiency and Renewable Energy section that contains excellent information about alternative and renewable energies.
  www.eere.energy.gov/RE/solar.html
- **Renewable Energy Policy Project and CREST, the Center for Renewable Energy and Sustainable Technology,** have a very detailed site with information about all renewable energies, including wind, solar, biomass, and so on. It's one-stop renewable energy information shopping.
  www.crest.org

Reading Handout Unit 5: FRENCH FRIES IN THE TANK?

In this reading, you'll learn about a sustainable fuel called "biodiesel," which is made from various vegetable oils. At this time, biodiesel is not a major source of energy in the United States, but it *is* used by pioneering individuals to power their cars, trucks, and boats, and to heat their homes. Answer the questions after the reading.

Imagine running your car engine with the icky sludge dumped out of the fryer of the nearest fast-food restaurant. That's exactly what many Americans are now doing. They're staunch believers in an alternative fuel called biodiesel. This fuel can be made from any plant-based oil, such as soy, canola, sunflower, rapeseed, or peanut. To work as a fuel, these oils have to be chemically modified to remove the naturally occurring glycerols—the chief ingredient in soap—from the oil. Most suppliers sell fuel that has been made from freshly made oil, but a few diehards use recycled cooking oil, which requires additional filtering to remove excess food bits.

The resulting liquid is safer to transport, handle, and breathe than petroleum-based diesel or even gasoline. It can be poured directly into any vehicle—car, truck, tractor, or boat—with a diesel engine. (The vehicle requires no mechanical changes at all!) A tank filled with biodiesel runs 78 percent cleaner than petroleum-filled engines and produces no carbon dioxide. The vehicle gets similar or better mileage and has a longer engine life because plant oils do a better job of lubricating and scrubbing the interiors of engines than ordinary "dino-fuel." Biodiesel mixes well with other fuels, so you can easily switch back to petrodiesel if you can't find your fuel of choice on a long car trip.

Biodiesel is popular in Europe because any nation that can grow soybeans or sunflowers can make its own fuel. The technology is slowly catching on in the United States as forward-thinking companies unveil new pumps and grassroots organizations form cooperatives to make and sell the stuff.

According to the website biodiesel.org, which features a clickable map showing all such retail outlets in the United States, Seattle alone is home to at least five different biodiesel retail locations. The Washington State Senate is considering a bill to help farmers grow enough "fuel crops" to produce 20 million gallons of biofuel by year 2020. Because this technology can help reduce American dependence on foreign oil and halt global warming, the bill also calls for generous tax breaks intended to bring the cost in line with ordinary gasoline.

One drawback: Biodiesel limits the make of cars you can drive. Currently, the only new passenger diesel vehicles sold in the U.S. are trucks or Volkswagens. A few older diesel brands are still available.

On a recent visit to Seattle, we spent time with a number of biodiesel entrepreneurs—those who brewed the stuff in their garages for their own use, as well as large companies. The day we stopped by Seattle Biodiesel, a railroad car filled with 26,000 gallons of soybean oil had just pulled into the facility, which can hold up to eight such railcars. Right now, Seattle Biodiesel uses the processed vegetable oil to crank out 5,000,000 gallons of biodiesel a year. At the moment, the closest source for all that vegetable oil is South Dakota. The railcars come in, pump oil into the tanks, and then head back east to the Great Plains. The owners of Seattle Biodiesel, John Plaza and Martin Tobias, dream of lining up enough vegetable oil producers closer to home.

Mr. Tobias's conversion to the new fuel came after he had spent considerable money supercharging his sport-utility vehicle to make it go faster. A few conversations with Mr. Plaza, and he was hooked. He sold his gas guzzler, switched to a couple of cars with diesel engines, and invested in Mr. Plaza's fledgling company. As Mr. Tobias says, his investment was more than a smart business move. It was a major reassessment of a lifetime spent in business.

"I've been investing in software companies for about five years myself," he says. "I started my own software company, and I really believe in the American spirit and the American ability to innovate our way out of any problem. Biodiesel is one way that we can do that. There's absolutely no reason why we should be so dependent on foreign oil when we can grow all the fuel that we need here. We can grow it, we can do it more efficiently, and there's no reason why we shouldn't. The industry just needs a bit of help from professional businesspeople with some capital. I looked at my own investment history, and I said, 'Do we really need another Internet security software company, or do we need to innovate and solve our problem of dependency on foreign oil?' I think we need to apply more capital to that problem, and then as Americans we can solve it. Biodiesel gives us the ability to solve a huge issue *today!* We don't have to invent a new technology, we don't have to change the infrastructure, we don't have to change our vehicles, and we don't have to change our behavior. We just have to change our source of energy."

His partner, Mr. Plaza, says their work is just one link in the chain toward a sustainable Seattle. "By starting this company, we can change our city from burning petroleum and only petroleum. We can start to see a change in behavior. Seattle has the largest population of personal biodiesel users in the nation. We have more than 1,500 people driving diesel cars for which they did no modifications. That's a huge change that we've seen in just the past two to three years. That's people empowering their community to effect change. What better hope can we have than to contribute to that movement? Two years ago, when we started Seattle Biodiesel, I'd wear my [Seattle Biodiesel] shirt and people didn't have any idea what biodiesel was. Now, when I walk down the street, everybody knows about biodiesel. They all want to talk about it."

When was the last time you hung out at your local gas station, chatting with your fellow gasoline buyers about your choice to switch to super unleaded? It never happens. But biodiesel converts are a different breed, and the places that supply their fuel have become a little like well-loved coffeehouses or organic co-ops where people gather to see friends and swap the latest news on their ever-growing circle. In his Seattle garage, Lyle Rudensey makes his own fuel and spreads the word by maintaining a website of biodiesel news, visiting classrooms, and running an organization called the Breathable Bus Coalition, which tries to convince school districts to switch their buses to biodiesel. Such a switch would not only reduce toxic emissions but also lessen the harmful impact of those emissions on children and bus drivers. As Mr. Rudensey points out, students have shown that children and bus drivers are exposed to up to 70 times as much cancer-causing, asthma-inducing petrodiesel emissions inside a bus as off the bus!

The emissions issue is an important one for families with young children. Seattle resident

Tom Kilroy switched his family to biodiesel a few years ago, after he was appalled by the low mileage of the new cars he was thinking of buying. When he heard about biodiesel, he couldn't sell his old cars fast enough. He and his 8-year-old son Dillon made a trip to San Diego to pick up their new diesel car, carefully planning their trip back to visit all the biodiesel filling stations from California to Washington State. "While we were fueling in Garberville in the redwood forest," he recalls, "a big 18-wheeler pulled up, and Dillon told the driver about biofuels. The trucker was clearly impressed and put in about 30 gallons of pure biodiesel. He thanked Dillon for his enthusiasm and info. He is quite the little spokesperson for the national biodiesel board! We now have three vehicles that run on biodiesel. We also use biodiesel as home heating oil for our furnace! Home heating oil is a huge untapped market!"

The switch thrilled Mr. Kilroy's wife, Sandy, who works in salmon restoration efforts. "Biodegradable, nontoxic fuels such as biodiesel fit into her goals of reducing water and air pollution," says Mr. Kilroy.

Questions

1. What is biodiesel made of, and how does it differ from the way regular gasoline is made?

____

____

2. According to the author, why is biodiesel better or gentler on cars than regular gasoline?

____

____

3. Why is being able to grow fuel more desirable than pumping it out of the ground?

____

____

4. What does Mr. Tobias mean when he says the biodiesel business needs help from businesspeople with capital? What is capital, and what role does it play in new businesses?

____

____

5. According to Mr. Tobias, the United States can easily switch over to biodiesel because only one thing needs to change in the way we use our vehicles. What is that one thing? What four things do not have to change?

____

____

6. Why do you think so many Americans are interested in reducing the nation's dependence on foreign oil?

____

____

7. Mr. Plaza says that by using biodiesel, people can "empower" their communities to bring about change. How do you think using a new type of fuel would "empower" a community, or make it stronger?

____

____

8. The reading does not mention any other types of new fuels or how biodiesel compares to them. What other types of information might you have liked to see in the story? Make a list of two to five questions you have that were not answered in the article that you wish had been.

____

____

REPRODUCIBLE 5.1A: Fossil Fuels

Fossil fuels—as their name suggests—are fuels whose formation dates back hundreds of millions of years, more specifically to the Carboniferous Period of the Paleozoic Era, around 360 to 280 million years ago. The main component in fossil fuels is carbon. All fossil fuels are nonrenewable—there's only so much available on the planet. They took millions of years to create and cannot be easily replenished. There are three types of fossil fuels: coal, petroleum or oil, and natural gas.

COAL

Coal is the most abundant fossil fuel produced in the United States. It is a rock made up of carbons and hydrocarbons. The rock was formed by the bodies of dead plants that lived hundreds of millions of years ago, even before dinosaurs. When the plants died, their remains sank into swamps and oceans and were buried, trapping their energy. Over hundreds of millions of years, Earth's pressure and heat compressed these plants, turning them into coal.

Coal is retrieved by mining into Earth, a labor-intensive, life-threatening task for miners. Once coal is retrieved, it is processed to remove dirt, unwanted rocks, and the like that reduce the coal's efficiency. After the coal has been processed, it must be shipped to its destination, which often costs more than mining it! Coal is burned to produce electricity at power plants. Burned fuel releases CO_2, or carbon dioxide, into the atmosphere.

OIL

Like coal, oil was formed underground millions of years ago. As many scientists believe, the formation of oil began with sea organisms called diatoms, which were able to convert sunlight into energy just as plants are able to do today. These miniscule animals, full of this converted energy, sank to the bottom of the sea when they died. They were buried by sea muck, and their bodies were transformed to a thick black liquid over hundreds of millions of years.

NATURAL GAS

Natural gas is almost always found near petroleum but is itself primarily made up of methane, or CH_4. Natural gas is highly flammable and can be used to cook food on your stove and heat your home, among many other things. Though petroleum products have a wide variety of uses, the vast majority are used for gasoline and transportation. Both oil and natural gas are obtained by drilling deep into Earth, where the deposits are found, sometimes within rocks themselves. Once pumped to the surface, natural gas also has to be refined and processed before being transported to its destination and used.

REPRODUCIBLE 5.1B: Sun and Wind: Renewable Energy Sources

The most popular forms of alternative energy are solar power and wind power. The biggest advantage of these forms of energy is that they are renewable. As far as we can tell, we will never run out of sun or wind.

SOLAR POWER

Anyone who has ever sat in a car on a hot, sunny day knows how powerful the sun's rays can be. The sun's energy travels 93 millions miles to Earth, carrying with it the power to generate heat and electricity for home and businesses, as well as providing necessary nutrients for photosynthesis in plants.

The sun can be used to generate and store electricity, and to heat water in pipes. You have probably seen solar batteries or photovoltaic cells on calculators and traffic signs. Using the sun to provide enough electricity to run a house, however, is more expensive and requires more equipment.

How does this happen? When light strikes a photovoltaic cell, it is absorbed by a semiconductor. Semiconductors contain silicon and are a vital part of all computer products. Light is absorbed by the semiconductor. Its electrons are knocked free and begin to flow. Flowing electrons form a current. That current can be drawn off and used as electricity.

Solar power is becoming more affordable all the time. Many power companies buy power from people with solar power cells on the roofs of their houses. If your house makes more electricity than you can use, you can send that solar power *from* your house *to* the power station, making your electric meter spin *backward*. Then the electric company subtracts from the money you owe them for the power you may have used when the sun wasn't shining!

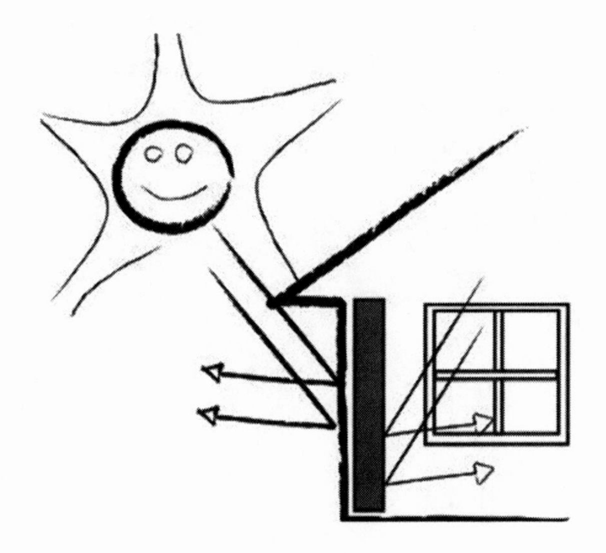

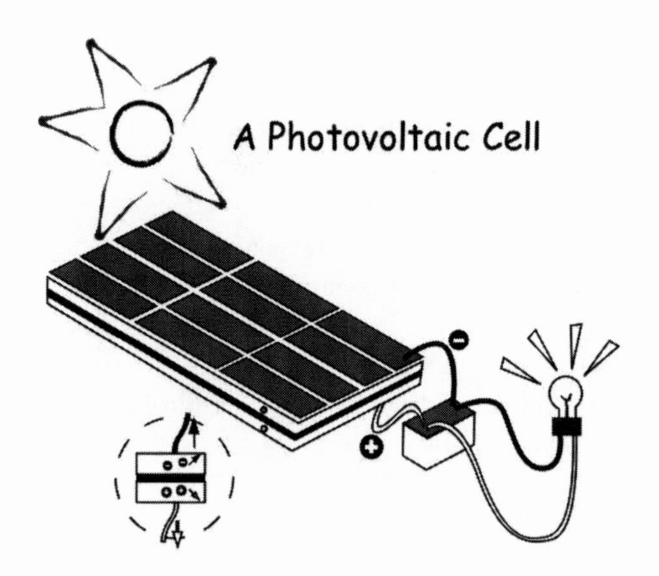

Wind Turbine

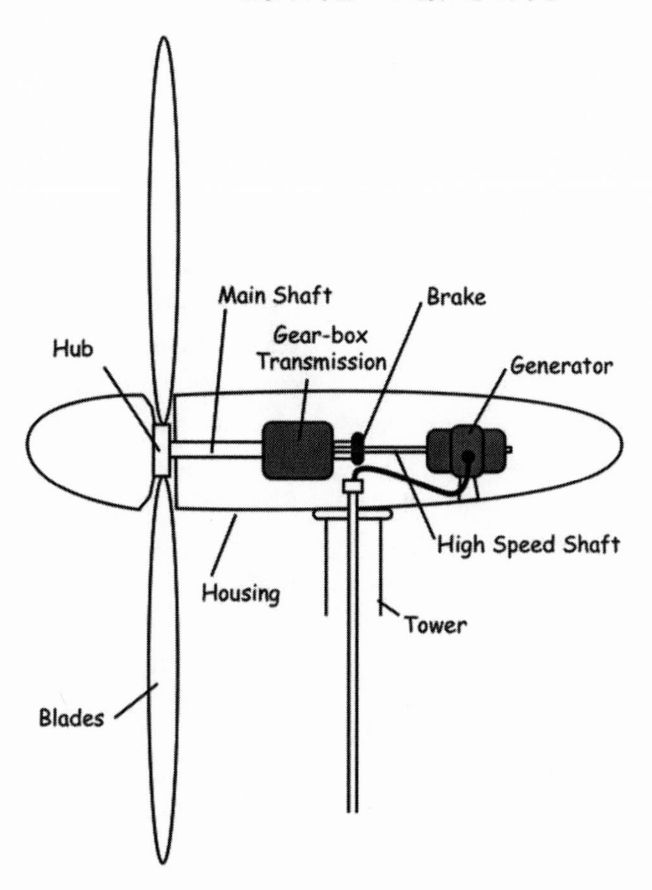

WIND ENERGY

Hundreds of years ago, windmills were a very common site on landscapes, where they helped pump water, for example. Though windmills are still used, an ever-increasing sight on the horizon is the sleek, futuristic-looking wind turbine. As wind turns the blades of the turbine, the shaft of the turbine turns. This is connected to a generator where the wind's *kinetic* energy (energy of movement) is then converted into mechanical energy (the energy of work). The resulting electricity is sent by wires to the nearest power station, where it is then sent to your house.

Name ______________________ Date ______________

HANDOUT 5.2: I Love My Car—I Just Hate Paying for Gas!

GASOLINE

Woman in minivan: I love my minivan, but I hate filling it! I get 19 miles to the gallon driving in town, and I have a 25-gallon gas tank. Filling it is sending me to the poorhouse! There's got to be a better way!

Man in huge SUV: Hey, if there is a better way, it's not worth driving! This baby can haul tons of cargo, tow a boat, and go off-road camping. Sure, it's expensive to drive. I've got a 44-gallon tank and only get about 14 miles to the gallon, but what choice is there? Nothing beats gasoline! You can buy it anywhere in the country. It runs every car made around the world. If I switch to some wacky new fuel and go on a trip, how will know if I'll find a place to fill up again when I run out? It doesn't seem worth switching. People are working on the problem. Sooner or later, they'll find more oil, and gas will be cheap again.

Woman scientist in lab coat with clipboard: You're right! Gas is very convenient and available all over the world. But oil is a finite resource, so some day we *will* run out. It will never be as cheap as it once was. The harder it gets to drill for oil, the costlier it will become. Sooner or later, other types of fueling options will become more visible at the pump near you. Don't worry: New fuels will be just as "powerful" as gas. Big strong trucks will work perfectly fine with them, too. I'm glad you enjoy your vehicle, but will you be happy driving it when it costs you $200 to fill the tank?

NATURAL GAS

Man and woman in small rental car: We're on vacation and rented this car at the airport. It wasn't until we opened the trunk that we found out that it runs on natural gas. A large natural gas tank takes up most of the trunk space! The car is much, much cheaper to fill up. But not every gas station sells natural gas, so we have to plan our trips in advance. (That's hard if you're a tourist and don't know your way around.) Because it needs a bigger tank, we have no room in the trunk. Why would anyone build such a silly car?

Female scientist in lab coat: Actually, it's not silly at all. The United States has more reserves of natural gas than of oil. That makes natural gas about 80 percent cheaper to buy than regular gasoline. Right now you'll find most natural gas cars in California, but soon they will be all over. For now, it's hard to find a natural gas station to fill up. But that will change. Soon, you'll be able to fill up U.S. cars in your own garage! The cars are great to drive to and from work, and natural gas emits 70 percent less carbon monoxide, 87 percent less nitrogen oxide, and 20 percent less carbon dioxide than ordinary gas!

BIODIESEL

Woman in old diesel car: My car runs on biodiesel, which is a mix of different types of vegetable oils. It's safer to handle than gasoline. It cleans out engines better than gasoline or ordinary diesel fuel, is biodegradable, burns 75 percent cleaner than ordinary gas, and emits no CO_2. The one hitch is that you need to buy a car with a diesel engine to drive with it, and only one car maker sells one in the United States anymore. (I bought mine used for about $2,500.) I love biodiesel. It's the greatest ever!

Trucker in a big rig: I like it too! But you can't buy it everywhere. There are only about 1,500 gas stations in the country that sell it. I buy it when I can. I like the way it makes my big ol' truck run. The exhaust smells like french fries. The good thing about biodiesel is you don't have to change your engine in any way to use it. When I run out of biodiesel, I fill the tank with ordinary diesel, and my truck runs just fine. Whoever invented the diesel engine must have been a genius!

Rudolf Diesel: Why, thank you! My name is Rudolf Diesel. In 1892, I designed my famous engine to run on peanut oil.

Female scientist: You're my hero!

HYBRID CARS

Woman in slick new hybrid: My family paid almost $22,305 for this little car, but I wouldn't drive anything else! It's a "hybrid-electric" car, which means it has a combined gasoline–electric engine. When I hit the brakes, the electric engine stores excess energy in a battery and automatically shuts off the gas engine. That way, I run on the battery much of the time and use less gas. Unlike regular cars, this one gets better mileage driving around town than on the highway. All that stop-and-go really charges the battery! I get 44 miles to the gallon of ordinary gas, on average!

Male accountant: I'm glad you love your new hybrid. They are indeed the hot new cars of the moment! But not all of them get great mileage, and even the good ones don't save you that much money in the long run. That's because they cost more to buy than other cars their size. You can find fuel-efficient small cars that get 38 miles to a gallon, and higher, that cost much less. You will consume less gas, and you are spewing less CO_2, but you are not "saving" more money for yourself. But who knows? If the cost of gas goes higher and higher, hybrids may turn out to be a better ride!

HYDROGEN FUEL-CELL CARS

Man in Test Car: I'm the luckiest man in the USA! I'm driving the car of the future! It runs on hydrogen gas. Inside my engine, hydrogen and oxygen are mixed together to make an electrochemical reaction. One by-product of this reaction is electricity, which runs the car. The other by-products are water and heat—*steam*. I'm driving a test vehicle because it could be another 10 or 20 years before hydrogen cars are everywhere, and all the gas stations in the world are switched over to hydrogen pumps!

Female scientist: That's right, you lucky dog! NASA has used hydrogen for decades to send rockets to the moon. Believe it or not, astronauts drank the pure water produced by those engines. The idea is too costly to sell in mass-market vehicles. Hydrogen is the most plentiful element in the universe, so we'll always have lots of fuel. Because water vapor is harmless, the engine would run cleanly.

Questions

1. How much would the man and woman driving ordinary gasoline cars pay, per gallon, to fill up their vehicles if they bought gas in your town today? (Consider checking out the website www.gaspricewatch.com)

____

____

2. What did you think of the man's arguments for why he liked gas-powered cars? Do you think he's got a point? Why or why not?

____

____

3. Which fuel did you find most interesting? What did you like about it, the technology or the rationale for using them?

____

____

4. Which fuel or transportation mode did you think was troublesome or foolish? Can you expand on some of the things already said about that fuel to explain why you feel that way?

____

____

5. Imagine you were in the market for a new car and intended to drive it in and around your neighborhood. Which fuel type do you think you would gravitate toward? Explain the reason for your choice.

____

____

6. In February 2006, a group of five high schoolers from Philadelphia presented a car that runs on soybean oil at the Philadelphia Auto Show. Now that you know a little about cars, fuels, and engines, can you guess which car in the handout might most resemble the soybean oil car in design?

____

____

Name ______________________________ Date ________________

HANDOUT 5.3A: How Much Energy Do We Really Use?

Look at the bar chart below. Then answer the questions on a separate piece of paper.

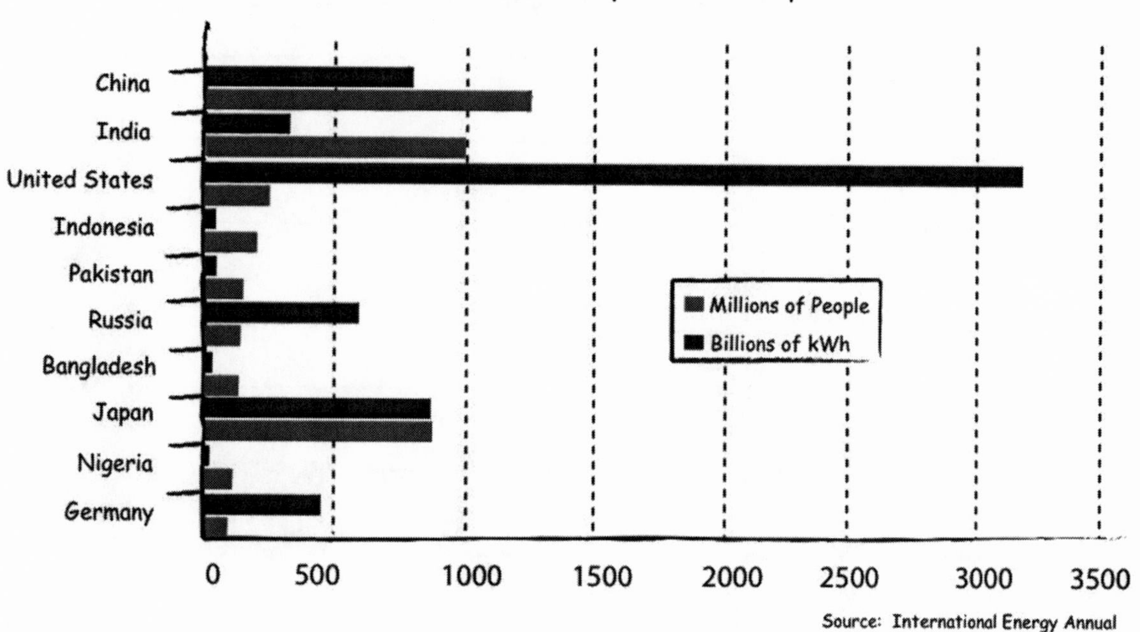

1. Which country has the most people? Approximately what is its population?

2. Does the country with the most people use the most kilowatts?

3. Which country has the least amount of people? Approximately how many kilowatt hours (kWh) does it use?

4. In which countries do the billions of kWh used exceed the millions of people living there?

5. In which countries are the billions of kWh used less than the millions of people living there?

6. Approximately how many more people live in India than in the United States?

7. Approximately how many more kWh does the United States consume than India?

8. Which is greater: The amount of kWh used by India, China, and Russia together, or the amount used by the United States?

9. What does this chart tell you about global energy consumption?

Name ______________________ Date ______________

HANDOUT 5.3B:

How Much Energy Do We Really Use?

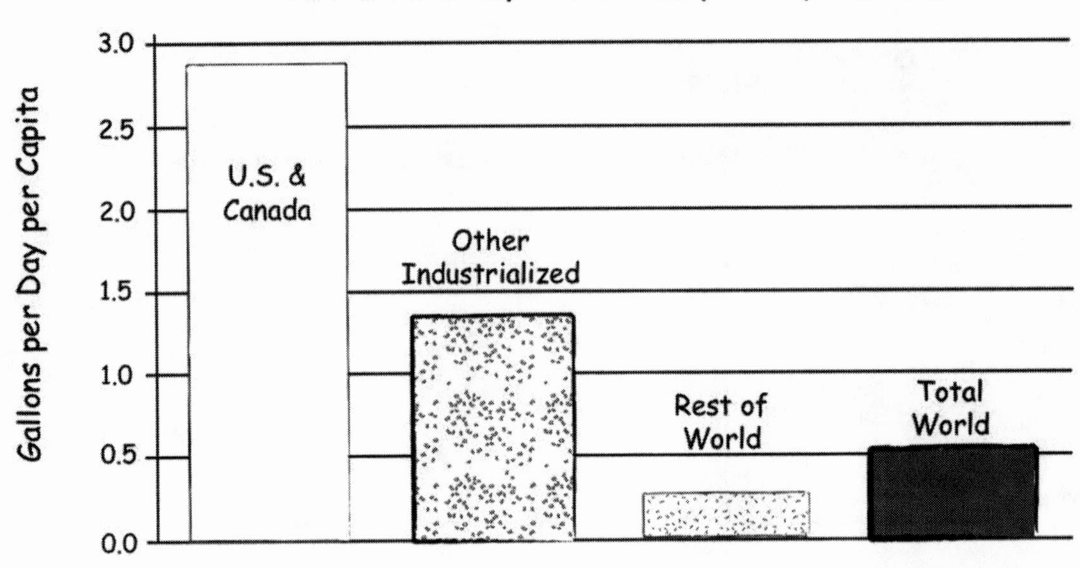

1. Approximately how many gallons per day per capita are consumed in the United States and Canada? __________
2. Approximately how many gallons per day per capita are consumed in the group of countries that would include *other industrialized* nations? __________
3. Approximately how much more oil per day is consumed in the United States and Canada than in the entire world, on average? __________
4. Which is greater, the amount of oil per capita consumed by the United States and Canada or the amount of oil consumed by other industrialized countries and nonindustrialized countries combined? __________ How much more? __________
5. Are you surprised by the information in this chart? __________Why or why not? Write your answer in complete sentences.

HANDOUT 5.4: HOME ENERGY SURVEY

Date Of Survey ______________________________

Home Of ______________________________

Survey Taken By ______________________________

IN THE KITCHEN

Is the freezer partially empty?
Fridges and freezers work best full. Fill empty spaces in your freezer with small plastic containers filled with water. Use the ice to water plants and cool drinks.

Is your refrigerator losing its cool?
Close the fridge door against a piece of paper. If the paper comes out easily, it might be time to change the seal. Buy new ones at a hardware store.

Is your fridge in the wrong spot?
Your fridge works harder if it is in full sun or next to your stove. Suggest to your parents that they move it to a better location, if possible.

Has your fridge been cleaned lately?
Twice a year, when you change your clocks, you should clean under your fridge and vacuum the coils behind it. (Clean its insides, too!)

Is your fridge at the right temperature?
The inside temperature of the fridge should be between 38 and 41 degrees Fahrenheit (3 and 6 degrees Celsius). The freezer should be between 0 and 5 degrees Fahrenheit (–18 and –15 degrees Celsius). If the refrigerator doesn't have a temperature gauge, the student will need to leave a small outdoor thermometer—the inexpensive ones used by hikers or skiers will do—in the closed refrigerator and freezer for about five minutes.

Does the kitchen faucet leak?
You could be wasting water. If you replace the faucet or change a washer, make sure the faucet has an aerator with a shutoff button.

Is the dishwasher usually full before running?
A dishwasher that is not completely full is wasting water. Also, it's not smart to rinse dishes before you place them in the dishwasher. You're only wasting more water. Simply scrape the leftover food scraps into the trash and load the dishwasher. Letting dishes air dry instead of heat-drying will also save energy.

IN THE BATHROOM(S)

Any leaky plumbing?
Check the sink and tub faucets. Check the showerhead. Check the toilet for excessive running. Visit a hardware store to get replacement parts.

Consider new heads for faucets and showers.
A "low-flow" showerhead uses only 2.5 gallons per minute. An aerator on your sink faucet lets you "pause" the water while you're brushing your teeth.

Make sure the bathroom fan is clean and working.
Remove excess dust or lint from the fan. Vacuum all heating vents.

Is the bathroom window drafty?
Make sure the window closes properly. Use weather stripping or other products to lock out drafts.

IN THE LIVING ROOM

Make sure electronic equipment is off when not in use.
TVs, VCR and DVD players, stereos, and other electronic equipment should be off when no one is using it. If you can live with the blinking "00:00" symbol, you might even consider unplugging them because they draw a charge even when on "standby."

Understand how much energy your computer uses.
It is usually better to shut off your computer when you are not using it. And screen savers often use more energy than they are worth. Check with your computer service provider if you are not sure of the efficiency of your model.

Check all windows for drafts.

Vacuum all heating vents.

Consider rechargeable batteries for small appliances.
Portable stereos, clock radios, and other small appliances work well with batteries that can be recharged again and again.

Check the fireplace.
When your fireplace is not in use, the flue should be closed tight. If you feel a draft, it might be time to have the damper replaced.

IN THE BEDROOM(S)

Check small appliances for rechargeable batteries.

Check that electronics are off when not in use.

Vacuum all heating vents.

Check windows for drafts.

IN THE BASEMENT

Check all windows for drafts.

Make sure doors leading outdoors are closed tight.

Check hot water pipes.
All hot water pipes should be insulated with foam tubing from the hardware store.

Insulate your hot water heater.
Water tanks should be "dressed" in an insulated blanket to keep excess heat from leaking out.

Replace or clean your furnace filter, if you have one.
Some furnaces are equipped with an air filter. All the dust in the house collects in this filter, which should be cleaned or replaced every three months.

Check the connections to your washing machine.
The water pipes leading to your washing machine may be leaking. Consider replacing them with antiburst tubing.

Check your clothes dryer exhaust line.
Clean out the lint trap on your clothes dryer and make sure that the exhaust line leading from the back of the dryer to the outside is free of lint, as well. Make sure the exhaust line is not squashed in any way. Hot air must be able to escape the dryer unimpeded.

THROUGHOUT THE HOUSE

Switch to compact fluorescent bulbs (CFLs).
Check every lightbulb in the house and consider switching it to a CFL bulb, which lasts longer and uses 75 percent less electricity. Bulbs that are used infrequently, such as ones in attics, closets, and basements, are less necessary to switch.

Check window air conditioners for drafts.
During winter, window air conditioners should be removed and stored someplace safe, so you can properly close the window. If that is not possible, then the outside part of the air conditioner should be wrapped with a waterproof tarp to keep cold air from blowing inside during the cold months. If you cannot reach the unit outdoors, do your best to cover the AC vents indoors.

OUTSIDE YOUR HOUSE

Check exterior lights.
Outside lights should be on a timer or on motion-detectors, so they go on only when someone passes by.

Check gaps around utility lines.
The spot where any outside power or utility lines enter your house might have holes around it where heat or cool air could escape. Seal any holes with caulk.

Questions

1. Did you find any evidence of energy waste in your home? Make a list of what you found.

   ____

   ____

2. What do you plan to fix in the near future? Why?

   ____

   ____

3. Visit a home center or hardware store and find prices for some common items needed to correct energy wasters on your list, such as insulating jackets for water heaters or hot water pipe insulating foam. What do these things cost?

   ____

   ____

4. Estimate the cost of the repairs you need to make at your own home. Do you think the repairs are worthwhile if they save your family money in heating, cooling, or electric bills? Explain your answer.

   ____

   ____

5. Listen to some of your classmates' reports of their home energy surveys. Did they uncover similar problems with their homes? What common problems were discovered?

   ____

   ____

6. How can you monitor the effects of the changes you plan to make?

   ____

   ____

UNIT 6

AIR QUALITY

TOPIC BACKGROUND

The air we breathe is one of the most precious resources on Earth, yet humans tend to treat it carelessly. Factories, power plants, and motor vehicles spew countless tons of harmful pollutants into the atmosphere every year. The chief issues in air pollution are the rise in greenhouse gases, the destruction of the ozone layer, and smog.

On their own, greenhouse gases such as methane, carbon dioxide, and ozone are neither bad nor good. Most occur naturally in the environment and are in fact necessary to sustain life on Earth. These gases have the ability to absorb the sun's infrared rays, or its "heat." Sunlight penetrates Earth's atmosphere, and these gases let the light escape but trap the heat, not unlike the function of glass windows in a greenhouse. Without our atmosphere to blanket Earth and retain some of the sun's warmth, our planet would be a very cold place indeed.

The trouble starts when human activity and industry pump many more greenhouse gases into the atmosphere than occur naturally. With a thicker "blanket" of gases around the planet, Earth retains more of the sun's heat. This is what scientists call the greenhouse effect, which has been implicated in climate change. (A diagram depicting this process is shown on page 85.)

Ozone is also a greenhouse gas that occurs naturally in the atmosphere. At higher altitudes—six to thirty miles up in Earth's "stratosphere"—ozone is one of the planet's chemical saviors. (For a detailed diagram of Earth's atmosphere, see Handout 6.1, page 92.) That's because the ozone layer is rather like a pair of sunglasses, shielding the planet from the sun's harmful ultraviolet rays. But stratospheric ozone is slowly being destroyed by the release of chemicals such as chlorofluorocarbons into the atmosphere. As the ozone layer thins out, more of the sun's harmful rays are getting through, harming living things.

At the same time, human activity is building up ozone in the troposphere (the part of the atmosphere nearest Earth), and this is not good. When tropospheric or *ground-level* ozone mixes with the fine soot, or particulates, from motor vehicles, factories, and power plants, it forms a dense, dirty layer of air called smog. Smog conditions are worse on warm, sunny days because sunlight triggers the chemical reaction that produces ground-level ozone. The production of ground-level ozone is Earth's vain attempt to rebalance the unstable mix of chemicals and oxygen in the air.

Unfortunately, that rebalancing act is unhealthy for humans. Taken together, these three air pollution issues are among the most challenging ones facing our planet. Smog exacerbates asthma and other ailments. The thinning ozone layer leaves us exposed to harmful radiation. And the buildup of greenhouse gases has heated up the planet about 1 degree Fahrenheit over the last century. Each new day brings news of melting ice in the world's coldest places.

Your class should keep in mind that all of these ideas are controversial. Some scientists insist that these issues are not nearly as bad as they have been portrayed, nor are they the result of human activity. They claim that these occurrences may be quite natural. However, the majority of scientists *are* concerned. Why else would 141 nations work to reduce their greenhouse gases through signing the Kyoto Protocol? (The Kyoto Protocol is an amendment to the international treaty on climate change, assigning mandatory emission limitations for the reduction of greenhouse gas emissions to the nations who sign the agreement.) As we go to press, the United States is not a signer of the Protocol, but many U.S. cities, led by Seattle, are working to reduce their greenhouse emissions. Nevertheless, the controversy behind climate change can make for spirited classroom debate.

What can be done locally about air pollution? Obviously, if a community identifies air pollution as a concern, it must do all it can to reduce harmful emissions. Municipal buildings and businesses would seek out less harmful sources for power. They may retrofit their structures with renewable energy. They may choose "green power" from their utility companies, if it is available in their area. They may encourage citizens to reduce their use of cars, to ride bikes, to walk, or to explore alternative-fuel vehicles. And they may plant more suitable trees to help filter the air of airborne pollutants.

Purpose
In this unit, students will learn the science behind air pollution, and how human actions and events are putting Earth's atmosphere in peril. They will learn about climate change, the thinning of the ozone layer, and how they can aid in cleaning the air where they live.

GROUP ICEBREAKERS

1. What is air pollution? What do you think pollutes the air? (Accept all reasonable answers at this stage, as students will not have had much instruction in this area.)
2. Can you think of three or four things each of us do every day that might contribute to air pollution? Think about how you get to school, how your home is heated, how different factories or businesses operate. (All acts of combustion produce CO_2, such as driving a school bus, running a furnace, or operating a factory.)
3. Do you think burning leaves, having a summer cookout, or having a nice fireplace fire on a cold winter night contribute to air pollution too? (They do! Some scientists say Earth's climate first began to change when humans began using fire. However, these three methods of combustion are far outstripped by CO_2 produced by the industrial world's machines.)
4. Do you think it is easier to clean polluted air or prevent air pollution in the first place? (The latter is easier and can be done by anyone—even young people.)
5. Can you think of some actions we can all take to lessen the amount of air pollution? (Simply choosing to walk or ride a bike to perform some errands would make a great difference. Lowering the thermostat in your home in winter helps, too, as does taking a shorter shower.)
6. Obviously, if we can think of ways to stop air pollution, it can be done. But why isn't it? Why do you think it's so hard to do some of these actions? (Accept all reasonable answers. In many cases, people want to help but simply do not know what they can do. The problem is one of education.)

ACTIVITY 6.1
The Air Up There

Materials
pen, paper; Handout 6.1

Primary Subject Areas and Skills
science, social studies, math, critical thinking

Purpose
Students use a chart depicting the layers of Earth's atmosphere to visualize the actual space above the planet that can be harmed by excessive CO_2.

To begin studying and discussing air quality, students first need to become familiar with Earth's atmosphere. The effects that different chemicals and emissions have on air quality often depend on where they're concentrated in the atmosphere. Ozone is important to Earth at upper levels of the atmosphere, yet very damaging closer to ground level. The objective of this activity is both to familiarize students with the different atmospheric levels and also to encourage them to think about the way that atmospheric conditions such as temperature and pressure can affect how different chemicals behave and interact.

1. Copy and distribute Handout 6.1, which depicts the layers of the atmosphere. Or you may wish to copy it for use with an overhead projector.
2. Instruct students to examine the diagram carefully and try to answer the questions that follow the chart. Explain to them that they may not know the answers, and that's OK. Make the best guess.
3. After they've finished answering the questions, go over the answers together and discuss any surprises.

Extensions
1. Have students choose a layer of Earth's atmosphere to research in more detail. Some levels are described by more than one name (ionosphere, for example). What do these names mean? What characterizes each atmospheric level?
2. Using sidewalk chalk outside on the pavement, students can create a replica of the layers of

Earth's atmosphere. Set the scale of the drawing based on the size of the pavement. For example: 1 inch = 1 mile. You may need to adjust (1/2 inch = 1 mile) after a trial run. Encourage students to do the math longhand, without a calculator.

ACTIVITY 6.2

Smog, Ozone, and the Greenhouse Effect

Materials

pen and paper; Reproducible 6.2 for overhead

Primary Subject Areas and Skills

science, social studies, math, writing, art, critical thinking

Purpose

Students will study three things—smog, ozone, and the greenhouse effect—with the goal of understanding the parts as well as the whole and the ways in which we can all work to reduce the detrimental effects on air quality.

Much more is going on over our heads than we think about. We hear about smog, the ozone layer, ozone pollution, and greenhouse gases—all important topics. But to truly understand the impact of these things, they must be viewed in relation to each other.

1. First, familiarize yourself with the following information on smog (section 1), ozone (section 2), and greenhouse gases (section 3). You may use this material as the basis of a class lecture.
2. Copy Reproducible 6.2 for use on an overhead projector.
3. Ask your students if they have ever seen smog hanging over a city, either from an airplane or off in the distance or even on TV or in the movies.
4. Remind students of the atmosphere chart they studied in the first activity. In which layer of the atmosphere do they think smog exists? (In the troposphere.)
5. Consider using the discussion questions or assigning some of the extension activities following the ozone and greenhouse gas sections.

1. Smog

A brownish haze hangs over Los Angeles, one of the cities profiled in *Edens Lost & Found*. All of us—no matter who we are, what we do or how much money we make—have to breathe the air. Air quality is a unifying need and cause, and one that is becoming increasingly more critical to our well-being. Two important components of smog are particulates and ozone. Particulates include dirt, dust, and smoke emitted into the air by fires, power plants, truck exhaust, factory smoke, and more. In the air these pollutants can irritate the respiratory system and hamper the body's natural immune defenses. Combined with ozone, discussed below, smog is not only annoying but may also contribute to life-threatening illnesses such as asthma, cardiovascular disease, and cancer.

2. Ozone

One of the biggest contributors to smog is ozone, and it can be very hazardous to the health of living things. "But wait," you might say, "I thought we needed the ozone layer?" True. We do. We need the ozone layer that exists in the stratosphere that protects us from the sun's harmful radiation. What we don't need is *ground-level ozone*—ozone that exists in the troposphere—which is a primary contributor to smog and a damaging irritant to the human lungs and eyes, as well as to other plants and animals. Simply put: Ozone up high, protection in the sky. Ozone down low . . . oh no!

Ozone is composed of three oxygen atoms (O_3), whereas breathable, atmospheric oxygen (O_2) is made up of two. Ozone is not emitted from anything we burn, like carbon dioxide (CO_2), nor is it the result of plant or animal respiration, like methane (CH_4). Ozone is formed in the atmosphere when volatile organic compounds (VOCs) combine with various oxides of nitrogen (NO_2 or NO_3, for example) in the presence of heat and sunlight. The role of heat and sunlight in the formation of ground-level ozone means that hot sunny days are more likely to increase "bad" ozone levels.

Ozone is also an important greenhouse gas that serves a vital purpose in the stratosphere. Beyond its role as a greenhouse gas, though, ozone in the stratosphere protects Earth from harmful ultraviolet radiation. The damage to the ozone layer in the stratosphere—which exists from approximately nine to 31 miles above Earth—means that more harmful ultraviolet (UV) rays are able to make it into our troposphere where they can cause skin cancer and damage our immune systems, as well as damage agricultural crops and even marine life.

What is depleting the ozone in the stratosphere? One of the main culprits is the group of pollutants known as cholorofluorocarbons, or CFCs. CFCs used to be much more commonly used as solvents, insulation, and refrigerants than they are now, thankfully, but they can last a long time, sometimes as much as a century, in the atmosphere. So the lasting damage of CFCs emitted five years ago can still be felt. In fact, just one chlorine atom can damage up to 100,000 good ozone molecules. CFCs aren't alone: Methyl bromide, hydrofluorocarbons, carbon tetrachloride, halons, and methyl chloroform are also responsible for depleting the ozone. These are found in pesticides, coolants, and aerosols, among other substances.

Extensions

Have students choose one of these topics and report more fully on it following some research:

1. What are chlorofluorocarbons? What is the chemical formula for some common CFCs? How are they released into the atmosphere?
2. Are CFCs regulated in the United States? How?
3. By what specific reaction does the chlorine atom in CFCs react to break down ozone in the stratosphere?

3. Greenhouse Gases

The greenhouse effect brings to mind climate change, unhealthy air, excessive emissions, and more. But the greenhouse effect is really a case of too much of a good thing turning into a very bad thing.

When the sun's rays penetrate the atmosphere and reach Earth's surface, some of that light is reflected back out of the atmosphere, while some of it is converted into heat in the form of infrared waves. This heat is kept within Earth's atmosphere by gases that have the ability to trap infrared waves. These are the so-called greenhouse gases, and this process warms Earth's atmosphere. The process is natural and desirable. If all of the sun's rays were simply reflected back into space, Earth would be a lot colder, and life forms wouldn't be able to survive. This very process is what has allowed plants and animals to flourish.

Now, however, human-made processes have drastically increased the amount of greenhouse gases in the atmosphere. One of the main greenhouse gases, CO_2, is produced by the burning of fossil fuels. Additionally, forests that help remove CO_2 from the atmosphere are being destroyed. As a result. Earth's average temperature is on the rise.

Carbon dioxide is not the only gas that results from human-made products and practices. CFCs also damage the ozone layer and are potent greenhouse gases. Nitrous oxide (N_2O), which can result from fertilizers, as well as methane (CH_4) can rise into the atmosphere from overburdened landfills, becoming greenhouse gases as well.

The following questions may require additional research. Students may work together or in pairs.

DISCUSSION

1. What is the link between greenhouse gases and excessive consumption and wasting of energy? (Wasted energy means more CO_2 in the atmosphere, which means that more of the sun's infrared rays will be trapped, which in turn warms up Earth.)
2. Which of these greenhouse gas sources do you come into contact with in your daily life? (The most common gas we come in contact with is CO_2, but our actions produce methane from landfills and other gases.)
3. How does each of the following activities help to send greenhouse gases into the air?
   a. Driving a car (CO_2 is released by the burning of gasoline.)
   b. Cooling your home (CO_2 is released to create electricity to power an air conditioner.)
   c. Using a clothes dryer (CO_2 is released to create electricity to power an appliance.)
   d. Taking a long, hot shower (CO_2 is released to create the heat to warm water, whether by gas, electric, or oil heat.)
4. Which of the greenhouse gases do you think would be easiest to remove or reduce from daily life? How would your new behavior reduce them? (By far, transportation contributes the most. We must figure out ways to reduce our

motor vehicle emissions. Walking or riding a bike helps enormously.)

ACTIVITY 6.3

How Much CO_2 Do You Contribute?

Materials

calculator

one year's worth of gas station bills or an estimation of the amount of gas used, based on miles driven over one year × the average gas price

one year's worth of household or school utility bills showing the amount of electric, gas, propane, or other fossil fuels used to heat and light a student's home

Reproducible 6.3 (optional) for overhead

Primary Subject Areas and Skills

science, social studies, math, research, critical thinking

Purpose

Students assemble real-life data from their families and analyze it to estimate how much CO_2 their family has released during a particular period of time.

As much as we'd like to help the planet, each of us adds greenhouse gases to the atmosphere, and CO_2 is unfortunately one of the easiest to generate. We can use math to calculate just how much our families are contributing and then think about ways to reduce that contribution.

1. Decide whether to analyze the class's home or school consumption. You may wish to divide the class and have different groups focus on just one of the different venues.
2. Estimate the gallons of gasoline bought in the year being studied. This can be determined from home receipts, online banking records, car mileage, or school maintenance records.
3. Have students check the home or school utility bills to determine how many kilowatt hours (kWh) were consumed.
4. If natural gas is used to generate electricity in the home or school, tally the therms used.
5. *Note:* If your local electricity provider uses some renewable energy, your calculation will not be perfectly accurate. However, this exercise still provides a good idea of just how quickly these emissions add up.
6. Parents may not want to contribute their bills, so you may wish to provide those from your own house or ask the school for its bills.
7. Have students use the data they've gathered to estimate the amount of CO_2 released into the atmosphere in the past year by their homes or by your school. Reproducible 6.3 shows some simple formulae that you may wish to place on an overhead.

Extension

Once students know how much CO_2 they have contributed, have them devise a plan for their families or for their school for reducing the amount of CO_2 that is released. Instruct them to use mathematical examples to support their suggestions, including charts or graphs that show how emissions could be reduced over a three-month, six-month, or one-year period. Consider presenting the students' writing to media outlets or civic leaders.

ACTIVITY 6.4

Understanding the Greenhouse Effect

Materials

Per experiment:

2 identical glass jars (e.g., large mason jars)

2 thermometers

1 plastic bag, clear, that can be sealed, large enough to contain jars

tap water and an evenly divisible number of ice cubes

a sunny spot (or sunlamp)

Primary Subject Areas and Skills

science, math, research, critical thinking

Purpose

In a simple classroom experiment, students duplicate the greenhouse effect on a much smaller scale. This experiment can be done as a class or in small groups.

1. Ask your students if any of them have ever made sun tea. How did it work? (Tea bags are left in a large, closed container of cold water for several hours in the sun.) Why did it work? (The sun warms up the water, which allows the tea to steep.) Do they think the results would be the same if the lid were not kept on? Why or why not? (The results would not be exactly the same. Without the lid, the water doesn't warm up as

fast. Colder or tepid water takes longer to make tea, and the taste is not always satisfactory.)

2. Divide the class into groups for multiple experiments or do as a group activity. Explain the instructions to the students.
   - Fill each of the two jars with exactly the same amount of water (two to three cups). The water should come from the same tap and be at roughly the same temperature. Take the initial temperature of the water and record the results.
   - Add several ice cubes to each jar (add the same number to each). The ice cubes should be uniform in size and shape.
   - Place one of the jars in the sealed, clear plastic bag.
   - Place both jars on a sunny windowsill, outside in the sun, or under a sunlamp for one hour.
   - When the hour is up, remove the plastic and the lid from the one jar and the lid from the other. Do this at exactly the same time. Measure the water temperature in each of the jars at exactly the same time. Record the results.
3. What were the results? Were students surprised by the results?
4. Ask students to explain how the results found in this experiment are analogous to the greenhouse effect. (The plastic traps the sun's heat, so the jar wrapped in plastic will tend to be warmer than the jar that isn't wrapped in plastic.)

EXTENSIONS

Greenhouse in the Garden

Consider having students build their own easy greenhouse to better comprehend the connection to so-called greenhouse gases. Greenhouses and cold frames are a familiar sight in many gardens and a fine way to keep plants protected from falling temperatures. Bear in mind that your local climate will affect what time of year you will be able to perform this experiment.

1. Get a pack of parsley seeds. Planting should take place in spring, when it's still quite cool, under 50° F—though not when the ground is frozen.
2. Pick a sunny location, perhaps south facing if possible. Mark off two 12-inch-by-12-inch squares. Place bricks around the edge of one of the 12-inch squares. Leave the other 12-inch square alone. Plant 10 seeds in each square, according to packet instructions.
3. Take a small piece of glass and gently lay it over the bricks so that it completely covers all the seeds planted in that square. Do nothing to the other group.
4. Use a classroom chart to record how well both sets of seedlings grow.
5. Ask students to interpret the results. Do the seeds exposed to cold air fare as well as those in the makeshift greenhouse? Why do they think the seedlings under glass did better?
6. Ideally, the lid of your cold frame should be removed when outdoor temperatures are solidly above 50 degrees. If possible, as a comparison, repeat the experiment in much warmer weather. Once the seedlings have sprouted, those that are growing under the glass will eventually become scorched and burnt if the glass is not removed. This extended experiment allows students to see, on a very small scale, how the greenhouse effect can be beneficial to plants up to a certain point and how it can become damaging when the warm air is not released.

Your Teenage Lungs

Scientists have always known that air pollution is detrimental to the health of infants and small children, who are in the developmental stages of growth. But new findings suggest that the lungs and health of teenagers may be likewise damaged by air pollution. Ask your students to investigate this aspect of the air quality issue and decide what they can do to bring this information to the attention of local leaders. This might be very important if your local high school is located near a factory, power plant, or major highway, for example.

Some links to investigate include:

- **New England Journal of Medicine**
  http://content.nejm.org/cgi/content/abstract/351/11/1057

- **U.S. News & World Report**
  www.usnews.com/usnews/health/briefs/childrenshealth/hb040908f.htm

- **CBS News**
  www.cbsnews.com/stories/2005/02/15/tech/main674353.shtml

Air Quality Near You

Have students use some web-based calculators and search engines to determine the relative health of their neighborhood's air quality. The website below will even tell them the name of the companies that are the largest polluters in their area! For comparison's sake, have them compare their zip code's air quality to neighborhoods where their friends or relatives live. The more scattered the zip codes are, the greater the opportunity for truly eye-popping differences. They may wish to expand their work to a full-fledged project, writing letters to major polluters, presenting their findings at local board meetings, and communicating their findings in the opinion (op-ed) pages of local newspapers: www.scorecard.org/

RESOURCES

- **Switch your home or school to "green" power**
  www.green-e.org

- **The EPA's air quality website**
  www.epa.gov/ebtpages/air.html

- **Andy Lipkis's organization, TreePeople**
  www.treepeople.org

- **Government site with news, maps, and features about air quality**
  http://airnow.gov/

- **2004 study showed that air pollution adversely affects the lungs of teens and preteens.** Listen to the NPR radio program.
  www.npr.org/templates/story/story.php?storyId=3905034

Reading Handout Unit 6:
TREE BOY

In this reading, you'll learn about a teenager growing up in the 1970s who decided to save an entire forest from the damaging effects of smog. After you read the story, answer the questions.

Andy Lipkis is a 50-year-old man who works to save trees. But his story begins back when he was only 15 years old, attending a summer camp in the San Bernardino Mountains outside Los Angeles. A naturalist told Andy and his young friends that the pine trees in the local forest were dying a slow death because of smog. In 25 years the forest would be entirely gone, the naturalist predicted.

Some trees are harmed by certain chemicals released into the atmosphere by humans. One such pollutant, sulfur dioxide, is released by power plants that generate electricity by burning coal and oil. Another pollutant is ozone, released by factories and automobile emissions. Both of these gaseous pollutants are major components of smog.

Andy and the other kids were horrified. Andy asked if anything could be done to save the forest. The naturalist mentioned that some trees had no problem breathing in smog. He theorized that these smog-tolerant trees could be planted in place of the dying trees, but that would take a superhuman effort. Thousands of these special trees would have to be planted by hand to offset the dying ones. Who could do such a thing?

Andy believed he could. After he left the camp, the plight of those trees stayed with him. He did some research and found that the trees that were dying off were ponderosa and Jeffrey pines. He needed to plant smog-resistant trees such as sugar pines, Coulter pines, and giant sequoias. To pull off such a massive project, he would need money. At first he tried to interest local companies and banks into giving him a donation, but none of them would. He almost gave up, but a few years later, when he was a little older and going to college, he hit upon the idea of getting other summer camp kids to help him plant trees. The camps he approached were willing to help, but when Andy phoned the California Division of Forestry to order some baby trees, he was surprised to discover that he had to pay for them. The amount wasn't much, only $600 for 20,000 seedlings, but every business he approached for a donation said "no." To make matters worse, the State Forestry Division had a policy of plowing under and killing each crop of unsold seedlings at the end of the season to make room for a new batch. (They weren't set up to take care of older trees.)

The situation infuriated young Andy. Here he was, volunteering to reforest entire swaths of the California forest system, and the state wouldn't give him the trees to do it! They wanted him to pay for them, and if he or no one else bought them by a certain date, they would discard all these trees.

Andy hit the phones. He called the offices of local newspapers, politicians, and famous actors who had taken a stand on the environment. The newspapers and one politician offered to look into the situation. As soon as they phoned the State Forestry Division, the state announced it was willing to donate the trees to Andy, provided he could plant them. Andy knew he had the manpower—all those kids—but he would need even more than $600 to pull off such a huge project. Buying those seedlings was only the first step. Each seedling would need to be transplanted into milk cartons and kept alive until they could be transported to the proper site and planted. He'd need dirt, shovels, and trucks, too. Where would he get the money to do all that?

One of the local newspapers wrote an article about Andy's encounters with the State Forestry Division. Andy had told the reporter that all he needed to plant one tree was 50 cents from a willing donor. This tiny, reasonable amount of money caught on with readers everywhere. After that, letters starting pouring in. Grade school kids, teenagers, and senior citizens all sent him a few bucks out of their pockets. As the cash started flowing in, the publicity about the adventures of California's Tree Boy grew. When donations reached $10,000—more than enough money to get those trees in the ground—the executives of all the companies and banks Andy had first approached now saw that he was a legitimate organizer and fundraiser. They started writing checks too. Andy didn't know it then, but he was well on his way to a life of activism. His organization, TreePeople, was born, and before 1980 they had planted more than 50,000 trees. Andy became so famous in California that he was invited to appear on a TV talk show, *The Tonight Show*, hosted by Johnny Carson.

"I heard about Andy in the early 70s and started sending him some money," says the actor and green proponent Ed Begley Jr. "I didn't have a lot of money back then, but I sent him $10, $15 when I had it, $25 when I had that, and very much liked what he was doing. Then in the late 1970s when I started to work more and be more successful, I sent a little more money, and I met him at some function. I thought to myself, 'Wow, this guy's incredible.' He's about my age, and he's done this incredible thing. Look at the contribution he's made within his lifetime. He's planted so many trees, he's affected so many people, his work will continue.'"

Andy's story is told in the pages of a children's book called *Tree Boy*, published in 1978 when Andy was only 24 years old. Today he's an energetic 50-year-old with boundless enthusiasm. TreePeople still plants trees, but its mission has matured greatly, along with the vision of its founder. Mr. Lipkis is one of the foremost environmental thinkers on the topic of trees in the urban landscape.

Questions

1. According to the story, how did Andy first find out about the plight of the California forests?

___

___

2. In your own words, what is it about the air surrounding California that harms trees? What pollutants are involved, and where do they come from?

___

___

3. Do you think California is alone in having a problem with air quality, smog, and airborne pollutants? How could you find out about air quality issues where you live?

4. Replanting a whole forest of trees is hard work. Can you think of a way to solve the problem of dying trees that would not require planting new ones? Is there a way to keep trees from dying in the first place?

5. Can you think of some adjectives that describe the young Andy? What word best describes how he was able to accomplish his goals?

6. Why do you think banks and businesses first said no to Andy's request for money? Why would they change their minds later and give him generous donations? What does their behavior tell you about the way to go about fundraising?

7. If you were raising money for a worthwhile goal, whom would you ask to contribute? Make a list of people, businesses, or organizations that you would approach.

Name ______________________________ Date ________________

HANDOUT 6.1: The Air Up There

Look at the diagram below and answer the questions.

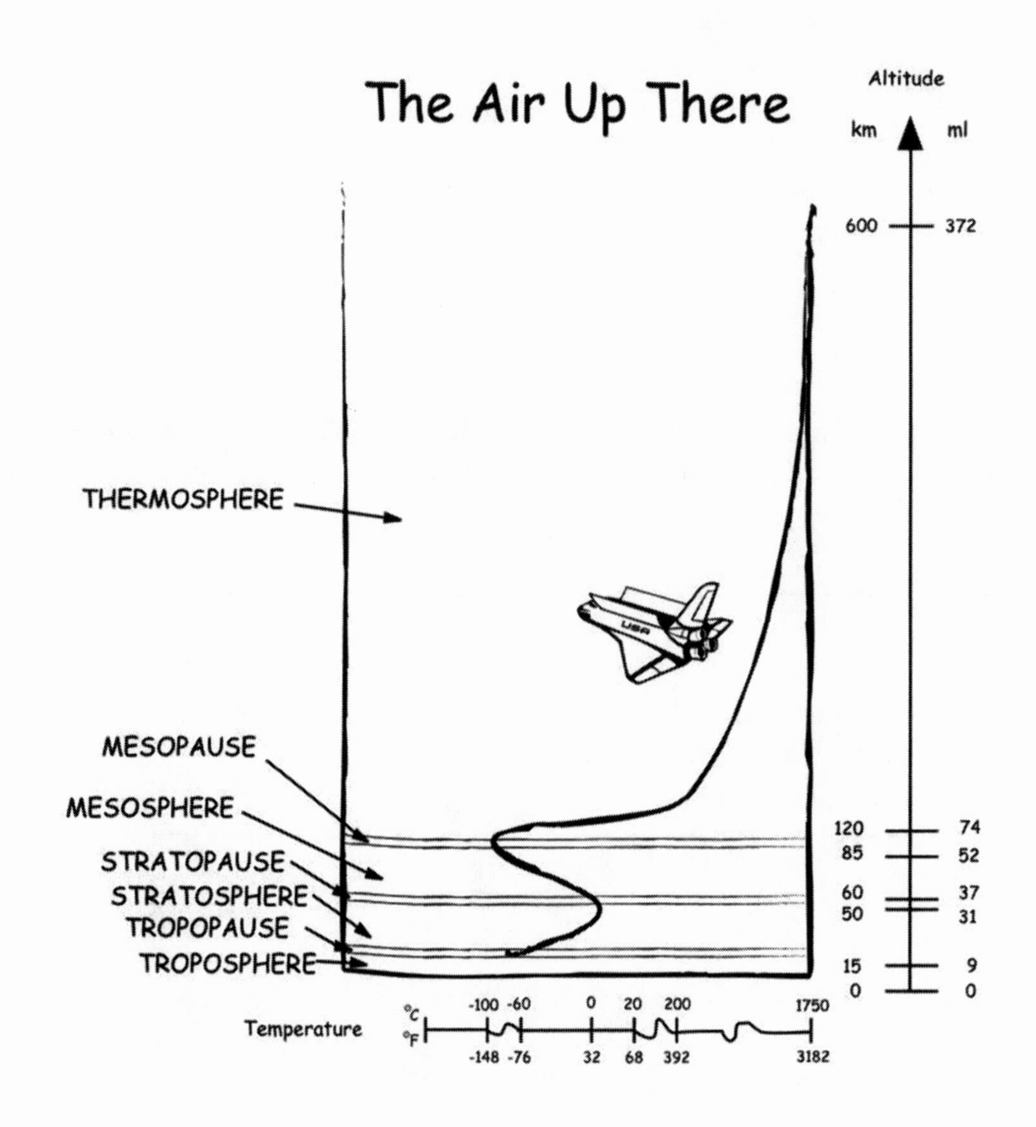

Source: NASA

1. In which section of the atmosphere do you think Earth's weather occurs?

2. The ozone layer is located 15 to 35 kilometers above Earth's surface. What layer of the atmosphere is it in?

3. About how many miles high does the stratosphere extend?

4. What is the average air temperature at 50 kilometers above ground level?

5. What separates the stratosphere from the mesosphere?

6. What do you think happens to the atmospheric pressure as altitude increases?

REPRODUCIBLE 6.2: Smog Diagram

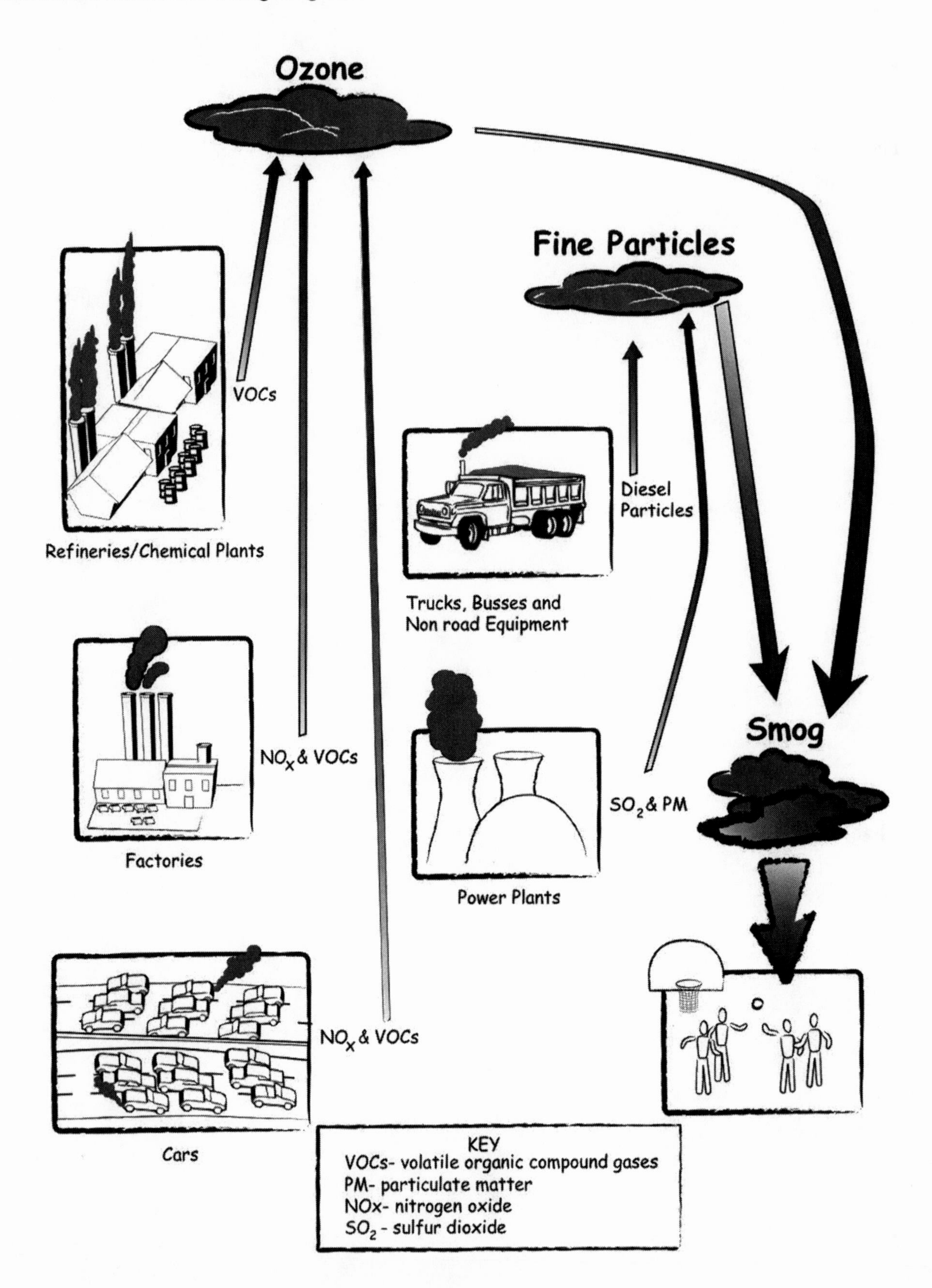

Name ______________________________ Date ____________________

REPRODUCIBLE 6.3: Calculating Your CO_2

Write your *monthly* totals for each type of fuel used in the blanks:

Gasoline	________	(gallons)	×	19	=	________	lb of CO_2
Propane	________	(gallons)	×	15	=	________	lb of CO_2
Electricity	________	(kWh)	×	1.8	=	________	lb of CO_2
Natural gas	________	(therms)	×	12	=	________	lb of CO_2
		Total per month			=	________	lb of CO_2

UNIT 7

WATER QUALITY

TOPIC BACKGROUND

If you have a chance to view the Chicago film in the *Edens Lost & Found* series, you'll see that, in the early days of that city, citizens and businesses simply dumped their trash or factory waste into the river. Unfortunately, the river was the city's main source for drinking water at the time. When the citizens recognized the error of their ways, they used engineering to reverse the river's flow, sending the trash away—down the Mississippi River. But this polluted every community along the river! Not a good solution.

In later years, Philadelphia's citizens realized that they had to do something to protect their drinking water supply. They set aside a massive plot of land, Fairmount Park, to act as a buffer between the city and the wilderness from which they drew their water supply. Today that park has grown to 9,000 acres in America's fifth-largest city. Many other towns and cities are recognizing now that they need to protect their "watershed," the geographical area on which rain falls and from which they collect their drinking water. To help matters, the U.S. Clean Water Act of 1972 cracked down on industrial polluters. As a result, dumping waste into waterways has gotten increasingly more difficult and expensive.

But that does not mean our problems are over—far from it! Corporations and small businesses may have cleaned up their act, but the rest of us have not. Today, each time it rains, rainwater falls on the hard surfaces in our neighborhoods (driveways, sidewalks, patios, parking lots, and the like) and washes away what we have leaked onto the ground, into storm drains and into our streams, rivers, lakes, reservoirs, and oceans. Individual citizens are not accustomed to thinking of themselves as polluters, but many of our household products can have damaging effects on wildlife, plants, and ultimately us if allowed into our waterways unchecked. For example, from a home in the suburbs, rain might wash away such chemicals as herbicides, pesticides, fertilizer, pet waste, and so on. Whether suburb or city, rain is constantly washing away remnants of our automotive age: motor oil, transmission fluid, gasoline, and windshield wiper fluids. When we wash our cars, the soapy water, window cleaners, and waxes trickle down the street and into a sewer. As our tires wear out, they shed tiny flecks of rubber dust. As our brake pads wear out, they shed ceramic and copper dust. This all forms a toxic soup that heads straight to nature or into our water supplies.

This toxic soup is causing major problems in the cities we profile in this project and likely where you live too. Waste from Los Angeles has been found in the remote regions of the Pacific Ocean. Whales that cavort at the North Pole carry chemicals in their blood that originated on the streets of Seattle. Salmon in Seattle are threatened by pollution from the city washing off into their habitats.

That's why so many towns, suburbs, and cities in America are taking steps to reduce or eliminate toxic runoff altogether. First, they are introducing more trees and planted nature areas into their local environments so polluted water can find a place to seep naturally into the ground. Trees and soil absorb and filter water naturally, performing for free the same work that storm water treatment facilities do at great cost. Second, cities and towns are trying hard to encourage people to reduce their reliance on toxic chemicals in and around their homes. This is slow going, because it means first educating the public about the dangers, then persuading them to switch to environmentally friendly products. If you could disinfect your kitchen countertops with a mixture of white vinegar, water, and biodegradable soap that you whipped up on your own at a cost of $1 a gallon, wouldn't you stop buying a caustic kitchen cleaner that costs $4 for a little can or bottle? If you could get your car clean with a sudsy soap made from harmless citrus acid, wouldn't you stop buying the toxic, heavy-duty cleaners? Most people would, but they need to know what's at stake. Your students, with their newfound knowledge of water quality issues, can be a powerful force for change in their communities.

Purpose

In this unit, students will explore a variety of water quality issues to learn just what is impacting the water we drink. In course of study, they will learn what they can do to improve local water conditions.

GROUP ICEBREAKERS

1. Chances are, your students don't know the source of their drinking, washing, and bathing water. Start this unit by asking them: What is the original, natural source of our local water? Is it pumped from a well in the ground? Is it collected in a local reservoir? Is it made up of rainwater or snow-melt water from local mountains, small streams, and underground springs? (Accept all reasonable answers, but assure them that they will know *exactly* where their water comes from by the time they are through with this unit.)
2. Next, help them to start thinking about the municipal and/or private agencies behind the water they drink. Do they know if their city has a water department? Is there a private water company? Is there an agency in charge of water quality? Is there a separate agency that handles the disposal or collection of rainwater? (Answers will vary. The agencies in charge of water differ from town to town, from state to state. In some cases, people are billed by a private company, in others they are billed by the city itself.)
3. If they don't know the answers to the two previous questions, talk about simple ways they can find out the answers. Do their families get a water bill, for example? If they flip through the "blue pages" of their local telephone book, do they find some water agencies or departments listed? How could they go about finding out who handles water and water issues in their city? Are there any governmental agencies that deal with water? (At this stage, simply talk about how they would do a search. You can table the phone book search for a later time.)
4. Ask if they know what a watershed is. If they don't, ask them to try to figure it out from the context of the word. (A watershed simply refers to the geographical area from which you get your water.)
5. Do they think, for example, that a year's worth of rain could fulfill their city's or town's water needs for a single year? Or must their city buy extra water from other places? If needed, where might a city buy such an important resource as water? (Larger metro areas cannot support themselves on the water that falls locally, so they must buy it from water authorities in other areas. Some U.S. states buy water from Canada.)
6. It's unpleasant to think of oneself as a water polluter, but the truth is, we all pollute our waterways in some way (adults more than young people, of course!). Can they think of some ways grown-ups might pollute rivers, streams, and ultimately drinking water without knowing they do? Using some of the descriptions given in the Topic Background, have them think about how some ordinary household items end up in our waterways. (Everyday items we use in the home can end up in the water supply, such as herbicides, pesticides, pet waste, automotive oil, transmission fluid, gasoline, windshield wiper fluid. In short, anything that we use that drips onto our driveways and streets can be washed into our water supplies when it rains. Indoor products and wastes enter the water supply when we wash them down sinks, flush them down toilets, or dispose of them in the trash.)
7. Have students brainstorm some ways to keep ordinary household waste from ending up in our waterways. (Accept all reasonable answers.)

ACTIVITY 7.1

Understanding Earth's Water Supply

Materials

pen and paper
Handout 7.1
Reproducible 7.1 for overhead

Primary Subject Areas and Skills

science, math, social studies, geology, critical thinking

Purpose

Students study a multibar graph, a pie chart, and a diagram of the water cycle to discover how Earth is supplied by water and how the cycle replenishes itself.

1. Where's the Freshwater?

Earth is covered with water, in the forms of oceans, seas, rivers, lakes, and so on. But not all of it can be used by humans. Additionally, the portion that can be used by humans is not always being cared for and preserved in the manner in which it should. In fact, many research and nonprofit agencies predict that by 2025 many countries throughout the world are going to be experiencing extreme water supply shortages. And even the wealthiest and most developed of countries—though not in dire straits—will need to begin to more seriously address and identify significant conservation methods.

The graphs and activities on Handout 7.1 are designed to get students to think about where their water comes from, where it is stored, and what happens to it as it passes through nature's water cycle.

1. Explain to students the difference between saltwater and freshwater. Ask them to give examples of saltwater (oceans, seas) and freshwater (lakes and rivers).
2. Tell students that they are going to create pie charts based on their assumptions of freshwater and saltwater distribution. Instruct them to draw a pie chart that shows the breakdown of the world's water supply as they believe it to be. How much of the pie is freshwater (the kind of water we need to survive), and how much is saltwater (seawater that we cannot drink)?
3. Tell students to draw and label their pie charts without researching first and to keep their charts to themselves.
4. After students have finished and without looking at each other's papers, students should answer where they think the majority of saltwater is located. Ask them where they think the majority of freshwater is located. (*Reminder:* You may wish to give them hints about groundwater, which is also a source of freshwater.)
5. Without giving students answers, copy and distribute Handout 7.1. Instruct them to interpret the information on the handout and answer the questions.
6. After the students have answered the questions, discuss the answers. Were they surprised by the answers? Why or why not? (Most people are astonished to learn that freshwater accounts for less than 1 percent of Earth's water.)

2. The Water Cycle

"You can't step into the same river twice . . ." the saying goes, and that's because rivers are constantly moving. But all sorts of water supplies—whether contained in a lake or underground—are constantly changing form: Water is evaporated, frozen, thawed, moving through groundwater reserves, and coming into contact with new substances, natural or other.

1. Reproducible 7.1 shows how water travels on Earth in various forms.
2. Copy Reproducible 7.1 and place it on an overhead projector.
3. Review the water cycle as depicted with your students. Prompt them with phrases such as, "Where does the water begin? What happens next? And then?"
4. Discuss the questions with the class, then move into a discussion of watersheds and the watershed discussion questions.

Water Cycle Discussion

5. In how many forms do you see water stored? (The diagram shows many forms, but there are three main forms: Water is stored as liquid, as a gas in the atmosphere, or on Earth as ice.)
6. Most people think of drinking water as coming from rainfall. Where does the rain come from in the first place? (This is a chicken-and-egg question. Water exists as a gas in the atmosphere at all times. It is rained down to Earth and is stored in ice, in oceans and waterways, and even in the ground. Humans collect water from variety of sources for drinking water, but rain is the easiest to capture and store.)
7. What is the relationship between surface freshwater storage and groundwater? (Water is constantly moving through Earth. Some water seeps into the ground and is stored there; some of it seeps out from the ground and comes out in springs, streams, and large lakes. Some groundwater flows out to sea.)
8. Explain, in simple words, two possible journeys for a water molecule from ocean to freshwater storage. (A water molecule might start as a gas above Earth, form rain and fall to Earth, then be frozen in ice and snow, until it melts and makes its way to where humans can harvest it. Or a water molecule might evaporate from the surface of the sea, rise into the air as gases, become condensed as water, and begin the cycle again.)
9. Looking at this chart, what kind of physical characteristics affect surface runoff? (Mountains, glaciers, and icecaps naturally capture and store water because they tend to be cold places. Because they are steep, they also allow melted

water to run down them and flow with gravity to the nearest body of water or groundwater supply. Wherever water goes, it adapts to the conditions around it.) What kind of atmospheric events affect surface runoff? (When it's cold, water may slow down, freeze, and be stored for long periods of time. If it's very warm, water may evaporate as steam and return into the upper atmosphere.)

3. Watersheds

Simply put, a watershed is a geographical area that catches rain, melted snow, and so on, which flows into places like rivers, lakes, and groundwater. It is an area of land that drains to a common point. A watershed can be as large as thousands of square miles, or a watershed can be as small as a few acres. There are 18 major drainage basins in the continental United States, and each of those is made up of hundreds of smaller watersheds, many of which are, in turn, composed of even smaller watersheds. A small watershed, for example, could be a land area that drains into the same stream. Whereas the land area that drains into a river—itself a larger watershed—can be composed of many smaller areas that drain into smaller brooks and streams. But they eventually converge in a common drainage point.

Because watersheds have geographical boundaries—not state or city borders—they of course include the towns and cities and farms and forests that fall within their limits. Wherever you go, you are in a watershed. You just may not have ever thought about it.

Watershed Discussion

1. What contributes to watershed pollution? (*Hint:* Encourage students to think beyond things that drain directly into a river: groundwater contamination, air pollution, and ground pollution.)
2. Rainfall in your watershed runs through farmland before it reaches the streams and rivers in your area. What kind of pollution might it gather along the way? (Herbicides, pesticides, animal wastes, and so on are all possible things that might enter the water supply in this way.)
3. Rivers such as the Chicago River discussed in *Edens Lost & Found* have served many purposes throughout history. What were some of the benefits achieved? (The Chicago River allowed the birth of a great city, which grew because it had a navigable waterway from which to ship goods.) How did some of these uses of the river damage the river itself and the watershed in which it exists? (Over time, humans spoiled the river by discharging wastes into it.)
4. Is it common for the role of rivers near cities to change with time? (Yes. You might challenge students to examine the roles of rivers in your area as well as famous rivers throughout the world. Some examples include: the Seine in Paris, the Thames in London, or the Tiber in Rome.)

ACTIVITY 7.2

Town Meeting: Watershed at Risk

Materials

pen, paper; research material (newspapers, magazines, maps)

Primary Subject Areas and Skills

science, math, writing, research, public speaking

Purpose

Students will conduct research in preparation for a debate on risks to a local watershed, posed by upcoming development. Though the debate is fictional in nature, students will be expect to cite real-life evidence to support their pro–con positions.

Debate Premise

A prominent developer is planning to build in the watershed area that supplies the vast amount of drinking water to your community. The development will bring housing and jobs to your area, which has been experiencing economic hardships in recent years. However, the development will be located in an area of critical environmental importance to your watershed. Trees will be cut down, and the area itself—including all the waste produced not only by the building but also by the new inhabitants—is in a major drainage area.

1. Divide the class into two teams and explain that the two sides in the debate will conduct research and prepare their findings to be presented in a town council meeting atmosphere. Give them at least one week to prepare. They will likely need time outside of class, if limited in-class preparation is possible.
2. Be clear about the ground rules: The team that supports the development must support their position with facts about how the benefits

outweigh the environmental risks or show how the development could be modified to decrease negative environmental impact. The team opposing the development must also suggest a viable alternative that would bring jobs and opportunity to the area.

3. Suggest to students that creating a "pros and cons" list may help them to organize their thoughts.
4. Role playing is welcome. Students may wish to take on the roles of developer, concerned citizen, and the like. Creativity is a plus! If students wish to dress the part, assume character names, and so on, encourage them to do so.
5. Mention to them that there are no easy answers to the debate, but this activity, if followed through, will help them appreciate the difficulties that face community members, businesses, and governments on a regular basis.

Extensions

1. Locate the primary sources of freshwater in your area. What condition are they in? What are the main concerns about these resources? Assign students—individually or in small groups or pairs—to report on your local watershed. Research where it is located, its name, where its water comes from, and what affects the water quality. Students can even draw diagrams that depict the watershed, the territory it covers, and its interaction with industry and pollution. What are the primary pollutants? What is the larger watershed and/or drainage basin of which your watershed is a part? Mapping is a wonderful part of this activity. Local and county maps should be gathered, as well as topographical information about your watershed area.
2. To help students tie the concept of watersheds to other water-quality issues, have them think about how the following events are connected:
   a. How does yard work affect the quality of the fish you eat? (Any chemicals applied to lawns will inevitably end up in the local water supply.)
   b. What is the relationship between automobile maintenance and groundwater quality? (Any chemicals used in car repair and maintenance will wash into local water supplies if they are allowed to drip onto an asphalt driveway or street.)
   c. What is the relationship between particulates and air pollution and the quality of your drinking water? (Soot and other airborne pollutants collect on windowsills and coat our vehicles, so they are inevitably landing in our water supplies.)
   d. How does farming affect groundwater? (Agricultural chemicals also end up in the water supply, which is a central argument used by organic farming proponents.)

ACTIVITY 7.3
The Rundown on Runoff

Materials

pen and paper; Handout 7.3

Primary Subject Areas and Skills

science, research, critical thinking

Purpose

Students perform a take-home audit to see what items in the home may present run-off problem if released into their local water supply.

Many contaminants in our water system didn't have to end up there. One way water pollutants arrive in rivers and groundwater is through runoff, a situation that has worsened with increased development. The more asphalt and cement that covers the earth, the less water can be absorbed into the ground. The planting of trees and plants helps drastically reduce this effect. Runoff and excess water running through streets carry toxins and trash down storm drains. Many of these will end up in the lakes, streams, or groundwater supply. This is sometimes referred to as nonpoint source pollution, meaning that it comes from many different places rather than one major source.

1. Ask your students if they've ever heard the word *runoff.* What is their understanding of it? (*Definition:* Water from precipitation or irrigation that flows into bodies of water, possibly carrying harmful pollutants.)
2. On the board or overhead projector, help students brainstorm a list of things that might be carried by runoff into the sewers in your neighborhood or community. Encourage students to think about the kinds of things that end up on the ground, such as candy wrappers, fallen leaves, animal droppings, paint thinner. Remind them that excess water running through streets

and gutters might increase the chance that these substances end up in the water system.

3. Copy Handout 7.3 and distribute it to students. Explain to students that they are to perform an audit at home to identify what they and their family may unknowingly contribute to runoff pollutants. Encourage them to be patient and thorough, as it is easy to overlook products that are common. They should then research the chemicals found in their household products to determine the possible effects they may have on living things. This is an ideal weekend assignment.
4. Once students have completed their audits, have them share their discoveries with the class. You may wish to keep a tally on the board or overhead projector that shows not only the products that students found but also the associated activities (washing the car, for example) that result in the pollution.
5. Finally, have students carefully report on the ingredients contained in some of the household products. (They will be surprised to find that manufacturers are not required by law to list the ingredients!) Remind them that merely buying these products is not the issue; disposing of them or letting them leak, drip, or trickle into the environment *is*. Challenge them to help their families go on a "tox-free shopping spree" once they have used up all the contents of the products they have identified.
6. Have them research good, tox-free substitutes for many of the products. (*Example:* Vinegar and water make a good countertop, window, and mirror cleaner.)

ACTIVITY 7.4
Understanding Water Quality

Materials
pen and paper; Handout 7.4

Primary Subject Areas and Skills
science, social studies, math, writing, critical thinking

Purpose
Students familiarize themselves with the information contained in a water quality report and use that information to analyze some of the components of water pollution.

Local water authorities are obligated to inform their customers of the quality of the water they're drinking. But when that information arrives in the mail, perhaps bundled in with our water bill, many of us don't give the valuable data a second look. All of the abbreviations, chemical names, and mathematical references can be overwhelming. Yet, what's contained in these reports is crucially important to our health and well-being.

1. Copy Handout 7.4 and distribute to students.
2. As they answer the questions, tell them to pay attention to the key and abbreviations listed at the bottom of the charts
3. To make the most of all the information on the chart, assign individual students or small groups of students to research some of the chemicals listed here. What products produce these? What other possible sources exist in your community that might produce these contaminants?

Extension
You may also wish to request similar information from your own local water authority so that students can analyze the data for their families or school authorities. A water quality task force can be formed through which students compare their local water quality information to recommended national standards and then devise a plan to reduce unwanted pollutants.

ACTIVITY 7.5
Swales and Cisterns

Materials
pen and paper; Handout 7.5

Primary Subject Areas and Skills
math, science, social studies, history, geology, critical thinking

Purpose
Students study a pair of diagrams to learn how swales and cisterns work. They answer a series of questions drawn from the diagrams.

Two methods for capturing water are swales and cisterns. This handout is designed to help your students understand how the two work. Have them work through the diagrams with you, in groups or alone.

1. Copy Handout 7.5 and distribute to the class.
2. Ask them if they have ever heard of a *swale* or a *cistern*. Both are important tools for preserving the environment through water management.

Explain to them that by the time rainwater reaches your local water company, it is probably contaminated with runoff. If your rainwater drains instead to a river or stream, it is doubtless carrying a toxic soup to wildlife. If we can somehow "hang" onto that water and let it seep *s-l-o-w-l-y* into the earth, some of those pollutants will be filtered out naturally, by soil and rock. Swales and cisterns help us hang onto that water, purify it, and—in the case of cisterns—reuse that water as soon as we wish.

3. When they've completed the handout, discuss things they can do to help capture precious rainwater.

Things You Can Do to Help Capture Rain

- Landscape your backyard to keep water from running into streets or sidewalks. Channel it to a specific spot where it will be naturally absorbed into the ground.
- If you can't build a swale or cistern, install rain barrels under rainspouts and use the water for your garden.
- Replace asphalt driveways with porous ones, such as gravel, which allow water to seep into the earth.
- Slope driveways into swales.
- Run all grass clippings, tree trimmings, and leaves through a wood chipper or mulcher and spread the resulting mulch around trees, shrubs, and flowerbeds. This absorbent layer will keep plants moist longer between rainfalls.

EXTENSIONS

Your Water Company

Visit your local water treatment facility and/or reservoir. Learn what steps water goes through on the way to your faucet. Create a class chart showing the typical purification process.

Storm Drain Stenciling Project

Many citizens mistakenly assume that anything dumped into a storm drain or street "sewer" is automatically filtered and purified. Nothing could be further from the truth. Your students can raise public awareness of this fact by painting a warning such as "DRAINS TO RIVER" on the sidewalk near a storm drain or sewer. Many stencils with charming artwork are available online, but your students can easily design a cute one on their own! Two caveats: Be sure you get permission for this project from local officials, and be sure to use a 100 percent acrylic latex (water-based) paint for exterior use. This should clean off hands and brushes easily with soap and water. Do not permit students to leak or dump paint down drains! Empty paint cans must be disposed of in accordance with your local hazardous waste ordinances.

Please follow EPA Guidelines:
www.epa.gov/adopt/patch/html/guidelines.html

Earthwater Stencils are available at:
www.earthwater-stencils.com/stencils/

School Swales

Using the information in the swales and cisterns handout, consider having students build their own swale on school grounds. (You may wish to enlist the help of a parent with access to construction tools.) You might consider the following experiment: Plant two gardens, one near a swale, one without a nearby swale, and monitor them during the year. Which garden appears to thrive best? (In general, gardens planted near swales do better than ones without a nearby swale—even when your growing season has been especially moist.)

Build a Rain Garden

A rain garden is simply an area of your garden that is designed to capture excess water and is planted with water-loving plats. Your backyard rain garden doesn't have to be that labor intensive. Start by studying the way water drains off your property. Do you have rainspouts leading down from your roof? Spots where the spouts discharge might be a good spot for a rain garden, provided they are about 30 feet away from your foundation. Also look at how your yard is sloped. Are there natural depressions that typically remain moist after a rain? This is also a perfect spot for a rain garden. Improve drainage in these areas by excavating a trench or square-shaped hole a few inches or even a few feet deep. Fill the hole with a layer of gravel, then mix a portion of the excavated soil with fast-draining materials such as compost, sand, or peat moss. On top of the filled hole, plant a selection of moisture-loving plants. A good nursery can help you select plants appropriate to your climate and hardiness zone.

Car Wash Coupon Drive

It is often more water-efficient to take your car to a commercial car wash than to wash it on your own in your driveway. Here's why: Businesses that generate contaminated runoff (such as soapy water) are often required by local ordinances to filter it before disposing of it. A person who washes her car in her driveway is under no such regulation. The dirty water often washes right down the street, into a sewer, and into a waterway. Some towns are encouraging citizens to make use of local car washes by distributing discounted car wash coupons—provided free by local car wash businesses—that are mailed out with local news on recycling, solid waste, and other matters. Challenge your students to broker a partnership between car wash businesses, city hall, and their classroom. If your students design and print up the coupon book, and city hall distributes it, the entire enterprise could result in an effective major public education campaign.

Watershed in the News

Water quality and availability are among the hottest political topics in the nation. Have students pore through local newspapers to put together a summary of the local water issues your community is facing. How can your community improve or expand your watershed? What threats exist to your watershed? Every water facility employs a water commissioner or water quality official. Invite this individual to speak to your class. What lessons can your students glean from this talk to share with their parents?

Protect Your Watershed

Before students can protect their watershed, they must know where and what it is. With a couple of online tools, they can locate their local watershed area by simply entering their zip code (or town or state). The EPA site listed below will give them the geographical location of their watershed, a list of environmental task forces working in the area, pollution lawsuits on file, and helpful statistics. The information can be used in reports, graphing activities, action plans, and as a starting point for service learning projects.

RESOURCES

- **EPA Surf Your Watershed**
  www.epa.gov/surf/
- **USDA water quality site**
  www.nal.usda.gov/wqic/
- **EPA monitoring water quality site**
  www.epa.gov/owow/monitoring/
- **Adopt a Watershed**
  www.adopt-a-watershed.org/
- **Map Your Watershed**
  http://water.usgs.gov/wsc/map_index.html
- **The River Network**
  www.rivernetwork.org

Check for green cleaning products at your local grocery store or market. If they don't carry them, you may want to request that they do.

Reading Handout Unit 7:
RESTORING THE CHICAGO RIVER

In this reading, you'll learn how a group of committed citizens brought a famous river back to life, using simple techniques that can be duplicated all over the nation. After you read the story, answer the questions.

Margaret Frisbie remembers the time she threw a barbecue for friends and colleagues who work with her. "I was setting up a picnic in a park, unloading bags of charcoal, and some people wandered by and asked me what I was doing. 'We're having a barbecue for our organization, Friends of the Chicago River,' I told them. And they looked at me and said, 'Oh, that's great, but where is the Chicago River?' I pointed over my shoulder. The river was flowing *right past us* in that park, and they didn't know it. That's the problem we face every day. People don't know that Chicago even *has* a river! And you can't save what you don't know."

Ms. Frisbie's job is to help change all that. She's fond of telling people that the river everyone takes for granted is the reason Chicago exists at all. In general humans have always been drawn to watery locations, intuitively grasping that these places best supported life. Chicago's original settlers selected the city's site because it was close to major waterways. But the river today looks nothing like the wilderness the explorers Jacques Marquette and Louis Joliet would have seen as they paddled by in their canoes in the 17th century. But the river doesn't look as it did in the 1970s either, and Friends of the River has had a lot to do with that.

The river's rebirth began in 1979, after a local magazine brought attention to the river's sorry state. "It was a complete dump," confesses Ms. Frisbie. "What you'd see was pretty despicable."

Even from the city's earliest days, people and businesses dumped garbage and sewage into the Chicago River, and the water whisked it out of sight and mind. At that time, the river flowed north and emptied into Lake Michigan, from which Chicago drew its drinking water. As the city grew in population, city officials realized that this practice was dangerous. They could not have Chicagoans drinking water contaminated by sewage. In 1900, a massive engineering project reversed the river's flow and sent it down the Mississippi River. This was fine for Chicago but bad for its neighbors downriver who complained about the refuse that now flowed past them on the way to the Gulf of Mexico.

As late as the 1970s, the river continued to be a site of illegal dumping. Rusty chain-link fences and concrete walls obscured river access. "Because it was so unpleasant, it was barred off," recalls Ms. Frisbie. Sewage treatment plants had cleaned up their act but still did damage. Runoff, treated with chlorine, destroyed plant and animal life. If you had been nutty enough to paddle down the river back then, you would have felt like you were coasting down a stinking no-man's-land littered with trash and overrun with weeds. Wildlife was exceedingly rare.

Friends of the Chicago River took a look at all this and dreamed of turning the river into a 156-mile park, flowing past 50 or so towns in the Chicago area. They tackled the work on two fronts: First, they physically cleaned up the river. Then they persuaded their local leaders to take action to stop people from dumping trash on the river's banks or pollutants into the water.

This was easier said than done. After decades of neglect, very tenacious exotic plants had colonized the banks of the river and were killing off the native Chicago plants. Friends of the River began inviting volunteers out to the river on weekends to help cut down invasive trees and replace them with native vegetation. It was slow, painstaking work. Local groups interested in wildlife, environment, and activities for children invited Friends to their meetings to talk about their work. Slowly, momentum built. Some groups or schools participated in an "adopt-a-river" campaign; they promised to clean up a specific spot on the river through all four seasons. If something was awry, they immediately reported it. Along the way, they mapped out walking trails, put up interpretive signs, and made that piece of the river part of their lives.

During that time, the federal government also got stricter. The Clean Water Act, signed in 1972 by President Nixon, put severe limits on waterway dumping. Previously, Americans thought nothing of dumping factory waste into rivers to be washed out to sea. As a result of this legacy, our waterways were dying in the most outrageous ways. The Cuyahoga River in northeastern Ohio was so polluted with toxic ooze that it caught fire several times between 1936 and 1969. Lake Erie, one of the five Great Lakes, was declared dead, unable to support marine life. In drier regions, the carcasses of dead wildlife littered the banks of small streams and arroyos, where creatures had drunk poisoned water and keeled over dead.

These terrible stories aroused a cry, and the U.S. government finally enacted the Clean Water Act. This law forced any business that habitually released its wastewater into a stream to get a permit and to treat the waste before releasing it at all. For many, the cost of such treatment was too much to bear, and it forced them to become more efficient about how they ran their operations. When sewage treatment plants stopped adding chlorine to the water they released, nature rebounded. Fish and wildlife became more noticeable; birds, for example, began to linger.

As the Chicago River began looking better, people suddenly wanted to get close to it, paddle in it, and fish in it. Some of the old fencing was torn down to provide access for recreational groups. Today, high school and college rowing teams use the river regularly. A canoe and kayak rental company has sprung up along the river path. The city's parks department is busy buying up parcels of land to incorporate into the grand vision of a river park. And some developers have actually built condos along the waterway, emphasizing the river view.

Today Friends of the River holds events during the year to introduce newcomers to the river and welcome back old friends. Chicago River Day, held in May, is a one-day clean-up blitz. The Chicago River Flatwater Classic, held in August, is a river race for paddlers. Another favorite is the "Voyageur" race, featuring huge, 25-foot canoes modeled after the ones used by Marquette and Joliet.

As someone who knows the river in all its seasons, Ms. Frisbie can tell you that there's more work to be done. Concrete walls and stands of buckthorn, a type of invasive exotic shrub, still block access in spots. The river teems with 68 species of fish, but some of them are exotics, and anglers are still warned not to eat their catch because the fish may be contaminated. "All in all," she says, "it's the beginning of what could be wonderful."

Today there are rumors that otters have returned to the river. If true, it would be a very good sign because otters require slightly purer water conditions than other mammals. For a while now, Friends of the River has had to teach people how to behave around beavers, which have returned to the river in startling numbers. "There were no beavers years ago," Ms. Frisbie gushes, "but to hear that they're back and to 'watch out, because if you plant some trees, a beaver might come along and chew on them,' is fantastic!"

Questions

1. After the city of Chicago reversed the flow of its river, in which compass direction did it flow? After the river's course was changed, do you think more people or fewer people felt the impact of its filth?

 ____

 ____

2. Why did Chicagoans turn their back on the river in the early days, putting up fences and walls to erase it from their sight?

 ____

 ____

3. What is one of the chemicals dumped in the river? What was it originally used for? How is it used in the home?

 ____

 ____

4. Why do you think biologists and others place such strong emphasis on saving an area's native plants, that is, plants that originated and evolved in a specific location?

 ____

 ____

5. From the tone of the story, do you think an exotic species of plant is a good thing, or a troubling thing, to have on the banks of the Chicago River? Why?

 ____

 ____

6. Why do you think it's important to get people involved in using the river for fun? If they enjoy themselves on the river, do you think they might be likely to support the river in the future?

____

____

____

7. Who were Marquette and Joliet, and about how many years ago did they travel the Chicago River? Why did they explore the area around Chicago in the first place? What were they looking for? (If you don't know already, do research to find out.)

____

____

Name ______________________________ Date ____________________

HANDOUT 7.1: The Distribution of Earth's Water

Look at the graphs below and answer the questions.

Distribution of Earth's Water

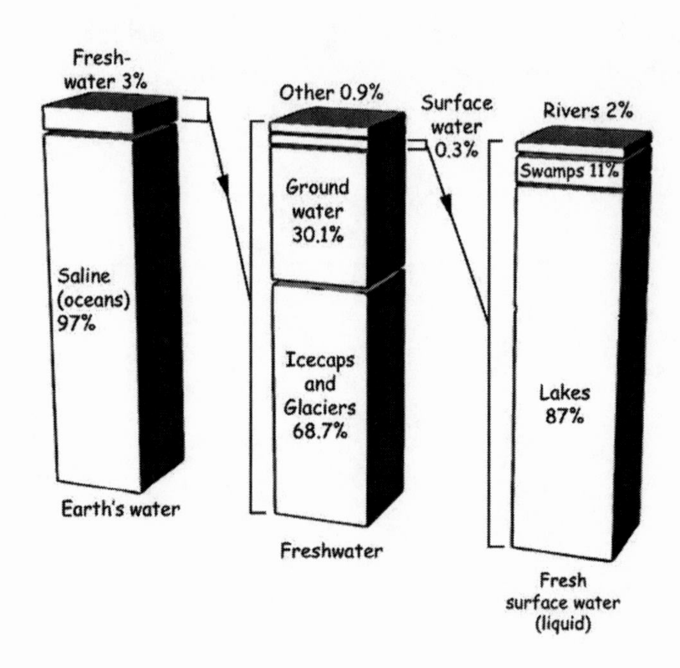

How much of Earth's water can humans use?

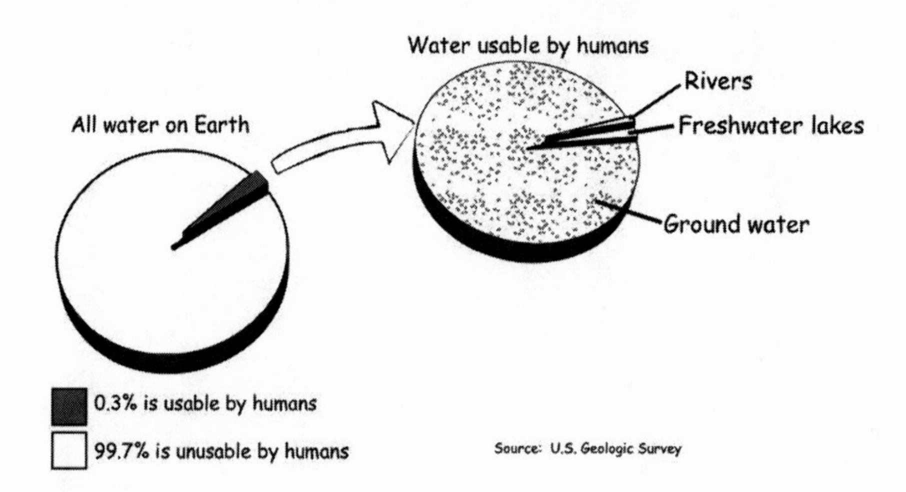

1. What percentage of Earth's water is freshwater?

2. What percentage of Earth's freshwater supply is groundwater?

3. What percentage of all Earth's water is usable by humans?

4. Of the water supply usable by humans, where is the majority located?

5. What percentage of freshwater is surface water?

6. What percentage of surface water is found in lakes and rivers?

7. Based on all the information presented here, on which water supply are humans most dependent?

8. How do these results affect your views about runoff? Ground pollution?

REPRODUCIBLE 7.1: The Water Cycle

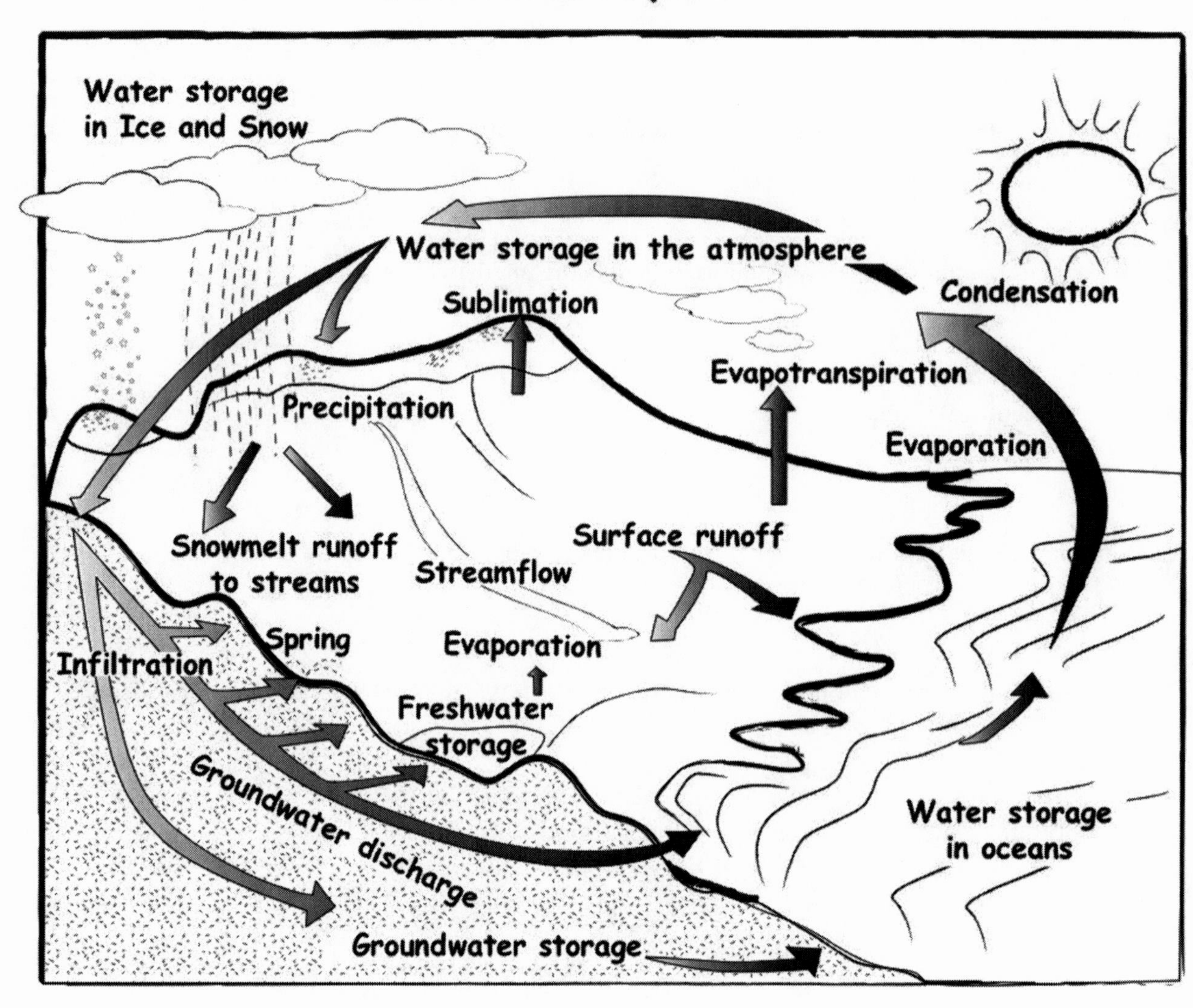

Source: The U.S. Geologic Survey

Name ______________________________ Date ____________________

HANDOUT 7.3: Take-Home Audit: Runoff Pollutants

Garage

Product	Chemicals Present	Possible Effect on Living Things	How It Could Get in the Water

Outdoor/Garden/Tool Shed

Product	Chemicals Present	Possible Effect on Living Things	How It Could Get in the Water

Kitchen

Product	Chemicals Present	Possible Effect on Living Things	How It Could Get in the Water

Bathrooms

Product	Chemicals Present	Possible Effect on Living Things	How It Could Get in the Water

Other Storage Areas (basement, laundry, pantry, closets, cabinets)

Product	Chemicals Present	Possible Effect on Living Things	How It Could Get in the Water

Name ______________________________ Date ________________

HANDOUT 7.4: Water Quality Analysis

What's in the water you drink? Created by a local water company, this chart is distributed to all residents who use, buy, and drink water from this water facility. All water agencies are required to send these reports to their customers on a seasonal basis. Use the chart to answer the questions below.

	Test Date	Unit	PHG	MCL	Average	Range	Source of Contaminants
MICROBIOLOGICAL CONTAMINANTS							
Turbidity[1]	2003	NTU	n/a	TT(5.0)	0.13	<0.01–0.16	Soil runoff
Arsenic	2003	ppb	10	50	12.50	11.–14.	Erosion of natural deposit runoff from orchards; glass and electronics production wastes
INORGANIC CHEMICALS							
Chromium	2003	ppb	2.5	50	11.25	<10.–14.	Discharge from steel and pulp mills and chrome plating; erosion of natural deposits
Fluoride	2003	ppm	1	2	1.13	0.50–1.9	Erosion of natural deposits; water additive, which promotes strong teeth; discharge from fertilizer and aluminum factories
Nitrate	2003	ppm	45	45	4.77	3.7–5.78	Runoff and leaching from fertilizer use: leaching from septic tanks, sewage; erosion of natural deposits
Chloride	2003	ppm	n/a	500	17.5	15.–21.	Runoff/ leaching from natural deposits; seawater influence
Foaming Agents (MBAS)	2003	ppb	n/a	500	0.05	<0.05–0.05	Municipal and industrial waste discharges
INORGANIC CHEMICALS: COPPER & LEAD							
Copper[2]	2003	ppm	0.17	AL = 1.3	<1.	n/a	Internal corrosion of household plumbing systems; erosion of natural deposits; leaching from wood preservatives
Lead[3]	2003	ppb	2	AL = 15	<10.	n/a	Internal corrosion of household water plumbing systems; discharges from industrial manufacturers
GENERAL PHYSICAL							
Color	2003	units	n/a	15	1.7	1.–2.	Naturally-occurring organic materials
Specific conductance	2003	umhos/cm	n/a	1600	408.	370.–468.	Substances that form ions when in water; seawater influence
Total dissolved solids	2003	ppm	n/a	1000	277.	232.–316.	Runoff/ leaching from natural deposits
GENERAL MINERAL							
Hardness	2003	ppm	n/a	n/a	101.	83.–140	Naturally-occurring polyvalent action present in the water, generally magnesium and calcium

	Test Date	Unit	PHG	MCL	Average	Range	Source of Contaminants
Sodium	2003	ppm	n/a	n/a	45.	41.–51.	Naturally-occurring salt; seawater influence
VOLATILE ORGANIC CONTAMINANTS							
Dichloromethane	2003	ppb	0.004	5	<1.	n/d–<1.	Discharge from pharmaceutical and chemical factories; insecticides
DISINFECTION BY-PRODUCTS							
TTHMS (Total trihalomethanes)	2004	ppb	n/d	80	52.6	n/d–86.	By-product of drinking water chlorination
Chlorine	2004	ppm	4	4.0	.6	.0–.9	Drinking water disinfectant added for treatment
HAA5 (Total Halocetic Acids)	2004	ppb	n/a	60	4.22	.0–.30	Drinking water disinfectant added for treatment

KEY TO TABLE
AL = Regulatory Action Level; n/d = none detected; PHG = Public Health Goal; MCL = Maximum Contaminant Level; NTU = Nephelometric Turbidity Units; SMCL = Secondary Maximum Contaminant Level; MCLG = Maximum Contaminant Level Goal; pCi/L = picocuries per liter (a measure of radioactivity); TT = Treatment Technique; n/a = not applicable; ppb = parts per billion, or micrograms per liter; ppm = parts per million, or micrograms per liter; umhos/cm = units of specific conductance
1. Turbidity is a measure of the cloudiness of the water. 2. None of the 44 samples tested had copper at a level that exceeded the Action Level of 1.3 ppm. 3. One of the 44 samples tested had lead at a level that exceeded the Action Level of 15 ppb.

1. Which volatile organic contaminants were found?

____

   a. What is the maximum contaminant level for each? ______________________________

   b. What are the sources of each contaminant? ______________________________

2. How many parts per billion of lead correspond to an "Action Level"? ______________

   a. How many samples tested exceeded the action level for lead? (Look closely.) __________

3. What does turbidity measure?

____

4. What is the average level of arsenic contamination? ______________________________

   a. What are the sources of arsenic in the water supply? ______________________________

5. What is the maximum contaminant level of copper? ______________________________

6. What are some of the primary sources of contaminants of nitrate in the water supply?

____

Name ______________________________ Date ____________________

HANDOUT 7.5: Swales and Cisterns

If rain washes down sidewalks and streets into a storm drain, chances are high that that water will contain contaminants. But if we can somehow "hang onto" that water and let it seep *s-l-o-w-l-y* into the earth, some of those pollutants will be filtered out naturally, by soil and rock. Two methods for capturing water are swales and cisterns.

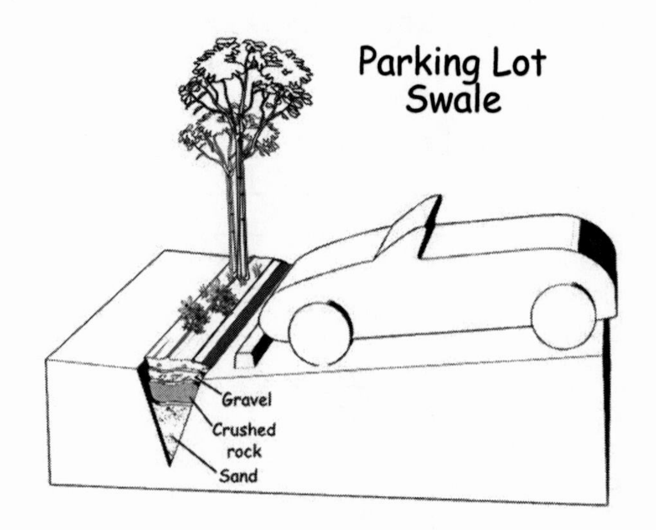

SWALE

A swale is a slice in the earth, filled with high-drainage materials that allow rainwater to more easily drip downward during storms and replenish groundwater resources. As the water drips down to lower layers of the earth, impurities are filtered out of it by the soil itself. If and when the water reaches the creek, it's that much cleaner than if it had simply washed down from asphalt to storm drain to creek. Swales are a natural drainage system.

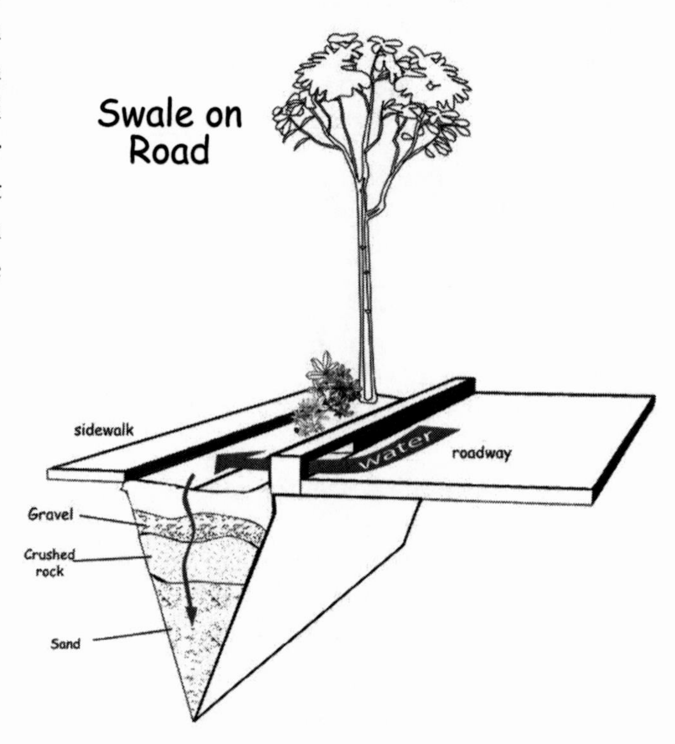

1. In the images above, what materials fill the V-shaped swales? Why do you think these materials are better than filling the swale with ordinary dirt?

____

____

2. Why is the parking lot sloped in the top drawing? During a heavy rain, in which direction do you think the water would go?

3. If there is a heavy rain and the swale can't absorb all the water right away, what do you think might happen to all that water? Will it wash away or pool up in the swale?

4. What kinds of plants do you think are best planted in a swale zone: ones whose roots thrive when wet, or ones whose roots prefer dry conditions?

5. In the road swale drawing, which spot would capture and hold the most water? Why?

6. Do you think rain that fell in the roadway would make its way into the swale? Why or why not?

7. Suppose you are an engineer who has been hired to redesign the road swale shown. What things might you change to encourage as much water as possible to flow into the swale?

8. What kind of maintenance do you think swales need? What kinds of things might need to be removed from a swale from time to time?

CISTERNS

More than 2,000 years ago, ancient peoples used underground tanks called cisterns to capture precious rainwater runoff in dry, parched environments. Modern cisterns can be enormous, capable of holding enough rainwater to fill a swimming pool. Cisterns are perfect for cities and towns where rain is rare and water is costly.

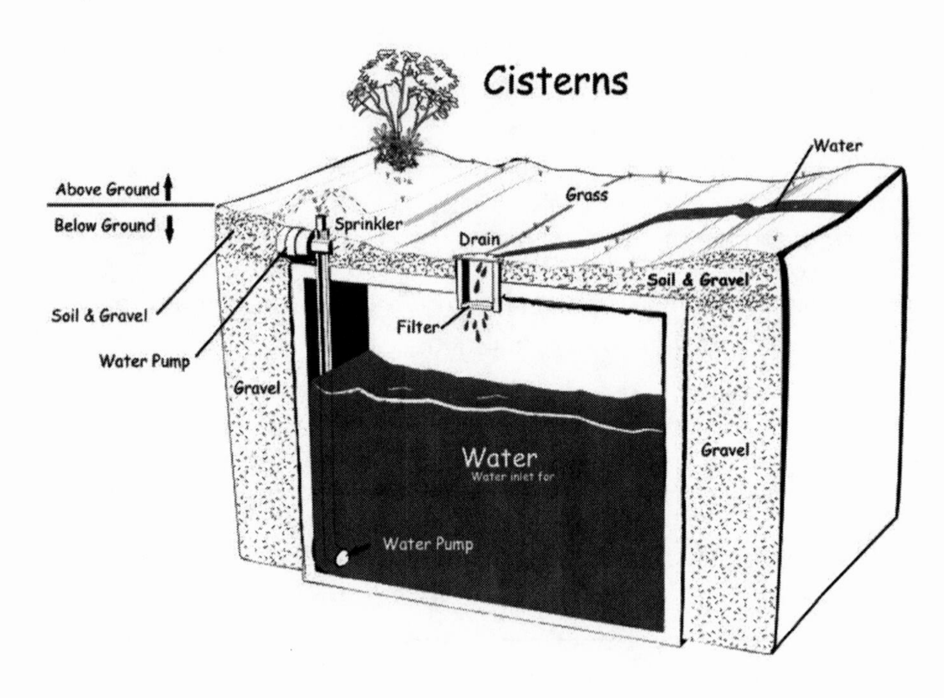

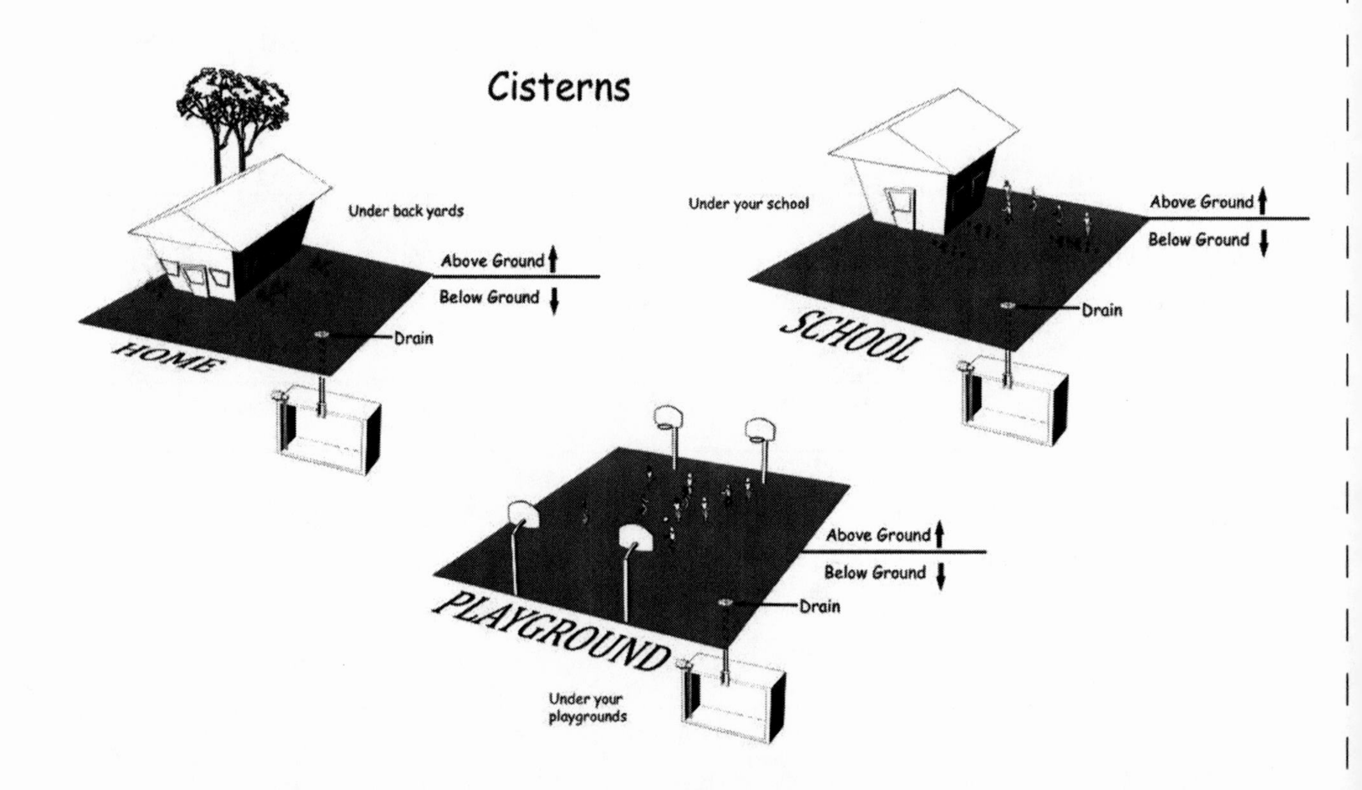

1. How do you think water gets into the cistern in the top drawing? How is the water purified?

____

____

2. How is the water taken out of the cistern so it can be reused?

____

____

3. Why do you think the soil is sloped on top of the cistern?

____

____

4. Cistern water is generally not used for drinking or bathing. How might a home owner use the water in his or her cistern? How might a school? A business? A park or golf course?

____

____

5. How or why might a cistern be easier, smarter, or cheaper than using purified water from your local water company?

____

____

6. Which do you think is cheaper to install: a cistern or a swale? Why?

____

____

UNIT 8

SOIL QUALITY

TOPIC BACKGROUND

The health of our soil is crucial to the health of our cities and towns. On the most basic level, we need soil because our food comes from it. Food crops grow in soil, and livestock graze on grass that grows in soil, or else they eat fodder that was raised for them from the soil. Soil surrounds our homes and parks and nourishes the trees that are so important to our environments.

Unfortunately, our soil is threatened by so many things humans do. Because Earth is dynamic, or constantly changing, what gets dumped on the ground has a good chance of being carried far afield by water and other natural forces. When factories carelessly dump waste on the ground or in waterways, contaminants can seep into the soil and carry toxins to underground water reserves. Such contaminants also can be absorbed by crops. Poorly designed landfills can leak trash residue into our communities. Home owners who dump paint or old bottles of pesticide in their backyards may not think they are endangering their neighbors. But those contaminants could later be washed by rain into the town's water supply.

When builders put up new structures in growing communities, their big machines tend to remove trees and other vegetation that leaves the soil vulnerable to the elements. Large-scale farming operations do the same. Stripped bare of turf, loose soil is easily washed away by rain. The land erodes, and once-healthy soil is wasted. It clogs waterways, muddies rivers, and increases the chance that floods will occur.

Your students may find this concern for soil erosion odd. After all, if some soil washes away, what's the big deal? There's always more underneath, isn't there? Well, that's not entirely true. Only the top few inches of a plot of land are the most productive for human use. This is the layer that is richest in nutrients, perfect for trees, grass, and crops. In the 1920s and 1930s, Americans witnessed firsthand what poor soil conservation practices could do to communities. When strong winds blew away the rich topsoil of farms in the American southwest, farmers could no longer grow crops. They were forced to abandon their farms and go elsewhere. This was the great "Dust Bowl" era, memorialized in *The Grapes of Wrath,* a classic novel by John Steinbeck.

Closer to our era, soil contamination has wreaked a similar kind of damage on communities. In one town in Pennsylvania, entire neighborhoods have been harmed by the toxic waste of an old battery-recycling plant. Lead levels were so high in some places that the owners of that now-closed plant were forced to buy the homes of residents closest to the danger zone and pay for them to move elsewhere. Further away from the site, federal workers in white protective suits removed the top layers of soil from many backyards, along with all landscaping and shrubs, and completely replaced it!

Now, we don't want to equate the dumping of toxic waste with cutting down a few trees to make room for a new home. But unless we take precautions, both of these behaviors can have serious effects on our communities. The destruction, erosion, and pollution of soil is not just bad for nature, it's bad for the health of all living things, not to mention costly to correct. The Chicago film in the *Edens Lost & Found* series shows how years of dumping the refuse of steel-making plants has left parts of Chicago looking like a brown moonscape. Elsewhere in the series, plenty of examples are given of bad corporate or individual behavior that has left a major cleanup job for the next person who comes along.

Purpose

In this unit, students will learn the meaning of soil quality, and why the health of a community depends on healthy, contaminant-free soil. Their studies will encourage students to be good stewards of the ground they walk on.

GROUP ICEBREAKERS

1. Ask students to describe what steps take place when a plot of land is cleared to build a house or a new strip mall. (Coax them to begin with the first step: the removal of trees and vegetation.)
2. Have them to imagine how the land looks at this point. (Without trees or grass, the ground would be completely bare, with plain dirt.) Have them picture what might happen if a major rainstorm occurred at this point. (Some of the soil would wash away and end up in streams and other waterways.)
3. If the soil washes into streams or rivers, why might that not be a good thing? (Help them to see that the more silt that builds up in a stream, the less room there is for carrying water. As a result, streams and rivers can overflow their banks, resulting in floods. Then, when water cuts away at a new point on a streambank or riverbank, it begins to loosen more soil, and the damaging cycle continues. Also, the habitat for fish and wildlife is forever damaged.)
4. Can they think of other times when soil might be left vulnerable to erosion? (Planting a new garden, installing a pool, and the like)
5. Can they think of some ways in which soil can become contaminated? (*Examples:* illegal dumping, overuse of pesticide, or a leak at an industrial site.)
6. How difficult do they think such contamination would be to clean up? Very difficult or not at all? (It's extremely difficult because such substances are virtually invisible once they are released.)
7. Can they think of a general reason why it is important to keep soil free of contamination? (In the long run, this behavior does not contribute to sustainability and displays poor stewardship on the part of all humans.)

ACTIVITY 8.1 ··························
Hydroponics to the Rescue

Materials (for each plant)

pen and paper for recording your results
seedling (nasturtiums and lettuce are popular, easy-to-grow choices; or let students choose for themselves)
empty 2-liter bottle, cap included. Teacher should punch a hole in all caps ahead of time, using an awl and hammer, knife, or other implement.
growing medium of your choice (or combination of media)*
nutrient solution (your own mixture)*
pH testing kit
wick or strip of cotton fabric
lemons or lemon juice, baking soda, vinegar
straw
sunny spot or sunlamp with timer
Handout 8.1

*See the Resources section for purchasing sources.

Primary Subject Areas and Skills

science, social studies, math, research, critical thinking

Purpose

Students conduct their own hydroponics experiment to see how plants and vegetables can grow without soil.

Ground pollution is one of the challenges facing not only urban gardeners and growers but anyone growing plants and vegetables in an area occupied by housing developments, businesses, or factories. Pollutants are rarely confined to any one space. The quality of the soil where you're living or planting may not be ideal, especially if you intend to eat the fruits and vegetables of your labors.

Some gardening situations are more serious than others. This is especially true if the ground has been contaminated with lead or other industrial pollutants that can work their way into any edible plants and vegetables you wish to grow. One way around this issue, and one that Philadelphia resident and *Edens Lost & Found* subject Mary Seton Corboy employs, is to use a technique called *hydroponics*.

Hydroponics is growing plants without soil. The phrase comes from the Greek words for water *(hydros)* and work or labor *(ponos)*. Water and the nutrients mixed into it do the work in this growing system, which enables the gardener to more strictly control what the plant "eats" every day. It's nothing new—the Hanging Gardens of Babylon used hydroponics in ancient times. Today, there are many different hydroponic techniques, ranging from quite simple methods that can be done at home (or in your classroom, as shown below) to very complex techniques employed by full-scale farms and other commercial operations.

"Passive" hydroponics is one of the simplest methods employed. It involves planting the plant in some sort of container, such as a regular pot or even old bottles, as seen below, that includes a growing medium.

Common media used in passive hydroponics include perlite, gravel, vermiculite, rockwool, clay, and many, many others, used either alone or combined. Media made specifically for home hydroponic systems can be purchased from various suppliers. These media allow oxygen to circulate to the roots. The pot then stands in, or is suspended over, a nutrient solution that will feed the plant. The liquid nutrient solution can also be transported up to the roots via a wick that is store bought or even homemade from leftover fabrics.

Nutrient solutions, or fertilizers, vary widely as well, and everyone seems to have her or his own recipe. Plants absorb nutrients through their roots. These nutrients, as they break down in water, are often in the form of ions. Plants specifically need nitrogen (N), phosphorus (P), potassium (K), calcium (Ca), magnesium (Mg), and sulfur (S). Popular nutrient "food" to add to the water includes calcium nitrate, potassium nitrate, magnesium sulfate, and potassium phosphate. As the plant interacts with the nutrient solution, depleting the solution of some of those nutrients, removing water, and so on, the pH of that solution may become more acidic (less than 7) or more basic (greater than 7). Adding acidic or basic substances (see below) to your solution can help bring the pH back into balance.

Here is an example of a passive hydroponics system.

Hydroponics to the Rescue: Putting It in Action

After conducting or viewing this experiment, students will have a better understanding of plant nutritional needs in general and hydroponics more specifically.

1. Decide if the experiment will be done as a class, in groups, or individually.
2. Go over the instructions with the class. Be sure that students are well equipped with any safety garb that they may need, such as gloves, safety glasses, and the like, and that they are well versed on safety precautions.
3. Instruct students to keep a journal to help them monitor the experiment.
4. Copy Handout 8.1 and have students answer the questions as they monitor their experiment.

Instructions

1. Cut off the top third of the empty 2-liter bottle. (If you are concerned about students handling cutting implement, do this yourself beforehand.)

Passive Hydroponic System

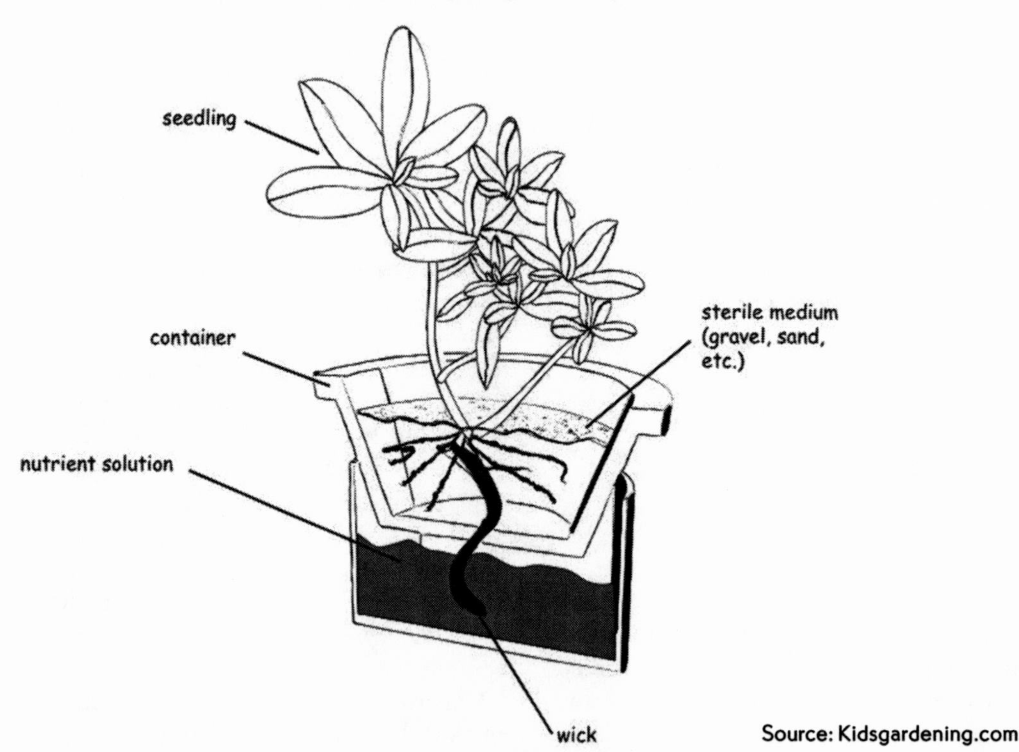

Source: Kidsgardening.com

2. Place the cap (with hole) on the bottle.
3. Cut a 10-inch section of wick and insert it through the bottle-cap opening.
4. Invert the top of the bottle and replace it in the bottom half of the bottle. (See above illustration.) You will now have the wick extending from the bottom area, or nutrient reservoir, through the bottle cap and into the seedling container. The wick will carry the nutrients into the container area.
5. Fill the container area with the medium of your choice and plant the seedling in it. As you do this, try to wind the wick through the medium (as shown in the diagram above) so that it will be exposed to the seedling roots.
6. You can then remove the container with medium and seedling temporarily to fill the reservoir with your nutrient solution. Your nutrient solution must come in contact with the wick so that the wick can carry the solution up to the plant. Nutrient solutions can be purchased through a hydroponic supply source and are usually diluted in water (often 1 to 2 teaspoons per gallon; follow the manufacturer's instructions). You can also add plant fertilizer to water to make your own nutrient solution.
7. Take the pH of the solution before inserting the container. Make a note of it.
8. The pH should be monitored on a regular basis and should generally remain approximately between 5 and 7. You can also research the optimum pH level for the seedling you have chosen to plant. Lemon juice can be added to increase the acidity of a solution that has become too basic, and baking soda can be added to a solution that has become too acid. Add them in small amounts until you achieve the result you want.

Ways to Vary the Experiment

1. If students wish to increase the CO_2 level in the nutrient solution, they can blow into the solution using a straw (being careful to blow out, not suck in). Also, baking soda and vinegar yield CO_2 and, under supervision, could be added to the mixture.
2. Changing the variables in the experiment will allow students to see effects of other factors on the plant growth. For example, change one of the following:
   - growth medium
   - nutrient mixture
   - exposure to light
   - kind of water used (tap versus bottled or distilled, for example)
3. Assign individual students or groups different variables on the same experiment and have them keep track of changes in growth and pH and then compare findings.

ACTIVITY 8.2
Alfalfa and the Importance of Cover Crops

Materials

pen and paper; Handout 8.2

Primary Subject Areas and Skills

science, social studies, critical thinking

Purpose

Students learn how cover crops are restoring wetlands damaged by steel factory waste.

In the Unit 8 Reading, as well as in the Chicago *Edens Lost & Found* film, evidence was shown as to how the state of Illinois is restoring the Calumet region with the help of a common crop called alfalfa. In this handout, students will work with some images that attempt to explain how the state of Illinois envisions the alfalfa planting will help the Calumet region.

ACTIVITY 8.3
Soil, Air, and Water

Materials

pen and paper; Reproducible 8.3 for overhead

Primary Subject Areas and Skills

science, social studies, math, writing, critical thinking, public speaking

Purpose

Students study a diagram to understand how soil contaminants may enter the environment and pollute water and air as well. Then they write a persuasive essay on the subject.

Spotting Soil Contamination

Soil contamination happens more often than we realize. It is not just the result of big companies dumping toxic materials on their lands but also the consequence of small businesses and many individuals who don't think it's a big deal to pour extra paint on the ground in their backyard or empty used motor oil under the bushes. Contamination can either be direct (meaning it's deposited directly on the soil) or indirect (meaning the contamination can travel, via soil or water, perhaps from a spill that occurred somewhere else).

The damage can be widespread: Plants take up the contaminants through their root systems. Contaminants can travel from soil to plant to animal to human, damaging each along the way. Coming into contact with contaminated soil can have unhealthy effects on humans, such as inhaling dust from contaminated soil.

The relationship among soil, air, and water means that if soil is contaminated, then groundwater that is exposed to that soil can be contaminated, too.

1. Copy Reproducible 8.3 to a transparency for use with an overhead projector.
2. Ask students to examine the image and think about the different ways that pollution travels between soil, water, and air.
3. Write students' suggestions on the board or label them on the transparency on the overhead projector. Some examples:
   - Factories dump toxic materials on their lands (direct contamination).
   - Small businesses and individuals drop paint or chemicals in their backyard or empty used motor oil on the ground (direct contamination).
   - Plants take up the contaminants through their root systems. Contaminants can travel from soil to plants to animal to human, damaging each along the way (indirect contamination).
   - Airborne pollutants from cars and factories can fall to Earth as rain and enter waterways or soil and be absorbed by plants, animals, and humans (indirect contamination).
4. Encourage students to think about relationships between elements and the effect of natural phenomena (rain, snow, tornadoes, and the like) on the transport of pollutants.

What to Do about It

In many cases, soil contamination is caused by the dumping of hazardous materials, often by companies. Some of this occurred long ago, before people were aware of ramifications of their actions and before there were regulations in place about how hazardous wastes should be disposed. No matter the cause, the legacy of toxic dumping is felt long afterward. So who's responsible for cleaning it up?

1. Copy several op-ed pieces from your local newspaper or from a more nationally focused newspaper such as *USA Today* or *The New York Times*. Ideally, copy both pro and con opinions relating to the same issue.
2. Distribute the articles to students and instruct them to read them.
3. Ask students what they noticed about the tone of the articles. How did the tone affect their opinion of the topic at hand? (Powerful writing can persuade readers.)
4. Divide the class into two sections, pro and con. Instruct them to write a 400-word op-ed piece about the responsibilities of cleaning up toxic dump sites that were abandoned long ago. In many cases, just as an example, if companies can show that they might need to use the land again in the future, they do not have to clean it up.
5. *Note:* The "pro" writers can suggest that companies clean up their hazardous dumping with their own funds, no matter what their future plans might be. The "con" writers can suggest that if they company owns the land, it is theirs to do with as they please, as long as they discontinue illegal dumping.
6. Have selected students on both sides share their articles with the class. This is a good opportunity for a class presentation.

EXTENSIONS

Hydroponic Headquarters

Start your own hydroponic garden at your school. Many schools have done this and the results can be not only educational for the long term, but also tasty and even profitable. A small greenhouse or underused shed can be converted; or, for smaller projects, a section of a science lab might be more appropriate. Experiment with different salad ingredients and vegetables. Offer some of the veggies to your school or even a community food bank.

Go Cover-Crop Crazy

Find an area on your school grounds where you can plant alfalfa. Not only will it look nice, but it will help the soil, too. Then seek out a spot in your community—a park or lot, perhaps—that could stand a little improvement. Take pH readings and test the soil (you can get a soil testing kit from most gardening and hardware stores) before you plant. Continue to check the readings at regular intervals to see how planting the alfalfa has affected the quality of the soil. If you keep the soil a little loose around a couple of the alfalfa plants, you can dig the plants up later (be patient and careful) to see for yourselves just how far and wide and deep those valuable roots go.

Pay a Visit to DEP . . . or Have Them Visit You

What are some of the major soil contaminants in *your* area? Ask the experts at your nearest Department of Environmental Protection. Also, what kind of natural events in your area (hurricanes, droughts, earthquakes) might affect your soil quality?

To Erode, or Not to Erode

Find a spot on your school grounds or in a local community garden to experiment with erosion. Create two identical mounds of dirt, roughly four feet in diameter and two feet high. Plant one with alfalfa and leave the other alone. Make sure the two mounds remain undisturbed. Water both mounds of dirt on a regular basis. Wait for the alfalfa to take root. Then, if you can afford to do so, wait for a heavy rain. If the weather does not cooperate, get a watering can and douse both mounds. Which one seems to hold up to the force of the water better? What other differences did you note in both mounds?

RESOURCES

For more information about hydroponics and hydroponic products:

- www.howtohydroponics.com/
- www.simplyhydro.com/whatis.htm
- www.hydroponics.com/articles/hydroatschool.html

The University of Minnesota has great information on the role of alfalfa in our future:

- www.misa.umn.edu/Putting_Alfalfa_to_Work_in_the_Environment.html

Check out this site from the folks at Discovery for more "dirt" on soil:

- school.discovery.com/schooladventures/soil/

For a great source of cover crops (alfalfa and others), have students do an Internet search to find seed companies that sell to individual growers or gardeners.

Reading Handout Unit 8:
RESTORING CHICAGO'S CALUMET REGION

In this reading, you'll learn how a group of committed people are bringing the contaminated soil in a Chicago neighborhood back to life, using simple farming techniques that can be duplicated all over the nation. After you read the story, answer the questions.

When tourists descend on Chicago, they tend to focus their excursions in the city center. Visitors come to check out the view from the Sears Tower, explore the fancy shops on the Magnificent Mile, take in the museums, catch a Cubs baseball game, and taste Chicago's deep-dish pizza. It is safe to say that few have put the Calumet region on their to-do list. This long-neglected region, pronounced "CAL-yoo-met," is situated on the city's southeast side and gets its name from the 700-acre Calumet Lake and the nearby Calumet River. However, historically its reputation is far from natural and pristine. Calumet was once home to Chicago's steel industry, and the landscape was forever altered by the mounds of molten residue, called "slag," that was dumped outside the factories. Hot slag was bright red, and as it cooled it hardened to thick slab as hard as concrete. Giant fields of slag now cover the ground outside these abandoned factories.

Times and economies change, and factory by factory the steel production moved out of Chicago. Once, this region employed 86,000 workers; today not a single one makes a living making steel. The old factories that brimmed with productivity are now silent, motionless giants. Refineries, steel mills, landfills, and railroads dot the area, most of them decrepit and abandoned. (A few are still in business, but they do not make steel.) "For Sale" signs, rusty pipes, Dumpsters, scattered slabs of concrete, and endless oceans of slag are all that remain. The highest percentage of Chicago's vacant industrial buildings is found in this region, and for years it was the city's biggest eyesore. It was a dumping ground—a place to forget or cover up. There was nothing here worth saving.

Or so it seemed. What few people realized or appreciated was that in the absence of humans, the original marshes and prairies of the lake and river were rebounding. Wildflowers sprang up in abandoned parking lots and in nooks and crannies of the tough slag. Herons, egrets, and cranes stalked fish in the Calumet's murky waters. People thought some citizens strange when they suggested the area was actually a rare jewel worth saving, a piece of Chicago's natural history that had managed to withstand more than a century of punishment.

Today, these people are not alone. The city of Chicago, state of Illinois, and the federal government have joined together to preserve nearly 5,000 acres of the region's natural spaces and attract less harmful businesses to the area. Some people, like engineer Victor Crivello, envision a day when the toxic dumps are cleaned up, the impact of slag is less important, waterways run pure again, and tourists come in droves to enjoy the water, wildlife, history, and views.

"The Calumet is less polluted than its reputation," says Mr. Crivello, who has spent his career inspecting and rehabilitating industrial areas. "There were nine steel mills within the city of Chicago that have closed, the major chemical plants and refineries have closed, so there are some historical problems left by those industries. But Lake Calumet never had any major industrial discharge into the lake. Calumet has pretty high water quality; the sites we have identified with problems are like any sites in the country with those same problems."

The Calumet is being rebuilt in two different ways. To transform the slag fields, officials from the state government are spreading sediment from Lake Peoria—largely topsoil that washed away during years of poor water management—on the slag fields to create a base for planting. Next, biologists are seeding these fields with alfalfa and rye grass, two crops that grow quickly, root deeply, and will help prevent erosion while sucking excess moisture from the wet soil. The roots can also penetrate existing cracks in the slag. Next year, biologists will shape and landscape the area and plant a large part of it with native plants and smaller areas with grass more suitable for foot traffic and other park activities. At the same time, volunteer organizations such as Mr. Crivello's group, Bold Chicago, invite citizens and students into the reclaimed areas to help them cut down invasive plants and plant new vegetation.

"I have a group of three high schools whom I work with," he says. "Eight Saturdays a year we'll put kids out in the fields to take water samples and do restoration work. They help us plant new native plants. They love it. They need a mission, and they get a sense of accomplishment. Bringing back the Calumet is very alluring."

Mr. Crivello expects that the region will grow on two levels. Eventually, he thinks, the Calumet will attract companies who are looking for real estate in pristine outdoor settings. Then, too, he thinks the area will lure birders, fishers, and boaters who see the place as the perfect recreational setting. Soon the Calumet will boast bicycle trails, public access to the wetlands area, and an active environmental center.

Mr. Crivello wants jobs to return to this area, too, but the kind that won't devastate the landscape and create unsafe work and living conditions. "The reason we're standing here right now—the reason we *can* stand here right now—is because we're on some solid ground. If industry never came through, if the railroad tracks never came through, there really would be no opportunity for residents living on the south side of Chicago to come here and enjoy this unless they could pay someone to come out here in a canoe."

The new strategy among the neighbors who live here is to visit the people who run the last remaining factories, to talk to them, and to include their viewpoints in their plans as the region moves forward. "I believe in ten years you can come back here, and it will be a different region." says Mr. Crivello, "You'll know it immediately when you get out of your car."

Questions

1. The author says it is safe to say that tourists do not want to visit the Calumet. Why is that? What is the Calumet known for? What industry was once there?

____

____

2. How did the old steel factories harm the soil? List at least one way.

____

____

3. What is slag, and how would you describe it? What do you think it feels like? How do you think it would affect the ability of plants to grow?

____

____

4. How is the state dealing with the slag problem?

____

____

5. What are some of Mr. Crivello's dreams for the Calumet area?

____

____

6. How optimistic are you that Mr. Crivello's ideas will work?

____

____

7. If you lived in Chicago, would you volunteer to help Mr. Crivello in his work? Why or why not?

____

____

8. The Calumet region is a wetlands area. Does it sound pretty? Could it be? What would it take to make it become more beautiful than it is now?

____

____

9. Would you pay money to rent a boat and paddle down the Calumet River—even if you had to look at a bunch of old factories? Would it be fun? Why or why not?

____

____

10. Imagine that you are a businessperson who wants to buy and fix up an old Calumet factory. What would you do about the slag? Do you think it's a big problem? Do you think it would be expensive or cheap to remove it?

____

____

Name ______________________________ Date ______________

HANDOUT 8.1: Questions for Hydroponics Experiment

Answer these questions as you monitor your experiment:

1. What are the chemical formulas for the following suggested plant nutrients?

Name of Nutrient	Chemical Formula
calcium nitrate	
potassium nitrate	
magnesium sulfate	
potassium phosphate	

2. How did the plant behave in the sun? Out of the sun?

____

____

3. How long did it take before a change in the pH was noticed?

____

____

4. How might increasing CO_2 affect the plant? (*Hint:* What role does CO_2 play in plant growth?)

____

____

5. What do each of the nutrients listed in the introductory material contribute to a plant's growth and development?

____

____

Name ______________________________ Date ____________________

HANDOUT 8.2: Alfalfa and the Importance of Cover Crops

In the Unit 8 Reading as well as in the Chicago film, you saw how the state of Illinois is restoring the Calumet region with the help of a common crop called alfalfa. Alfalfa is a legume, a member of the bean family that is famous for being a natural soil fertilizer. Farmers plant legumes, or cover crops, to protect loose topsoil from blowing or washing away.

The next year, when farmers want to plant on this field, they plow the alfalfa underground and let it rot, releasing its nutrients into the soil. Alfalfa and other cover crops have a special ability to capture nitrogen gas from the air and in the ground and convert it into a form that is easily usable by other plants. When a plant does this, it is said to be "fixing" nitrogen. When the plant dies, its body decays, releasing nitrogen into the ground where more plants can reach it.

Study the diagram below and use it to answer the questions on the next page.

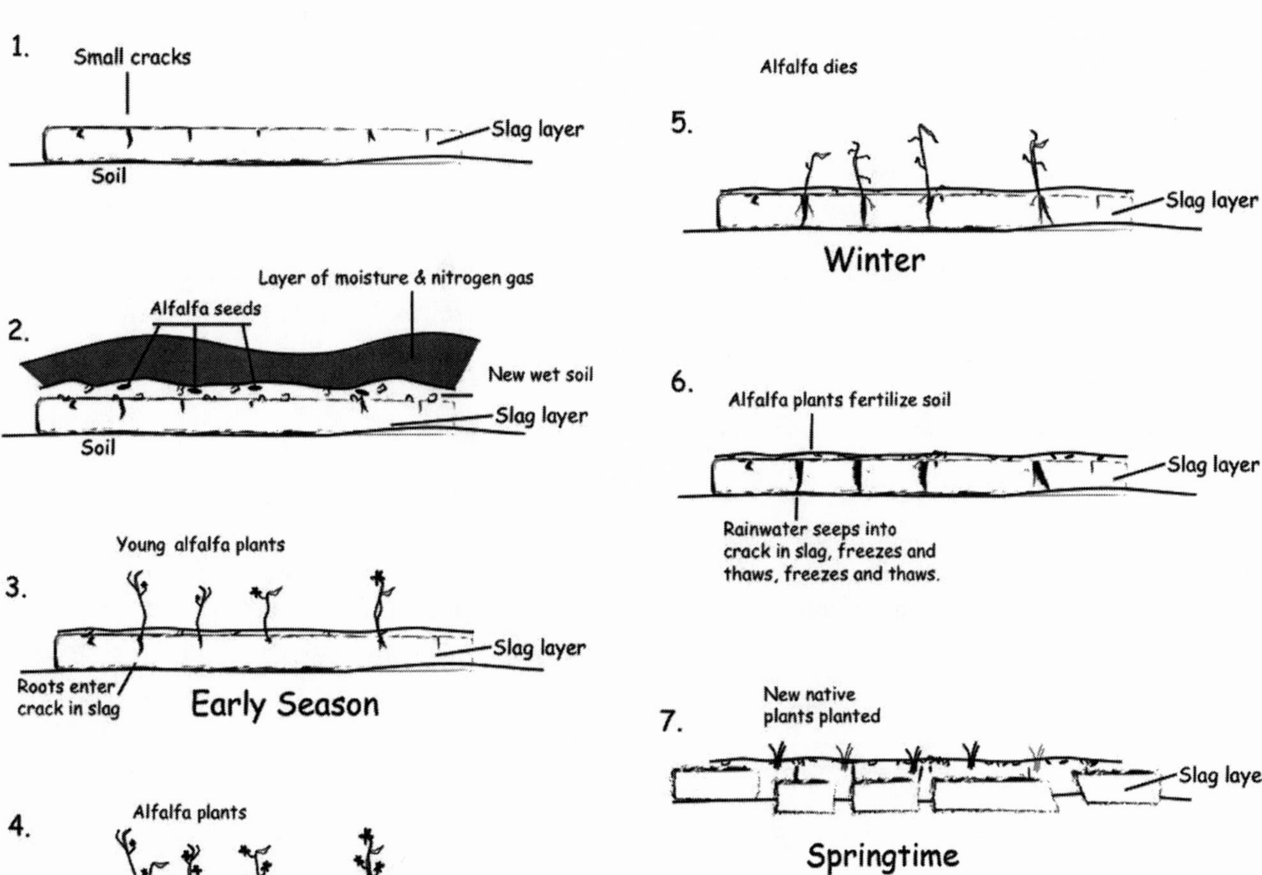

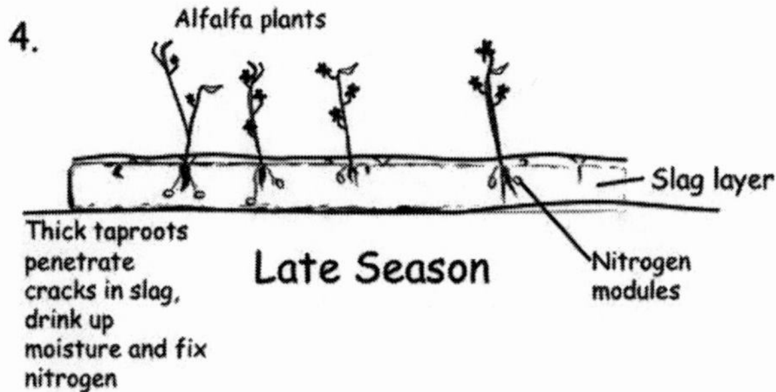

1. In Figure 1, a layer of slag, a steel industry waste product, covers the ground. When it cools and dries, slag is solid, like concrete. Would you expect plants to grow through a substance like this? Yes or no?

   ____

   ____

2. How would you describe the progress of the plants between Figures 2 and 4? How are the alfalfa seedlings getting the nutrients they need?

   ____

   ____

3. What happens to the slag during the winter months? What role do you think the alfalfa plays in this process?

   ____

   ____

4. Alfalfa has three important traits: It has a thick central root that reaches down into the earth for many feet, it requires lots of moisture, and it is a natural fertilizer. How do you think these two characteristics made it a good start-off crop to plant in the Calumet?

   ____

   ____

5. The Calumet's restorers decided to use natural methods to break up the slag and fertilize the ground. Do you think they could have broken up the slag with bulldozers and applied commercial fertilizer out of bags? Which method do you think would be more time consuming, difficult, and expensive?

   ____

   ____

REPRODUCIBLE 8.3: How Pollution Travels

UNIT 9

PARKS AND OPEN SPACES

TOPIC BACKGROUND

Parks and open spaces are important parts of any community, but they are especially important in large, growing cities. Biologically speaking, cities can't "breathe" because hard surfaces such as concrete and asphalt encase so much ground. Impermeable surfaces are unforgiving in the summertime. Asphalt, glass, steel, and concrete absorb heat and pump it back into the air, heating up the city even more.

Parks and open spaces, on the other hand, ventilate cities with their trees and permeable surfaces, such as grass, gravel, mulched landscaping, and even leaf-covered soil. Green surfaces absorb rain, which reduces the likelihood of floods and polluted waterways and builds up precious drinking water reserves. And, as we've seen in previous units, open ground helps filter contaminants from water. In this way, parks and open spaces meet the goals of urban sustainability.

They also serve a very real human need as well. Human beings are social creatures. We like to congregate, see friends, and make new acquaintances. Parks give us a setting for those planned and chance meetings to occur. When people pitch in to clean a park, plant trees, or just hang out in a public place, they are obliged to cross paths with other people. Again, they are lifted out of their smaller world and exposed to different people from different backgrounds and cultures. This gives them a chance to explain themselves and what they are doing. Little by little, others become intrigued and may join in. Before long, a powerful feeling of community takes root, and people feel more connected to their city than ever before. This is important at time when many adults worry that their young people are spending too much time in front of computers or else text-messaging their friends from a distance.

American culture does not always make room for a public space. We have parks, but we do not always appreciate the value of a public square, an open spot in our downtowns where people can hang out and meet each other. Malls are apparently expected to fill this void. Yet, squares were important features of older American towns and European cities.

Purpose

Today's movement toward sustainability recognizes green, public spaces as having a definite role to play in the health of a town or city. In this unit, students will start thinking about the reasons that parks and open spaces are so valuable.

GROUP ICEBREAKERS

1. Start by asking your students to jot down on a piece of scrap paper the kinds of activities they like to do in their community, such as skateboard, play sports, and go to movies. (The longer their lists, the better for your purposes.) Next, have them go down their list and check off how many of these activities occur in a park. How many are conducted in public spaces? How many are on school grounds? Some other place?
2. Write some of these locations up on the board and ask students to describe the attributes of each. (Many of your students may simply say they hang out at a mall, a specific street corner, or a coffeehouse.)
3. If their choices are *not* parks or downtown open spaces, get them to explain why such places are not included. Do such places not exist in town? Or if they do, are they dull or uninteresting? Are they poorly situated? Unwelcoming to young people? (You may be surprised: In some towns, public parks are locked up when sports events are not taking place, ostensibly to prevent young people from gathering in them.)
4. If parks and open spaces are among their favorite spots, ask them to consider why. Can they offer any criticisms for improving them? (Accept all reasonable answers.)
5. Based on their answers to questions 2 and 3,

what characteristics might a new park or open space need to have to be considered desirable? (*Examples:* The need to be open, clean, skateboard friendly, and so on.)

6. Because they would not be the only users of such a public space, how must they go about making sure the interests of others are represented? (*Examples:* Do a survey, talk to current park users to see what they would like to see in their park, and the like.)

ACTIVITY 9.1

Thinking about Parks

Materials

pen and paper; large tablet, overhead, or board

Primary Subject Area

social studies, history, critical thinking, research

Purpose

The instructor will use a series of prompts designed to have student engage in a conversation about the value of public parks.

Parks are a key to life all over the world, though they are often either overlooked, undervalued, or even mistreated. Yet parks can provide the much-needed space for exercising, socializing, or just plain relaxing outdoors that is not always available in cities, towns, or suburbs. They can be a place where members of different communities come together and allow for the kind of activity and interaction that does not normally take place on the streets of a town or even neighborhoods.

Some of the parks highlighted in *Edens Lost & Found*—Fairmount Park in Philadelphia and Millennium Park in Chicago, for example—have played a central role in the rejuvenation of these cities. But what are the features that really make a "good" park?

Use the questions below to spark a class discussion that will get kids thinking about the value of parks as part of the greater community. This activity and discussion will also lay the groundwork for the "Design a Park/Piazza" in Activity 9.4 later in this unit.

1. Begin this discussion by asking students, in a general sense, why they think parks are necessary. Write students' answers on the board. (Accept all at this point.)
2. Instruct students to take a blank piece of paper and list three different parks with which they are familiar, whether they are parks in your area or parks they have visited while on vacation or living elsewhere.
3. Have students make a chart at their seats that compares the following characteristics about the parks they listed: How are these parks the same? How are these parks different? How do individuals benefit from the parks? How do groups benefit from the parks?
4. Ask students to share their lists verbally with the class, describing what it is they noticed that was different or similar in the structure and use of the parks they named. See if students who listed the same parks had similar impressions.
5. Ask if they think people in cities, suburbs, smaller towns, or rural areas use the parks more. Why or why not?
6. On the board or overhead projector create a list of all the things that make a good park. Encourage students to think about facilities, activities, and signage, as well as sport fields and bike paths.

Extensions

As a lengthier assignment, students can pick a world-famous park in the United States or abroad and write a short report on it. This report should include not only the history and layout of the park but also how the park has historically been used and whether that usage has changed over time. Were there once horse-and-buggies where now there are bike paths? Was it a place of commerce where now it is used simply for recreation? You may also assign this activity for homework before coming to class, asking students to think about the different types of communities that exist in the world.

ACTIVITY 9.2

Open Space Around the World

Materials

pen and paper; Reproducible 9.2

Primary Subject Areas and Skills

social studies, writing, critical thinking

Purpose

Students are invited to compare and contrast the

design and uses of public spaces in the United States and in great European cities.

If you've traveled to Europe or even seen images of the large open *piazzas* in Italy or *places* in France or *plazas* elsewhere in the world, you've seen that many Europeans like to congregate in city centers as much as they do in parks. People sit quietly sipping coffee, reading newspapers, or people watching as others go about their business, ducking into shops and offices. All day and into the night, these centers of social activity and business are the heart of these cities and are gathering places for all.

With the expansion of suburbs, much of the "heart of downtown" has been lost for many towns and cities across the United States. Yet some towns are working to restore this feeling. Many cities and towns, large and small, have adapted some of the more popular aspects of the so-called "café lifestyle," such as outdoor seating at restaurants and benches and vendors dotting the parks that run alongside rivers. But open spaces in urban centers are much more than that. They offer an opportunity for commerce and community to come together and for people of all backgrounds to interact with each other as they go about their daily lives, whether working, playing, relaxing, or running errands. The Bunker Hill project featured in the *Edens Lost & Found* Los Angeles DVD is an example of a targeted return to this approach.

This activity gives students the opportunity to analyze the features and history of urban open spaces here in the United States and abroad. It will give them insight into what brings communities together in urban settings and the benefits and challenges of including urban open spaces in towns and cities across the country.

Copy Reproducible 9.2 onto a transparency for use with an overhead projector. Then discuss the questions that follow.

1. Ask students to describe the different kinds of activities that take place in this piazza. Which are work related? Who do they think gathers here and why? Write these responses on the board. (*Examples:* Bikers, artists, shoppers, diners, mimes, gymnasts.)
2. What do students *not* see here? What probably would *not* take place here? (Some examples might include sports games, skateboarding, picnics, cookouts, cars driving through.)
3. What kind of environmental and health benefits might arise from having more open public spaces in towns and cities? (The more open spaces there are in a city, the more residents have a place to congregate and relax in a venue where automobiles are not present.)
4. Instruct students, as individuals or in small groups, to research some of the more famous open spaces in the world. They may suggest their own, based on their own experiences and travel, or you may have them choose from the following:
   a. Piazza del Campo, Siena, Italy
   b. Trafalgar Square, London, England
   c. Piazza San Marco, Venice, Italy

Their reports should discuss the following:

5. What are the chief characteristics of these areas? (All are surrounded by businesses or residential buildings. They are meeting places, tourist attractions, and destinations for people who just want to hang out.) What do they offer? What businesses are located there? (Some have food venues, nearby shopping, tourist sites, and so on.) How were they used historically as compared to how they are used today? (Many European sites were once markets for tradesmen and are now regarded as tourist sites.)
6. What purpose do they think the area serves to the greater community? (They provide open spaces in very large, crowded cities.)
7. How do these places compare to American parks, such as Millennium Park in Chicago or Central Park in New York City? (American parks are designed to be places containing nature. They are meeting places, but one does not go there to shop or enjoy fine dining. The closest American comparison to a European square or piazza is a shopping mall.)
8. Does their town or city have a town square? How is it used? How would students like to see it used? (Many smaller American cities do not have town squares in the European sense. Students may be attracted to the rather grown-up notion of hanging out in a lively town center. A large park does not provide the same type of atmosphere.)

ACTIVITY 9.3

Mapping the Millennium

Materials
pen and paper; Handout 9.3

Primary Subject Areas and Skills
social studies, math, critical thinking

Purpose
Students analyze a map of Chicago's new Millennium Park to determine what features modern Americans regard as essential to new parks.

Chicago's new Millennium Park, featured in *Edens Lost & Found,* looks like it's always been there. Seemingly perfectly situated in downtown Chicago, flanked by businesses and galleries, abuzz with signs of Chicago's busy daily life, you would think that the city grew up around the park—not the other way around. But that's exactly what happened. Where once was a train yard and rundown parking lot is now a brand-new green space with art, gardens, performing spaces, and more.

1. Copy Handout 9.3 and have students read the map and answer the questions. Alternatively, you may wish to read the map with the students on overhead.
2. When students are finished, ask them where they think the names for these areas of the park came from? Using the Millennium Park website (www.millenniumpark.org), have them research the background of each of the park's features and their namesakes.
3. What do they think of corporations donating money to parks in exchange for having their name on prominent features? (Some will have no problem with it; others may object on grounds of commercialization. Have them explain their stances.)
4. Now that they have learned a little about the park's various amenities, which one(s) do they think would be a good idea for a park or open space in their community?

ACTIVITY 9.4

Design a Park or Piazza

Materials
pen and paper

Primary Subject Areas and Skills
science, social studies, math, art, research, public speaking

Purpose
Students use what they now know about parks and open space to design one of their own.

By now your students know enough about parks and town squares (or piazzas) to try their hand at designing their own. They may think it is a simple endeavor, but once they get into it, they'll see that a number of choices need to be made, and not everyone will agree on them. Depending on the number of students in the class, you may wish to break them up into a couple of different teams. One team could design a park, the other a town square. Alternatively, you could have the entire class design one park but break them up into teams responsible for different design aspects: one team to design the signage, another to select and lay out the plants and trees, still another to design placement of seating, sporting sites, and so on. If you go this route, stress that students from each committee must meet regularly with representatives of other committees to compare notes.

Above all, once the park is completed, all the design elements must come together in a unified way. All the signs, for example, must use a similar style of typeface or font, and the feeling they engender must be a good fit for the rest of the park, not to mention the town. (A "Goth" look, for example, would be inappropriate for a quaint New England village.) We envision this as a long-term project, ideally based on real locations in your own town. To give a sense of seriousness to the proceedings, you might schedule a day for a final "crit," or critique. Real architecture students use crits to present their ideas to their superiors and get feedback and criticism. You might invite your principal and other teachers to your crit, or, if you truly want to go big-time, try to schedule a presentation to a local committee of your town council. Below are some guidelines for this project.

Team Breakdown
Teams can be broken down into the following design committees: sporting areas, storage and maintenance

areas, seating, landscaping, artwork, cultural, and historic or military monuments.

Site Assessment

If they will be designing or redesigning a real space, they should view it with an eye toward preparing a critique of the place. What would students like to see changed? Are there any eyesores or obvious problems they feel need to be addressed?

1. Some topics to investigate at a park include: Are exits and entrances well marked and secure? Are there sporting facilities? Areas for small children? Bathrooms and water fountains? Flags and monuments?
2. Some topics to investigate in a potential town square site are: What types of shops could there be? What kinds of cafés? Will car traffic flow through the area or not at all? Will the area be open to bicycles? Will street vendors and performers be allowed? What kind of paving stones will be used?

Flora and Fauna

If you are using a real site, students must be mindful of the existing biodiversity: plants, shrubs, trees, birds, wildlife, and more. (This will require repeat visits during different seasons because some wildlife and plants may only be active during certain times of the year.)

Regulations

You'll need to establish rules of behavior for your setting. In some parks, for example, playing loud music is not permitted. Some mandate a curfew after a certain time of night, which tends to vary seasonally. Some places do not allow pets, or if they do, require pet cleanup and that animals be leashed. Their work must include a list of rules, subject to approval by your town council.

Human Diversity

Teams should be mindful of the different types of people who might visit a park or square. Their locations should be handicapped or stroller accessible. They should be easily accessible to the elderly. Parents with small children may have a need for a diaper changing station.

Presentation

Allow students to use all their talents in bringing their designs to life. Sketches, dioramas, and scale models may all be part of the final presentation, but they should also be able to produce a thoughtful written component, such as essays, journals, or brochures. The written piece should spell out their design challenges, research, and the choices they considered before arriving at their final designs. Expect that these would contain evidence of math and science tie-ins. Be sure to stipulate that they precisely spell out all measurements and the scientific processes underlying why, for example, they decided to choose one tree over another.

Outside Help and Materials

It would help if your students could review some drawings done by building architects, landscape architects, landscape designers, and others. You may also want to present these plans to local council members or other civic leaders.

ACTIVITY 9.5
One Space, Two Plans

Materials

pen and paper
library or Internet research materials
poster board or other display material (optional)

Primary Subject Areas and Skills

science, math, writing, art, research, critical thinking, public speaking

Purpose

Two teams of students develop two different plans for the same fictional or real-life parcel of undeveloped land.

This activity will encourage students to think both clearly and creatively about a challenge that often faces communities: what to do with an undeveloped parcel of land.

In this exercise, students will try to understand what motivates us as a society, as a community, and as individuals. When competing intentions collide, it can be very trying for everyone involved and, more often than not, counterproductive.

It can be valuable for students to work and research on behalf of a project that they don't necessarily support. This helps them gain insight and perspective that can help them become better communicators and savvy negotiators.

1. Divide the class into two teams and read aloud the Premise below.
2. Each team will create a presentation that details their plan for an undeveloped parcel of land.
3. Students should show that they have answered the following questions about the project:
   a. How will their project be financed?
   b. What purpose does it serve for the community, and whom will it benefit?
   c. What will the site look like?
   d. Why is this plan needed?
4. Students should include visual guides in the form of posters, transparencies, or dioramas in support of their plan. Graphs, charts, or other representations of their project plan are strongly encouraged. Students may wish to create names for the different features or commercial venues in their plans. They should make their design as real as possible.
5. Each side will be given a chance to present their entire project plan.
6. After the presentations, each side will choose one person to represent their plan in a formal debate.

Premise

Read aloud: A large bus depot is closing in your town, leaving a three-acre parcel of land up for grabs. (*Note:* Make sure students understand what will fit on three acres.) The location is an ideal one, and many different organizations are vying for the land so that they can put forth their various plans and projects. Team 1 wants to use the land for commercial development, and Team 2 wants to use the land for large open space or a park.

State Your Case

After students have completed their project plans and presented them to the class, a few students from each team will represent their side in a formal debate:

1. In this case the teacher will act as judge of the debate because all of the students had a larger stake and level of participation in the project presentations.
2. Flip a coin to determine who will speak first. The first team will speak for three to five minutes to state their argument.
3. The second team will then have three to five minutes to present their argument.
4. Each team will then be given an additional three to five minutes to respond to the other team's initial argument.
5. The teacher rates the arguments of each group and presents their findings to the classroom. *Remember:* The decision must be made based on who gave the most thorough, well-researched, logical and, most importantly, convincing argument.
6. Discuss among yourselves the emotions and ideas that came up during the debate. Did anyone change his or her mind about which plan he or she thought would better serve the community?

Extension

The word *compromise* is a three-syllable word that can be very hard to swallow. Unlike most debate activities that end once a winner is decided, this activity goes a step further. The two sides must devise a compromise. Using the research they have already done, ask the teams to now work together to arrive at one design incorporating elements of both original designs.

EXTENSIONS

Take a Park Survey

Choose a park in your area or region and plan a day trip there to examine how the park is used and who is using it. Before going out, ask the students who they think uses the park near you most and what activities do they think are the most popular. Do they think they will find more adults or kids? What neighborhoods do they think most of the users hail from? Travel to your park on a weekend day, if possible. Have students interview some park passersby. They should strive to speak to a variety of people of different ages and ethnic backgrounds, as well as families, individuals, couples, even police officers, if they're around. Students should explain that they are taking a survey for a class project and looking at ways that the park could be improved. One student can take notes, while the other asks the questions. Students should take turns doing each. Some questions to get students started:

- How often do you come here?
- Do you live nearby? How far do you travel to use this park?
- Whom do you come here with?
- What are your three favorite things to do while you're here?

- What would you like to change about the park?
  - see more of?
  - see less of?
- Do you think the park is safe?
- In your opinion, is the park well maintained?
- Are there enough parks in this community?

Compile the information you gathered, analyze it, and create a report that can be presented to the official in charge of the Parks Department in your town or city. Create a report that can be left behind or distributed at a council or commission meeting. The report should describe the people with whom you spoke, where they're from, and what their concerns are. Include your own critique—a map of the park with suggestions for additional signage or parking, for example. Create a volunteer squad from your school or community to work on keeping the park clean, starting an outdoor art project, garden, and so on. Students may even wish to record the experience for a podcast or a school radio or television broadcast.

RESOURCES

For more information on parks, piazzas, and squares, check these websites:

- **Chicago's Millennium Park**
  www.millenniumpark.org
- **New York City's Central Park**
  www.centralpark.org/
- **Philadelphia's Fairmount Park**
  www.fairmountpark.org/
- **Paris's Jardin du Luxembourg**
  www.v1.paris.fr/EN/Visiting/gardens/jardin_luxembourg.asp
- **Siena's Piazza del Campo**
  www.greatbuildings.com/buildings/Piazza_del_Campo.html

Reading Handout Unit 9:
THE GREAT PUBLIC SPACE

In this reading, you'll learn how the city of Chicago turned a public eyesore into a beautiful, new public park. To see what Millennium Park looks like, watch the Chicago movie on the *Edens Lost & Found* DVDs. After you read the story, answer the questions.

For decades, Chicago wrestled with major issues: How can we add more public space when space is already tight downtown? How do we get rid of ugly parking garages and train tracks without banning cars and trains entirely? How can we encourage people to start riding their bikes to town instead? For Chicago, the answer lay in a single project, 25-acre Millennium Park.

One of the city's biggest eyesores used to sit smack in the middle of downtown, on Michigan Avenue just north of the Art Institute. Here, in a 20-foot-deep canyon, commuter trains and an 800-car parking lot occupied a patch of real estate that was among the choicest in town, the perfect place for a beautification project. But how should it be done? The train tracks and parking were too important to eliminate, yet too unattractive and noisy to be tolerated forever.

City officials had talked about the problem since the 1970s, but little got done until Mayor Richard M. Daley arrived on the scene. The task of turning the site into something practical, functional, and beautiful was assumed to be too expensive to be done. "The genius of the mayor," says John Bryan, the chief fundraiser of the project, "was realizing that he could build an underground parking garage first and start earning some revenue. Trains also would be tucked below ground. He called in the architects and had them build a 4,500-space garage, which helped him raise $175 million."

Building a park this big required even more money than that. One day, out of the blue, John Bryan—who was then the head of Sara Lee, the famous company that makes baked goods—got a call from the mayor.

"Think you can raise some money for me?" said the Mayor.

Mr. Bryan had done his fair share of fund-raising over the years, and now that he was preparing to retire, he thought he'd have plenty of time on his hands to devote to the cause. "How much money are we talking about?" he asked.

Daley didn't flinch: "Oh, $30 million."

"No problem," replied Mr. Bryan, and off he went.

"Of course," he confides now, four years later, "at first we were just talking about building a nice little outdoor pavilion. We had no idea—not the mayor and not me—how much we would actually bring in, and how much we would be able to do with it."

Mr. Bryan got to work phoning major corporations in town and some of the better-known philanthropic Chicago families. People were interested, but they wanted to know what their money would pay for. The plans got more and more ambitious as the pledges poured in. Donors got more excited when they saw that top-notch artists and architects, such as Frank Gehry, had been invited to design buildings for the park. "No one wanted to give their money for something dull and uninteresting," says Mr. Bryan. "The selling point was being able to say we will bring in the world's best designers and give Chicago to the world. The best of this generation will stand in this park forever."

The $30 million goal grew—to $210 million!

As construction went forward, Mayor Daley opted to enlarge the vision, adding eight additional acres to the park and upping the price tag considerably. The press hounded him when one of the earlier parking garages had to be torn down and rebuilt. "They went after the mayor on those overruns," recalls Bryan. "But they left me alone. It was as if no one was interested in what the private sector was doing to contribute to this park."

Building a 25-acre park was bound to be expensive. But building a smaller or less grand park would send the signal that the golden age of parks was dead. Instead, the city went to work rounding up the money, and its effort became a good example of public and private sectors working together to achieve a dream. In the end, the final price tag was more than $475 million, with everything above ground—the trees, the grass, the structures and sculptures—made possible by contributions by private donors or corporations such as Wrigley, the chewing gum makers.

Builders rushed to get Millennium Park done in time for the year 2000, "the Millennium," while the local media continued to predict a financial fiasco. Some newspaper writers thought it ridiculous to spend so much on a park when so much of the city's economy seemed at a standstill. Finally, the park opened in 2004, and the populace was overwhelmed. The result was a harmonious complex that gives back to Earth almost as much as gives to the people of Chicago. What is Millennium Park? Well, depending upon your interests, it's a band shell, a home to perennial gardens, fountains, restaurants, sculptures and monuments, an indoor theater, an ice rink, four parking garages, a commuter train platform, and a world-class 300-bicycle station where you can park your bike, rent a locker, take a shower, and head off to work.

Because Millennium Park is built on the roof of a parking garage, it could be considered a "green roof," a type of garden that is constructed on top of a building as a way to cool it in the summertime and insulate it in the wintertime. Many of Chicago's air and water problems are tackled by the 900 elms, maples, hawthorns, pear, crabapple, white fir, and red bud trees planted in Millennium Park. These trees help the park function as a miniwatershed. Because trees shade buildings and cool the air, they reduce the demand for air conditioning. A mature tree reduces the amount of carbon dioxide (CO_2) by about 115 pounds per year. It does this in two ways: by drinking in CO_2 to make food for itself, a process called photosynthesis, and by lowering the amount of CO_2 pumped into the atmosphere by power plants. The California Energy Commission has calculated that the CO_2 reduction achieved by a single tree has a dollar value of $920 per ton per year. Shade trees only cost a few dollars to plant. "They absorb pollution, add oxygen to the air, and temper the heat gain that the city gets every summer," says Ed Uhlir, the park's chief engineer. "But they're also important emotionally to people, for them to come and safely enjoy themselves in the outdoors."

People not only recreate, they appropriate. Take Crown Fountain, for instance. The monumental fountain, designed by the Spanish sculptor Jaume Plensa, was conceived as a giant work of art. Every five minutes, the two towers flash an image on their 50-foot-high screens, while torrents of water cascade over them. Students from Chicago's School of the Art Institute took 1,000 photos of Chicago residents, which the fountain's computer selects from its database. But the first summer the fountain opened, local kids used the fountain another way. Decked out in swim trunks, they huddled at the base of the fountain, counting down the seconds until the floodgates opened and torrents of water crashed down upon them. Beats a yard sprinkler any day.

Small or large, parks build community. A city is enriched when its citizens have more places to hang out, visit with friends, and meet new people. Chance encounters in a park can lead to lifelong friendships. Such spaces are even more necessary today than they were decades ago because more and more Americans are starved for unscripted interaction with their neighbors.

"In a sprawling area we just have to spend more time in metal boxes taking us from one area to another so we use up a lot more time," says David Putnam, a Harvard professor of public policy and author of *Bowling Alone: The Collapse and Revival of American Community.* "Our lives are lived in large triangles, one point being where we sleep, one point being where we work, one point being where we shop. When we used to live in villages, the distance between those three corners used to be measured in feet or hundreds of yards, but now in any large city, the distance between those points for anyone might be 30 or 40 or 50 miles. In that kind of world it's not clear where home is. It's not clear where you should be basing your community."

Critics who wondered if Millennium Park would be worth the expense appear to have gone mute. Within the first six months of the park's opening, more than 5,000,000 people visited—a record for any public space in the city. "Statistics are difficult to come by with precision," says Mr. Bryan. "What we have is mostly anecdotal. Every place I go, people are telling me that they've gone, they took the relatives who have come from out of town. 'I went, I saw, I loved it,' is the reaction. It's amazing how good art and architecture moves people."

If you build it, they will come. If you build it beautifully, they'll *keep* coming.

Questions

1. What was on the site of Millennium Park before it was a park? Did people like what was there originally? Why didn't they do something about it?

____

____

2. What creative steps did the mayor take to earn some money for the park?

____

____

3. Using the numbers in this article, could you argue that trees do far more work than needed to recoup the money it costs to buy and plant them? How do you know?

___

___

4. In what way can Millennium Park be considered a miniwatershed? What work do the trees do for the city of Chicago?

___

___

5. The park's budget of $475 million is *huge*. Scour a newspaper for other big numbers like this. What kinds of items cost this much?

___

___

6. Say you were a newspaper reporter writing at the time of the building of Millennium Park. How would you describe it, as a fiasco or a dream?

___

___

7. What do you think of Dr. Putnam's theory that people today don't have time to interact? What are the "metal boxes" he's talking about? Does his description sound like the lives of grown-ups you know?

___

___

8. The author says Americans are "starved for unscripted interaction with their neighbors." Do you agree or disagree? How can parks help meet that need?

___

___

REPRODUCIBLE 9.2: Diagram of a Piazza

Diagram of a Piazza

Name ______________________________ Date ____________________

HANDOUT 9.3: Mapping Millennium Park

WHAT'S IN A PARK?

Everyone who came together to build Millennium Park put a lot of thought and imagination into its design. How do you think they did?

Look at the map and keys below and then answer the questions in full sentences.

Millennium Park

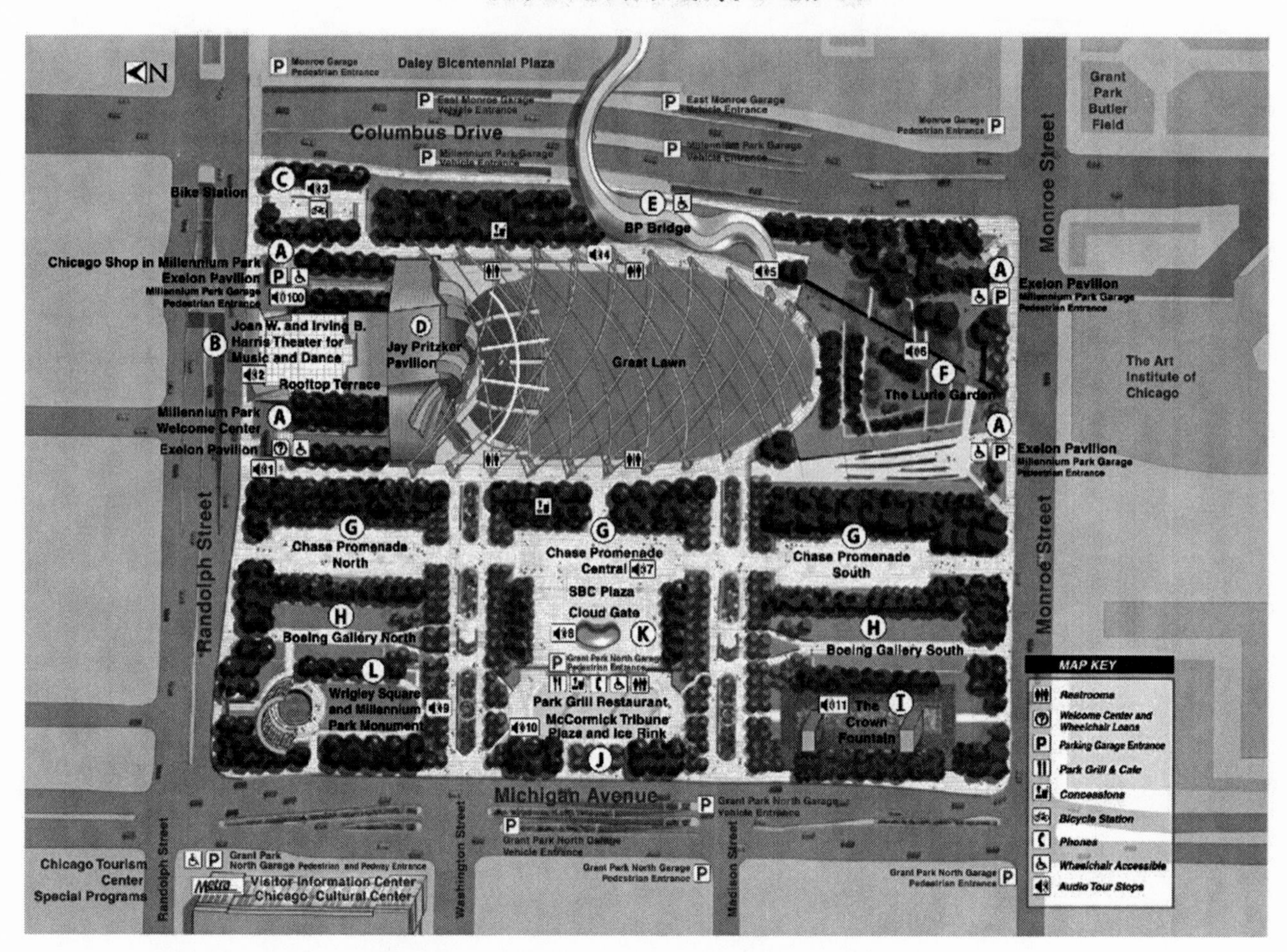

A. Exelon Pavilions: Millennium Park Welcome Center and the Chicago Shop in Millennium Park
B. Joan W. and Irving B. Harris Theater for Music and Dance
C. Bike Station
D. Jay Pritzker Pavilion
E. BP Bridge
F. The Lurie Garden
G. Chase Promenade
H. Boeing Galleries
I. The Crown Fountain
J. McCormick Tribune Plaza and Ice Rink
K. SBC Plaza/Cloud Gate
L. Wrigley Square and Millennium Monument

1. What do you consider the best feature of this park? Why?

2. How many food venues are there in the park?

3. Which one of the park's features were you not expecting? Do you think it is a good attribute? If yes, why? If no, why not?

4. How is this park similar to other famous parks and open spaces you have heard of or discussed in class? How is it different?

5. Which of the park's features can earn money to pay for the park and manage ongoing upkeep and maintenance?

6. At Millennium Park, you can park your bike, take a shower, and leave your things in a locker when you go off to work. Does this sound like something you would do? Why or why not? Do you think a lot of people do this?

7. How might this park improve the overall environmental health and well-being of the city of Chicago?

UNIT 10

TRANSPORTATION

TOPIC BACKGROUND

One of the important links in the chain of sustainability is the availability of quick, cheap, easy public transportation. If people know that they can get to work and run errands reliably by bus, train, ferry, or some other means, they will leave their cars at home. Ultimately, that plan is best for everyone because individual cars, trucks, and sport-utility vehicles consume the most fuel, spew the most carbon dioxide, and generate the most traffic. Cars, not factories, are the biggest single contributor of CO_2 on the planet. Cars and small trucks gobble up 40 percent of all the oil-based fuels used in the United States and produce 20 percent of the nation's CO_2. In its lifetime, the average car will cough out 70 tons of CO_2. An SUV will spew 100 tons! Besides CO_2 pollution, cars leak fluids and shed toxic dust that can wash off streets and into our waterways. They are also annoyingly expensive to operate: You must insure them, repair them, keep refilling them with gas, and you never know when they might break down.

By comparison, buses, trains, subways, ferries, and so on, leave the cost, hassles, and driving to someone else. For a fair price, you get where you want to go. Admittedly, mass transit isn't always comfortable, and it isn't for all people or for all places, but modern American cities cannot consider themselves sustainable, or world class, unless they have a good transportation system. It's the only way to move large numbers of people around without chaos and uncontrolled emissions.

Unfortunately, as shown in our four films, some American cities built up without incorporating mass transit into their plans. Los Angeles was conceived as a city where the average person could zip in and out of town on his or her own set of wheels. Now, choked by smog and traffic, the city is expanding its mass transit choices. Seattle has buses, ferries, and some trains, but it is still largely a driver's city. Today that city is exploring biodiesel fuels and light-rail trains. Light rail is one of the hot modes of transportation in the United States. Such trains are lighter than ordinary passenger trains and thus cheaper to build. Like the trolley cars of old, they run on electric power and can potentially be run on renewable energy.

Some cities, notably Seattle, have experimented with a program in which two-car families volunteer to go without one of their cars for five weeks. During this trial period, families carpool, use mass transit, or simply use one car more intelligently. At the end of the program, some families sell off a car, realizing that they can indeed get by without it and save money on insurance and maintenance costs.

Trains, subways, and so on, cost millions, sometimes billions of dollars and are always a sore subject with voters. Cities contemplating major construction must convince voters that the expense is worthwhile and then raise the money and keep the project on budget. If citizens can get through these hurdles, they will likely never regret it. In the cities we visited, when new transit projects finally opened for business, people flocked to them in droves, and months or years of ire seemed to be forgotten.

Purpose

By the end of the unit, students should understand the essential role public transportation plays in modern life. They will also understand why many cities resist making a commitment to public transportation.

GROUP ICEBREAKERS

1. Ask what kinds of mass transit are available where your students live. Write all the choices on the board as students mention them.
2. Ask them if they have ever ridden any of these modes of transportation. Have them generate a list of adjectives that describe the experience. Encourage them to be as frank as possible. If they didn't like riding a bus or train, they should

say why. If they thought the ride was uncomfortable but nonetheless convenient, they should say that too.

3. Ask them to consider how most adults they know get around. Do they drive their own cars? Do they take buses or trains? (For comparison's sake, you might ask them to list all the ways people their age get around town. They will likely find that their methods of transportation—walking, biking, skateboarding—are healthier for themselves and better for the planet.)
4. Ask them how popular they think mass transit is in their community. If so, why? If not, why not? (They may want to review some of the attributes they compiled earlier. Chances are, if buses are crowded, dirty, and infrequent and people have access to cars, those may be compelling reasons why mass transit is not popular.)
5. Some buses and trains do not go where many people need or want to go. This is another reason why mass transit is unpopular in some cities and regions of the country. Do your students know, for example, how to find a local bus, train, or ferry schedule for their area? (They probably will not. Ask them to find and study one of these schedules; most are available online these days.)
6. Why do they think people prefer to drive most places in their own cars? (They may simply say that driving your own car is just better. Press them to articulate why.) Ask students to think what their own dream car would be like. Then, ask them to imagine that they are able to get that car. How often would they use mass transit, then? (Most would drop mass transit if they could suddenly have their dream car.)

ACTIVITY 10.1

Passenger Costs

Materials

pen and paper; Handout 10.1

Primary Subject Areas and Skills

science, math, critical thinking

Purpose

Students work with math to determine how many BTUs are produced or saved in several different driving scenarios.

Transportation affects both our environment and our lives. The resistance to mass transportation is often based on a desire for convenience, freedom, and autonomy and a feeling that one who can pay for her or his own transport should be able to use it as she or he sees fit. And it's true that no mode of transportation—aside from those powered by the human body alone—is without drawbacks. But one thing is certain: Mass transportation can often significantly reduce the amount of pollution released into the atmosphere and the amount of energy consumed.

The charts that follow will demonstrate to students just some of the issues to take into consideration when considering mass transportation.

1. Copy Handout 10.1 and distribute to students.
2. After students have answered the questions, discuss their reactions.
3. For an additional math extension activity, have students convert the BTUs into American therms, kilowatt hours, joules, or another energy unit of your choice.
4. After students have completed the questions related to the "Energy Efficiency" chart, ask them to think about the ramifications of increasingly longer commutes for Americans. Each year, millions of Americans commute over 50 miles *each way* to get to and from work. Some commute as far as 100 miles each way, and a smaller percentage travel even further. Discuss the following with your students:
   a. Do any of your students have family members who commute more than 50 miles one way to work? If so, how many times per week?
   b. Why do they think longer commutes are becoming more prevalent? (As cities grow, real estate closest to cities becomes more expensive, and people are forced to move farther away and make longer commutes.)
   c. What do students think the effects of these supercommutes have on the environment? (More people and more vehicles mean more pollution and more resources consumed.)
   d. What effects do they think supercommutes have on the family? The community? (Long commutes are difficult for families because parents and children have less time to spend together. This can undermine neighborhoods.)
5. According to national statistics, more than 96

percent of those who have "stretch commutes," as they're called, use individual vehicles. Using the "Energy Efficiency" chart, have students compute how much energy one person would use in one week traveling to and from work. (Remind them that there are five business days to every week, not seven. *Answers:* 31,740 BTUs/vehicle mile; 26,275 BTUs/passenger mile.)

6. How much CO_2 ends up in the atmosphere if one car, getting 28 miles per gallon, travels 50 miles each way to work? (1 gallon of gas = 19 lb CO_2.)
   a. In a day? (About 70 lb)
   b. In a week? (About 350 lb)
   c. In a month? (About 1,400 lb)
7. If three people decide to carpool (using the same mileage figure above), how much less CO_2 would be emitted by the three people all together (rather than individually) in . . .
   a. One week? (About 700 lb)
   b. One month? (About 2,800 lb)
   c. One year? (About 33,600 lb)

ACTIVITY 10.2
Pros, Cons, and in Between

Materials

pen and paper
library or Internet research materials
Handout 10.2

Primary Subject Areas and Skills

science, social studies, math, writing, research, critical thinking

Purpose

Students research the attributes of four different types of transportation options to learn which are the best options for various cities.

In preparation for this activity, you may wish to watch the Seattle DVD of the *Edens Lost & Found* series, in which transportation choices for that city are hotly debated. Decisions of this sort affect different members of the community in very different ways. Transportation is an integral part of our daily lives, and people may respond very passionately if they feel that their day-to-day lifestyle—and pocketbooks—will be adversely affected.

1. As preparation, copy Handout 10.2 and distribute to students. Discuss the different types of transportation with students. Inquire as to whether students have had any experience with all four. Did they like them? Which was their favorite? Why?
2. As extra preparation or as an extension activity, have students research each of the four main types of public transportation presented in the handout. Using the information they gather, have the students help to finish completing the chart. How do costs compare? What are the pros? What are the cons? (Researching may be difficult, though many online sources will quote train costs in dollars per mile of track.)
3. You may also wish to consult newspapers or their websites in order to study how other cities—Austin, Texas, for example—have struggled to decide whether or not to install a light-rail or monorail system. What are the arguments? Which groups seem to prefer which kind of system, and why?
4. After reading aloud the Premise, assign different groups of students to research and think about the points of view that each of these demographics might have. What is each group's concerns? What are their priorities? This can all serve as valuable research for the role-playing activity that will follow.
5. Have students write a first-person essay from the viewpoint of one of the individuals listed in the Premise section. Then ask them to role-play a "town meeting," during which all of these groups will state their views and concerns and maybe come to a compromise.

Premise

Read aloud: The town of Greener Pastures is desperate to improve its public transportation. Traffic is a nightmare, air quality is worsening, and commuters' lives are miserable. But not everyone has the same idea. Some want light rail, some want monorail, and some want to simply improve the existing bus service. Whatever change is made will require a big hike in taxes on items like food, clothing, gasoline, and income. But how much tax will depend on the choices that are made. Do the necessary research and make the best possible choice not for *yourself* but for each of the following individuals. Are you ready? Put yourself in the frame of mind of each of the following citizens:

- A 20-year-old college student
- A business owner along the new line
- A politician seeking reelection
- A resident living along the new transportation line
- A senior citizen on a fixed income

ACTIVITY 10.3

One-Less-Car Challenge

Materials

pen and paper; many copies of Handouts 10.3A and 10.3B

Primary Subject Areas and Skills

math, science, social studies

Purpose

Students challenge their families to live with one less car for the duration of the project to see what life would be like if they did not rely so much on car transportation.

In this activity, you will challenge students and their families to use one less car for the duration of the project. Next, students will analyze the data they have collected, make recommendations to their parents, and examine opportunities for a wider, townwide study. This is a powerful subject, but you must undertake it with the cooperation of your students' parents and with some sensitivity. (Some families may not have cars for financial reasons, or you may live in an area where cars are superfluous. Judge accordingly. Variations on the project are discussed below.)

1. If you can watch the Seattle film with your students, pay close attention to the segment on the family who volunteered to go without a car for five weeks. Then assign the unit reading and have students answer the questions.
2. Ask how the class feels about undertaking their own one-less-car study. Explain that you are asking them and their families to monitor their car and noncar transportation usage for a period of time. Ideally, you should do this for eight weeks: The first three weeks they'll collect data on how their family uses the family car; the next five weeks they'll see what life is like without a car. But because this is a long commitment, it's best to have them work as a group to write a letter to their parents explaining the purpose and scope of the study. Based on parental reaction, they can adjust the time period accordingly.
3. *Variations:* It's entirely possible to modify the data collection to study all modes of transportation, not just car use. For example, if you live in an urban area where cars are already somewhat superfluous, you might challenge students to reduce their public transportation usage for a certain number of days a week. Tread carefully: You do not want to affect the ability of any student to attend school or get to school in a timely fashion. In general, we have found that there is wisdom in the class voice: Students will let you know what they consider to be acceptable boundaries for the study. You could conduct multiple studies at the same time, allowing some students to collect data and report back on their car-using families. Allow another set of students to report on their non–car-using families. If parents are resistant to the idea, maybe students will consider reducing their car usage or ridership. We are confident that suitable parameters can be found if the class exercises some creativity. What follows are our instructions for the full, eight-week study.
4. Have students decide the start and end date of the study, eight weeks apart. Some days before the start date, they should gather data to complete Handout 10.3A, the "Car Cost Worksheet." This will tell them and their families how much it costs to own, lease, and operate their car. The numbers will be quite surprising, so be sure to discuss the results in class. The one item with which most people are unfamiliar is their automobile's "depreciation." Simply put, it's the amount by which a car's value drops year to year. We all know a new car costs more than a used car of the same make and model. The auto industry uses depreciation to calculate that drop in value.
5. Copy sufficient diary sheets, Handout 10.3B. Each family should have enough sheets on hand for every family member to complete for each day of the eight-week period.
6. On the first day of the "driving" period, students should record the odometer readings of each family vehicle, identifying which car it is if there is more than one. The family should decide on Day 1 which car(s) they will not drive during the nondriving period. The car they intend to drive will be "CAR 1."
7. For the first three weeks, each member of a

participating household should record *all* methods of transportation they used to make *all* trips during their day. The family should request receipts and keep a record of all costs *other* than car costs, such as bus, cab, train, ferry, and subway fares. As a reminder, you have already calculated your monthly car costs with some accuracy. Hold any car receipts till the end of the study for recalculation and comparison.

8. *Note:* A "trip" is any movement between one location and another. Example, a student walking from home to school in the morning counts as one trip. A parent traveling from work to a supermarket and then home counts as two trips. Any walking a family member does while in a school building or at the office or job does not count as a trip.
9. At the beginning of the fourth week, the families begin employing strategies that allow them to go without one of their family vehicles. They should brainstorm ways to do this ahead of time, so they are not stymied in their efforts to get around. They can carpool, for example. They can walk. They can plan their days more efficiently so that they only need to use Car 1, and so on.
10. At the end of the fifth non-driving week, have students compile the data for every person in their home. They may wish to generate a spreadsheet to analyze their data. They should carefully analyze the car costs during the eight-week time period to see if they need to make an adjustment to their Monthly Car Cost. For example, the family may have taken a long trip, incurring more gas and toll expenses. Students need to decide if this plus/minus is actually quite typical and well represented by the Car Cost Summary.
11. Have them present their findings to the group, answering all questions below.
12. *Margin of error:* This project is large in scope and vulnerable to errors in data collection. As long as you and your students are aware of the trouble spots, you should be able to correct for them. Some families may not keep accurate receipts for all car expenses, so the student will have trouble calculating annual gas and toll expenses. Family members may not be scrupulously attentive when charting their daily movements. This is to be expected. Work with students to brainstorm ways to compensate for these gaps in the data.

Questions

(Use after challenge has been completed.)

1. Which costs more per week: Driving all the cars your family owns or driving one less car?
2. What did you save per week? Per month? Per year?
3. One gallon of gas burned by a car generates 19 pounds of CO_2. About how many pounds of CO_2 did your family not pump into the atmosphere during the five nondriving weeks?
4. What strategies did you and your family employ to get around when a car was not available?
5. Did you notice any difference in the way family members felt before and during the nondriving weeks? Was going without a car more stressful or less stressful?

ACTIVITY 10.4

Writing about It and Making Connections

Materials

pen and paper; Handout 10.4

Primary Subject Areas and Skills

writing, critical thinking, social studies

Purpose

Students summarize what they have learned by writing some short essays making connections between transportation in the greater issue of sustainability.

The automobile has long been central to the American way of life from coast to coast. With a few notable exceptions—such as New York City—public transportation is not the norm. Sharing a ride with others is often viewed as an inconvenience, or something you do only when you have to, such as when your own car is in the shop. Even within families there is a reluctance to share the ride: Everyone has her or his own schedule and errands. The further away we live from urban centers, the more likely we are to have to commute. This also means that we're further removed from, and may have less social contact with, others outside of our immediate family, circle of friends, or schoolmates. Public transportation not only brings people to a geographic location, it can also bring together people who would have had little chance of interacting otherwise.

Have students consider the following questions as writing prompts. You may wish to assign all or only several of them. You may also choose to use some of these for group work and discussion, while assigning others for individual exploration. Many of these discussions also lay wonderful groundwork for intriguing debates.

After discussing the introductory material above, instruct students to answer the following questions in full sentences. The best answers will be reflective and well thought-out—not knee-jerk responses with little consideration. Honesty is the best policy here. No "right" answers exist.

EXTENSIONS

Transportation Beyond Our Shores

Tokyo, Sydney, Dortmund . . . All over the world there are monorails, light rails, electric buses, trams, trolleys, and more! Use the Internet or library to do a survey of the kinds of public transportation that are popular throughout the world. You may wish to limit this to one type—monorail, for example—or simply assign different students to different major metropolitan cities around the world. Students can do a presentation about the city and its transport. You may wish to even host a Worldwide Transit Day display, during which traditional fare from each of the countries profiled is served.

Traffic Watch

Pick a particularly busy intersection in your area. At the busiest time of day, count or estimate the number of cars that pass through this intersection. Stationing students on each corner and assigning them a particular task is the best way to handle this. Much will depend on the intersection you choose. Perform this exercise on more than one day, if possible, and take an average of the number of cars students counted. If this exercise is not feasible, request traffic data from your local department of transportation. Use this information to calculate emissions. If possible, note whether cars had one or more passengers. If the number of cars was reduced by just 20 percent, how many pounds of CO_2 might be removed from the atmosphere? Assume one gallon for every vehicle. Clearly, this is just one kind of estimate and doesn't account for the kind of vehicle and the distance it is traveling. However, it will give students an idea of just how many cars—especially single-passenger cars—pass through an intersection on any given day.

Ask the Experts

Arrange for a class visit to the local transportation authority. Interview your speaker or other workers there about the transportation challenges your area is facing. What are some solutions that have been suggested? Are there any current plans in the making or up for a vote? How is transportation paid for in your area?

City Hall Car Audit

Offer to perform a transportation audit for members of your town or city government. How many cars do they have? How often do they use them? Do they carpool or take public transportation?

Research Your Dream Car

Have kids research and calculate the costs of different cars. They could study gas mileage of different sizes and types of cars as well as alternative-fuel cars. For those students with sophisticated math skills, it would be interesting to compare the costs (new car price, fuel per year, estimated repairs) of a gas engine car and a hybrid and/or electric car. High school students may be considering purchasing their first car, and this would be a valuable exercise.

RESOURCES

- **Rapid Transit, Light-Rail, and Monorail Index**
  www.publicpurpose.com/utx-usr.htm

- **Light Rail Transit Association**
  www.lrta.org/

To investigate how Austin, TX, has dealt with their transit decisions, see:

- www.lightrailnow.org/

- **The Monorail Society**
  www.monorails.org/

- **Bureau of Transportation Statistics**
  www.bts.gov/

- **Zipcar**
  www.zipcar.com

- **Flexcar**
  www.flexcar.com

Reading Handout Unit 10:
LIVING WITH ONE LESS CAR

Many families in the United States have more than one car. They believe that having these many vehicles is absolutely necessary. But are they, really? Cars are expensive to own and maintain. They require insurance, gasoline, and mechanical repairs, and they are the chief pollutant of our American landscape. Could some families get by on one car alone? In this reading, you'll find out how the city of Seattle experimented with a program to show people that they can indeed give up one of their cars and still run all their errands and get their families around town. If you have the *Edens Lost & Found* DVD set, watch this segment in the Seattle movie.

Automobiles are the biggest single contributor of carbon dioxide (CO_2) in the world. Cars and small trucks gobble up 40 percent of all the oil-based fuels used in the United States and generate 20 percent of the nation's CO_2. In the American Northwest, such as Seattle, cars contribute 60 percent of the CO_2 emissions. In its lifetime, the average car will cough out 70 tons of CO_2. A sport-utility vehicle will spit out 100 tons!

Automobiles can contribute to the waste stream in unexpected ways when driven in large numbers. As rubber tires wear down, rubber dust washes into our waterways. Ceramic and copper dust are shed when brake pads wear down. To be accessible to cars, cities are forced to shell out millions in roads, parking lots, garages, and so on, when that money might be better spent elsewhere.

Today city planners are mindful of this, but it often takes superhuman effort to wean a city off its cars. That's because many people have used only cars to get around their whole lives. In some places, mass transit has never been an option. Even when it is, drivers are often too attached to their ride to give it up.

Seattle, which has buses, ferries, and a new train system on the way, has come up with an interesting way to help its citizens give up their extra car, as well as to collect some excellent research for the city. The "Way to Go Seattle" program, promoted through various news outlets, invited Seattleites to volunteer for a study on car use. For the first week of the study, families were instructed to drive their family vehicles as they normally do, keeping tabs on where they drove, how far they drove, how they felt after their drives, as well as the cost of operating the vehicle. Then, for the next five weeks, families with two cars were instructed to use one less car and use alternate means of transportation such as walking, biking, and mass transit instead. Families with one car were asked to leave it at home as well.

Jemae Hoffman, Mobility Manager of Seattle's Department of Transportation, dreamed up the project. It's her job to come up with ways to make it easier to walk, bike, take transit, and move freight through the growing city. She herself was curious to see what would happen at the end of those five weeks.

At the end of the project, many participants were stunned to find out how much money they had saved by leaving their cars at home. "People made hugely different choices when they discovered they were saving $70 a week," says Ms. Hoffman. The small group of test families drove 42,000 fewer miles and reported feeling less stressed. "That's 15 swimming pools worth of climate change gases not put into the air and a lot less neighborhood traffic and safer places for our kids to play," she says. Participants reduced the number of miles they drove by 22 percent, increased their mass transit usage by 125 percent, upped their walking mileage by 38 percent, and boosted their biking mileage by 30 percent.

The program does have its critics who say the One-Less-Car program, as it is called, preaches to the converted. That is, the only people who would challenge their behavior and give up their car voluntarily are those who are already committed to environmental issues. If Seattle is going to clean up its act, they say, it must do what it takes to stop encouraging people to drive everywhere they want to go. That means investing more in mass transit than in roads and parking facilities and not relying on voluntary reductions in car use.

Ms. Hoffman counters by saying that the city *is* headed in that direction. There is no one solution, she says. "We need to create communities where people don't have to be reliant on the car and provide more frequent and reliable transit service and facilities to bike and walk," she says. When planning the location of future light-rail stations, the city chose to prohibit the building of new parking lots. The city would rather see free parcels used to develop multiuse structures—a building that features, say, a supermarket on the ground floor and apartments upstairs. It's that kind of construction that will ultimately create livable, walkable city neighborhoods.

If anything, the One-Less-Car program was a perfect illustration of a "feedback loop," meaning that when you show people concrete results of their behavior change, they'll opt to make that change permanent or, at the very least, a more frequent habit. "Now," says Ms. Hoffman, "every time they walk down their driveways, these people think about whether or not they really need their cars for that trip and start to change. If you can live in a place where you don't have to own a car, or if you can save four to six thousand dollars a year by not driving it, bells go off in people's heads. They feel less stress, get more exercise, get to know their neighbors better, and have more of a sense of community."

It worked in the case of David and Bobbi Martin, a couple with two sons who took the challenge for six weeks and ended up giving up their second car. They haven't needed it since February 2002. They learned that Seattle's bus system could get them where they needed to go. Mr. Martin, for example, works only four or five miles from home. A bus got him there quickly. He learned that he didn't need to have an extra car, and he didn't need to be spending extra money each year on insurance, fuel, and maintenance.

The couple now says that they hardly miss the extra car. "It's no big deal," says Mr. Martin. "It's a matter of getting used to it more than anything else. We found that it really wasn't a hardship. It was just a matter of changing the way you think about things, planning a little bit more." For example, the couple found that if they communicated each morning about who really needed the car that day, who was driving the kids to school or to various activities, they could easily map out their daily driving strategy. "Whoever is dropping off the kids somewhere usually gets the car," says Mr. Martin, "and the other person takes the bus. Either one of us can take the bus pretty much as easily."

"I don't think it works for everyone," adds Mrs. Martin. "It works for our family. We live and work within the city of Seattle. If we had to make a longer commute, it would be difficult. The effort is worth it. We find that we spend more time together as a family as a result of this. It's really been beneficial for all of us, and we are contributing to the environment that we live in in a positive way, and we're happy to do that."

The realization that you can get by without a car is not the only benefit," says Ms. Hoffman. "One woman said to me that before she tried this program, she always drove to [a distant discount retailer]. But now she walks more to neighborhood stores and sees her neighbors more. She sees flowers in her neighbors' gardens more often. She gets to know local people who work at the neighborhood stores and is developing more of a sense of community. She's been really enjoying herself."

"I'd like to think of it as a responsible thing to do for the city," says Mr. Martin of his family's choice. "It's also a responsible thing for us to do for one another, and it teaches our kids good values. You don't just drive a car because you can. There are other ways to get around the city and it can be better for you. You can be burning less gas and spending less money."

Questions

1. In what ways do cars pollute? From this article, identify at least three ways cars pollute.

   ____

   ____

2. What is the One-Less-Car program, and how does it work?

   ____

   ____

3. What were some of the results of the Seattle study? How did people get around? Would you say that the study participants increased their alternative transportation options more than they decreased their car usage? What evidence in the story backs this up?

   ____

   ____

4. Mrs. Martin says the One-Less-Car program is not for everyone. Can you describe the kind of family that might really need more than one car?

   ____

   ____

5. What is gained when people can walk to their destinations? How could a city or town make more of their neighborhoods more walkable?

____

____

6. In what ways do cities have to spend money to make themselves accessible to cars?

____

____

7. What message is Seattle sending to its inhabitants by deciding to build fewer parking lots?

____

____

8. Why do you think some people find driving stressful? How would mass transportation help alleviate this feeling?

____

____

Name ______________________________ Date ____________________

HANDOUT 10.1: Energy Efficiency Chart

Energy Efficiency of Public Transportation and Personal Vehicles, 1998

MODE	BRITISH THERMAL UNITS/ VEHICLE MILE	BRITISH THERMAL UNITS/ PASSENGER MILE
Bus	41,338	4,415
Commuter Rail	54,071	1,612
Heavy Rail	19,789	911
Light Rail	29,688	1,152
AVERAGE	38,251	2,741
Automobiles, Sport-Utility Vehicles, and Light Trucks	6,348	5,255

Source: *Conserving Energy and Preserving the Environment: The Role of Public Transportation*, Robert J. Shapiro, Kevin A. Hassett, and Frank S. Arnold, 2002.

1. Which mode of transportation uses the most BTUs per vehicle mile?

    ____

2. Which mode of transportation uses the most BTUs per passenger per vehicle mile?

    ____

3. How many BTUs are consumed by a bus traveling 5 miles?

    ____

4. How many BTUs are consumed per passenger, per mile, traveling via heavy rail?

    ____

5. The average BTUs used per vehicle mile of public transportation is

    ____

    a. The average BTUs used per passenger mile of public transportation is ____________

    b. Explain briefly the difference between these two numbers: ____________

    ____

6. Explain how the BTU passenger mile for autos could be affected by carpooling.

____

____

8. Which uses more BTUs per vehicle mile: One bus carrying 40 people or 40 people driving individual cars? ____________ How much more?______________

9. What does the answer to question 5 tell you about the benefits of mass transportation in regard to reducing energy use? What does it tell you about reducing pollution?

____

____

Name ______________________________ Date ____________________

HANDOUT 10.2: Transportation Chart

Type of Transportation	Light Rail	Monorail	Heavy Rail	Bus
How It Works	Usually runs at street level, but not always. Runs on the same roads as cars, other times in its own right-of-way lane. Runs on standard dual rails.	Usually runs above ground, but not necessarily. Can be suspended or straddle the beam on which it runs. Runs on a single (mono) rail; the train is often wider than the rail on which it runs.	Trains running on standard dual rails. Larger and heavier than light rails. Includes commuter trains and subways.	Runs on fossil fuels, occasionally electricity. Operates at street level. Almost always uses the common roads.
Fuel Source	Electricity, often supplied by overhead lines	Electricity	Electricity and/ or fossil fuels	Fossil fuels—sometimes electricity
Subject to Traffic?	Sometimes	No	No	Yes
Require a Driver?	Usually	No	Yes	Yes
Costs				
Pros				
Cons				

HANDOUT 10.3A: Car Cost Worksheet

WHAT IT COSTS TO DRIVE YOUR CAR

Monthly payment: ________________
(What your family pays each month to have this car)

Monthly depreciation: ____________
(How much, per month, the value of your car goes down.
Follow the steps shown below, then divide by 12.)

Monthly insurance: ________________

Monthly gas costs: ________________

Monthly maintenance costs: ________
(Add up all maintenance bills from the past year and divide by 12.)

Monthly parking/tolls: ____________
(Ask for receipts and add them up.)

Monthly car fees: ________________
(Add annual car registration, license, and property taxes
from past year and divide by 12.)

Monthly Car Cost Total: ____________

Annual Car Cost Total: ____________
(Multiply monthly total by 12.)

Daily Car Cost Total: ______________
(Divide monthly total by 30)

CALCULATING DEPRECIATION

What is depreciation?
Cars lose value the older they get, the more miles they have been driven, and the more damage they sustain. In general, you can calculate how much value your car loses each year with these few easy steps.

1. Find the "Blue Book Value" of your car at www.kbb.com
2. Find an Online Depreciation Calculator, such as the one at http://calculators.aol.com/tools/aol/auto02/tool.fcs
3. For "purchase price," type in the current Blue Book Value of your car. For "Vehicle Age (years)," type in the current age of the car. Lastly, for "Years You Will Own Vehicle," type in how many years more you expect to own the car. Calculate for "Average" depreciation. Write the "First Year's Depreciation" on your sheet.

Name ______________________________ Date ____________________

HANDOUT 10.3B:

Daily Transportation Diary

Is this a **DRIVING** or **NONDRIVING** diary?
(circle one)

	Trips You Made and Why	How You Got There*** Who was with You	How Far You Traveled*	Costs**	How You Felt
1					
2					
3					
4					
5					

*If driving, note odometer before/after
**If driving,compare to your weekly car cost
***Write all ways you used to get around—bike, skateboard, family car, friend's car

Name ______________________________ Date ____________________

HANDOUT 10.4: Writing About It and Making Connections

THINKING ABOUT TRANSPORTATION

1. If you were an adult with a full-time job, would you like to be able to walk or bike to work? Why or why not? Would the possibility of saving money affect your decision? Would the possibility of helping the environment affect your decision? If yes, would you be willing even to change where you live in order to not need a car to get back and forth to work?

____

____

2. What do you consider to be the personal benefits of walking or biking as a means of transportation? What would you consider to be the personal benefits of taking mass transportation?

____

____

3. What would you be willing to do, right now, to alter the way you get around: to and from school, to and from activities and friends' houses? Would your family be willing to go with one less car? What would you *not* be willing to do?

____

____

4. What would you take into consideration when thinking about public transportation for your area? If you already have public transportation, what would you change about it? If you don't have it, what kind of things would you look for in a new transportation system? Taking all of these into consideration, describe the ideal public transportation situation for your area (light rail, monorail, bus, train, and so on).

____

____

5. How does or how could mass transportation benefit different communities in your area?

____

____

6. What is the relationship between mass transportation and air pollution? What is the relationship between mass transportation and water quality? What is the relationship between carpooling and soil quality?

____

____

UNIT 11

BIODIVERSITY

TOPIC BACKGROUND

The word *biodiversity* refers to the variety of life found on Earth. *Bio* means "life," and *diversity* means "a variety of attributes within a given group." Scientists generally refer to three different types of biodiversity: *genetic* diversity, or the variety of genes in a given species; *species* diversity, or the variety of species in a given region; and *ecosystem* diversity, or the variety of different habitats found around the world that are home to a collection of different animal species.

Biodiversity is a major issue in modern science because so many species are becoming extinct. You may wish to share these statistics with your students:

- Three species disappear every hour.
- Twenty to 30 species disappear off the face of Earth each night while you sleep.
- Twenty thousand species disappear every year.

Now, it's true that species have gone extinct for billions of years, and that this process is nothing new. The cause of the most famous mass extinction, the disappearance of the dinosaurs, is still hotly debated. However, what worries many scientists today is that the current extinctions are caused by the impact of humans on the planet. Humans consume land and resources to survive, pushing further and further into territory once occupied solely by wild animals. Humans pollute environments with their waste, destroying habitat.

Why does this matter? Why should we care? Well, there are many reasons, but three make the most sense:

1. *Every species on the planet has a role to play in its ecosystem.* When an ecosystem is rich with life, it's better able to withstand various types of stress, and it can be more productive. For example: Everyone knows worms and other bugs consume the bodies of other dead animals. Imagine if the worms and bugs suddenly disappeared from an ecosystem? The resulting chaos is too horrible to envision!
2. *Every species on Earth is a potential aid to humans.* Humans rely on animals and plants for food and other goods, to perform services, to use in the creation of beneficial medicines, and so on. If we lose an important species, we're doomed. If we lose a species before we even learn how it can benefit us, we're missing out.
3. *Every species on Earth contributes to knowledge.* The more species we have to study, the more we can learn about the evolution of life on this planet. Why would we ever want to lose out on more opportunities to learn?

Purpose

In this unit, you will explore some issues of biodiversity with your students. We'll ask you to start by exploring the diversity of your classroom. Students will see that a diverse population is a rich one that can call upon the skills of all its inhabitants to help the group. As they work though the unit, challenge them to keep answering the central question for themselves: *Why should we care about biodiversity?* They are sure to come up with answers that surpass or expand the three we've listed here.

GROUP ICEBREAKERS

1. Ask students to define the word *biodiversity* without looking it up in a dictionary or online. How can they express the term in their own words? (Accept all reasonable answers. For guidance, refer to the definition given above in the Topic Background.)
2. After explaining the three different types of diversity, challenge them to verbally express all the different ecosystems they can think of. (Their answers can be as simple as "desert,"

"rainforest," "coral reef," and the like.) Write each on the board as they call them out.

3. Have them guess which animal species on the planet is the most abundant and diverse. (The answer—insects—will probably stun them. Scientists have identified between 1 and 2 million insect species so far, and they think there could be as many as 6 million species.) At the rate humans discover and classify insects (about 10,000 new ones a year), it would take four more centuries to identify all the insects on Earth. Of course, by then, they may have evolved into different species or gone extinct!
4. Use the previous question to launch the question they will be studying throughout this unit: Why is biodiversity important? Why should we care if some of the 6 million species of insects go extinct? Aren't 1 million of them enough? (Help them to see that if an insect species goes extinct, it may be a sign of a larger problem, such as the disappearance of that creature's favorite food source, a change in climate, or destruction of its habitat.)

ACTIVITY 11.1

Classmate Diversity Scavenger Hunt

Materials

pen and paper; Handout 11.1

Primary Subject Areas and Skills

science, social studies, writing, research

Purpose

To develop an understanding of the term *biodiversity,* students conduct an in-class scavenger hunt to discover what skills, talents, knowledge, or other attributes their classmates possess.

Biodiversity is a broad and descriptive term and one that your students will strive to better understand as they work through this unit.

Discuss with your students the word *biodiversity* and its root words, *bio* and *diversity.* What do the two word parts tell you about the meaning of the whole word?

The following activity is designed to encourage your students to think about their very own classroom community as a diverse and varied one. They should value the experiences and attributes of each classmate, as well as highlight the similarities that they share with each other.

1. Copy and distribute Handout 11.1 to your students. Explain to them that they are going to further explore the meaning of *biodiversity* by first examining the diversity in their own classroom.
2. Instruct students that they have 15 minutes to gather information from their classmates and complete their research form.
3. Be sure to tell students to talk to as many classmates as possible. Just because they find one person who fits the bill does not mean that they should stop. They should strive to find as many classmates as possible who have the same experiences and so broaden their understanding of the makeup of their classroom community. After a classmate is interviewed, he or she should sign the handout. Students can have no more than one signature from any one student.
4. Once time is up, have students return to their desks. Write each description on the board or an overhead. Have students share their experiences. (*For example:* If there is a student who was born in another country, where were they born? How and when did they come to the United States?)
5. Remind your students that humans are animals, too. Discuss what makes it possible for so many humans to live in the same place: What do they need to survive? Then ask them to think about what other living species need to survive in a constantly changing world. What can they adapt to? What makes it dangerous for them to live there?

ACTIVITY 11.2

Getting to Know Your Own Backyard: An Urban Survey

Materials

pen; notebook or sketchbook; Handout 11.2

Primary Subject Areas and Skills

science, writing, art, critical thinking

Purpose

Students conduct an audit of the types of plants they see growing in their neighborhood, cultivated or otherwise, to gain an understanding of the diversity of flora and fauna in their surroundings.

Many students believe that "nature" and the "environment" are things that exist only "out there" somewhere, detached from their everyday surroundings. A key part of understanding the essence of biodiversity is to begin to examine your immediate surroundings more closely and to appreciate how the diversity of nature and all living things is around us all day, every day. Many of the stories told in the *Edens Lost & Found* series are about kids living in urban centers who learn to recognize, appreciate, and preserve the biodiversity around them.

1. Review the meaning of biodiversity with your students. Ask students how familiar they are with the plants and animals of their own region. If you have studied this in your class, review those topics. If students have very limited knowledge in this area, you may wish to set aside the first 20 minutes of class for research. You may also wish to assign the questions that follow to individuals or groups.
2. There are two parts to this activity: The first part is done in the classroom, while the other is conducted outside. Before distributing Handout 11.2, discuss the following questions with students. You may wish to write the answers on the board or an overhead projector. (Because answers to these questions vary regionally, we cannot provide answers here. We refer you to your nearest state park office, cooperative extension office, or other nature and wildlife authority. They may even have a website where all of these items can be easily found.)
   a. What is the name of the major habitat you live in?
   b. Name as many native animals as you can.
   c. Name three native trees.
   d. Name any animals that migrate to your area. When do they come and why?
   e. Name any edible plants that are found in your area.
   f. Name three birds native to your area.
   g. What species of plants in your area are endangered?
   h. What species of animals in your area are endangered?
   i. What is the growing season in your area?
3. When you are ready to do the survey, you can choose to either:
   a. Go outside on your school grounds.
   b. Take a field trip to a nearby park.
   c. Send the Handout 11.2 with students and have them observe plants and animals in their own yards or neighborhoods. Encourage them to observe even plants that would normally be termed "weeds."
4. Students may wish to sketch or photograph plants for follow-up class discussion.

Extension

After the students have completed their surveys, either as a class or at home, discuss their findings as a class. Then instruct students to write an essay that answers the following questions. (Answers will vary greatly. Accept all that are reasonably argued.)

- What is a weed? (Simply put, a weed is an undesirable plant growing where it is not wanted.)
- Does the word *weed* conjure positive or negative images? (Negative.)
- How is a weed different from a flower? (This is a tricky question: Most of us agree that flowers are desirable plants, but many weeds flower as well.)
- Do weeds grow in the forest? (Certainly, but again, this is a tough question: Because humans are not cultivating plants in a forest, anything that grows there is wild vegetation with a right to be there. It is hard to term it a weed.)
- How do you decide what's a weed and what isn't a weed? (Only humans can decide what a weed is. There is no such designation in nature.)
- The American writer Ralph Waldo Emerson once said: "What is a weed? A plant whose virtues have not yet been discovered." What did he mean by that? What are virtues, and why would discovery of them change the way you viewed a plant? (Emerson was pointing out that humans call plants "weeds" when they have no use for them. As soon as a plant's virtues, or usefulness, is discovered, it is no longer a weed but a plant worthy of planting.)
- Ask students to think about how the term *weed* is used metaphorically in the English language. (For example, "growing like a weed," or "weeding out unnecessary items.")

ACTIVITY 11.3

Overfished and Contaminated: Crisis in Our Seas

Materials

pen and paper; Handout 11.3

Primary Subject Areas and Skills

science, social studies, math, critical thinking

Purpose

Students create a Venn diagram to pinpoint a list of fish that share the following two attributes: They are both plentiful and safe to eat.

Fish are among the healthiest foods humans can eat. At the same time, many fish species are contaminated with dangerous chemicals such as PCBs or dangerous elements such as mercury. Pregnant women and young children are at the greatest risk because fetuses and young children are still growing, and their bodies absorb nutrients and contaminants at a faster rate. When choosing fish in restaurants and supermarkets, consumers need to be mindful of which fish species are safe to choose and which should be avoided. *At the same time,* however, many species are being overfished—that is, too many of these fish are being plucked from waters for our dining pleasure, leaving too few in a particular area to reproduce in large numbers.

The conflict between fish as healthy food versus contamination and overfishing can leave many consumers feeling like fish out of water! What are we supposed to eat? How can we keep these important concerns in mind and yet make the choice that is healthy and environmentally sound? In this activity, students will use a Venn diagram to audition some good choices. Depending on class time and interest, you can extend the activity in a couple of different ways, as discussed below.

After introducing students to the issues, distribute Handout 11.3 and challenge them to create a *Venn diagram* interpreting the fish data. Venn diagrams illustrate the relationships between different groups of items that share things in common. One circle should contain all the fish considered "SAFE TO EAT." The other circle should contain all fish that are considered "PLENTIFUL" in oceans or do not have other serious environmental issues surrounding them. The intersection of the two will yield a good working list of fish suitable for eating. (*Note:* Lists of fish vary greatly from organization to organization. See the "Extensions" section, below.)

Extensions

1. Challenge students to research and design a wallet-sized card that they could give to their parents to help with their fish shopping. They may need to do additional research online to inspect how some existing cards look (consult the "Resources" section on page 168). If time does not permit designing your own class card, consider having students print up a few cards they find online and critique them: Does the information on one card conflict with the others? (Yes.) Why is this so? Do they think the information is debatable or simply inaccurate? (Every organization uses different resources to gauge the health of fisheries. Information varies from place to place.) How can they find out the truth? (The only way students can know the truth is to stay informed, keep researching, and check back with sources from time to time.)
2. Students can also research the safety of fish specific to their state. Every U.S. state has issued advisories on the consumption of fish caught in local waters. Have students study the advisory from your state and report on local fish. Are there any fish caught in your state that you should never eat? (This may well be the case.) How often can you eat others? (Answers will vary, depending on local resources and information.) Can students suggest things they, their families, and their communities can do to improve the quality of local waters? (Limiting toxic household chemicals and stopping dangerous runoff are two things we can all do to improve the health of local waters.)

ACTIVITY 11.4

Ecosystem Swap Meet

Materials

encyclopedias or Internet access
library or Internet research materials
old magazines
scissors
markers
pens
papers
colored pencils
glue

Primary Subject Areas and Skills

science, social studies, writing, art, critical thinking

Purpose

Students create trading cards and posters depicting various ecosystems and the creatures that inhabit them.

Many ecosystems are highlighted in the *Edens Lost & Found* series, including prairies and wetlands. This activity is designed to help students to further develop their understanding of the vast biodiversity and ecosystems present in their communities and in the world.

Explain to students that they will be creating trading cards and posters for various ecosystems. The cards will be for animals living in a particular ecosystem, and the posters will depict the homes for these species. *For example:* If the ecosystem is a tropical rainforest, the poster might depict layers of lush tree canopies and large ferns with mushrooms and mosses carpeting the forest floor. The animals featured on the cards might include monkeys and sloths.

1. Review the definition of an ecosystem with your students and brainstorm different kinds of ecosystems together. Write the names of each on the board. (Some examples include mountains, tundra, temperate forest, desert, tropical dry forest, cold climate forest, grassland, savanna, tropical rainforest. Freshwater ecoystems include rivers and streams, ponds and lakes, and wetlands. Marine ecosystems include shorelines, temperate oceans, and tropical oceans. You may find others in books or online. Be sure that they are not redundant before sharing them with the class.)
2. Divide students into six equal groups. Assign each group two different ecosystems to research. Keep a list of which groups are responsible for which ecosystems. Be sure to include a variety of ecosystems, including aquatic ecosystems, and the like.
3. Instruct each group to research the plants and animals that live in their ecosystems. They should create posters or collages that depict their two assigned ecosystems, highlighting and labeling their key components. Each group will also create 10 "creature cards"—five for each ecosystem—depicting information about different creatures that live in the ecosystems. Each card should include a picture or drawing of the animal, its name, and descriptive information about the animal, such as what it likes to eat or hunt, whether it's warm- or cold-blooded, or any other unique characteristic that will help link it to its ecosystem. Students should get adequate time—in or out of class—to research their ecosystem and make their posters and cards. Remind students that insects, fish, and birds are animals, as well as larger creatures and mammals. Students should also remember that nonliving things—such as rocks or water—are part of an ecosystem.
4. Once the creature cards and posters have been made, you will have a total of 60 cards and 12 different ecosystems.
5. Have students from groups 1 through 3 exchange their creature cards and posters with the students from groups 4 through 6. Students must then match each of the creature cards to the appropriate ecosystem by placing the creature cards on, near, or under the correct poster. Students should write down the names of their creature cards and to which ecosystem they believe each belongs.
6. Check everyone's answers and check that creature cards were matched to the correct ecosystem.

Extension

- Laminate cards for use with future classes. This way, if you do not have time to do the entire activity in future classes, students can still match the cards to the correct poster.
- Create classroom incentives: Encourage students to make more creature cards representing more ecosystems. For every ecosystem studied and cards created, students get points toward a prize or extra credit.
- Creature cards can be made into biodiversity holiday cards and sold to raise money for a particular local environmental cause or charity.
- Combine posters for an art exhibit in your classroom or school hallways.

EXTENSIONS

Grocery Store Field Trip

Where do the fish in your grocery store come from? Are they wild caught or farm raised? Are they overfished? Take a look at what's on sale and investigate whether it's a species that is in danger of being depleted. Consider sharing the information with the store manager.

Fish and Wildlife Services

Contact your local Department of Parks and Recreation, U.S. Fish and Wildlife Service, or other similar organization. Schedule a class talk about your local ecosystems. What are the indicator species in your area? (An indicator species is a one whose well-being provides information about the overall health of an ecosystem.) Which species, if any, are endangered?

Taking It to the Streets

Develop a public awareness campaign about overfishing in North American waters. Research which fish are the most in danger and create posters, diagrams, and charts that depict the data your class has gathered. Hold a talk in the school library, on Earth Day, or at a PTA or town council meeting. Write a report about the experience for the school or local paper.

RESOURCES

- **Interactive map**
  www.biodiversityhotspots.org/xp/Hotspots
  This wonderful interactive map shows some of the world's most amazing and endangered communities.

- **National Wildlife Federation**
  www.enature.com
  This National Wildlife Federation website features incredibly useful field guides for virtually all habitats in the United States and even features a ZIP code finder that will highlight all of the species in your area. This is an excellent resource for the "Urban Survey" activity.

- **The American Museum of Natural History**
  www.amnh.org/nationalcenter/it_takes_all_kinds/
  The American Museum of Natural History's site on biodiversity has maps, interviews, and kid-friendly information. It also features a wonderful interactive diagram of a kitchen, showing the origins of some of the foods you have at home.

- **Fish safety guide**
  www.nrdc.org/health/effects/mercury/guide.asp
  NRDC's guide tells which fish are safe to eat and which are being overfished. Use the information here and the links available to discuss with students what's healthy for their bodies and what's healthy for the future of aquatic biodiversity. Are they always the same? How does one affect the other?

- **Indicator species**
  www.nwf.org/wildlife/pdfs/speciesatrisk.pdf
  National Wildlife Federation's guide to species at risk. Also information about "Frogwatch," their program for monitoring endangered amphibians and important wetland and other indicator species.

- **The Biodiversity Project**
  www.biodiversityproject.org/
  The Biodiversity Project has a wide variety of links, publications, and other resources about biodiversity worldwide.

- **Endangered wild places**
  www.savebiogems.org/
  The NRDC profiles a selection of endangered wild places and what you can do to help save them. You can use this information to develop an action plan that might involve letter writing or research, for example. Are any of these endangered wild places near your area? What can you do as a class, school, or community to help protect these areas and spread the word to others?

- **Wallet cards for fish species**
  Wallet cards resources abound on the Internet. A simple search such as "wallet card + fish" will yield numerous results. Here are a few to get you started:
  http://seafood.audubon.org/seafood_wallet.pdf
  http://a1410.g.akamai.net/f/1410/1633/7d/images.enature.com/newsletter/fish_card.pdf
  www.nrdc.org/health/effects/mercury/walletcard.pdf
  www.mercuryaction.org/fish/images/wallet_card.pdf

- **State fish advisories**
  http://epa.gov/waterscience/fish/states.htm

Reading Handout Unit 11:
RESTORING THE PRAIRIE

Who cares about grass? Can you imagine spending your weekends pulling up one species of grass, only to plant another one? In this reading, you'll find out about one man who devoted his life to restoring the American prairie in Illinois. Additionally, you can watch Steve Packard talk about his work if you have the Chicago movie in the *Edens Lost & Found* DVD set. After you read the story, answer the questions.

Illinois calls itself the Prairie State, though many of its citizens have never seen a prairie or know what a prairie looks like. Prairies are open fields populated by special, resilient grasses, wildflowers, and other plants—but few trees. Many well-known garden plants, such as purple coneflower (Echinacea), got their start in the American prairie. Because a prairie is treeless, there's no shade, and the grasses are constantly exposed to the harsh rays of the sun. They have adapted to their environment, growing rapidly even in the hottest summers. If struck by lightning, a prairie will catch fire, and flames will whip through the dry grasses, turning the fragile-looking stalks and flowers to cinders. A scorched prairie is a terrible sight, but it doesn't last long. Shaped by evolution over thousands of years, prairie plants grow back quickly when burned, faster than most forests.

If you rode the elevator to the top of Chicago's famous Sears Tower, you would see a vast urban landscape, punctuated by factories, cut by waterways, and obscured at times by billowing steam and smoke. Where in this landscape is the prairie, and how far would you have to go to find it?

That's the question a man named Steve Packard asked himself nearly 30 years ago. Mr. Packard grew up in Massachusetts, where he spent a lot of time outdoors. A young man during the 1960s, he spent most of his time as a peace worker, and worked at a number of odd jobs. In the late 1960s, he came to Chicago to work for a film company. Inspired by a 1972 book called *The Prairie: Swell and Swale*, which depicted prairie lands along Lake Michigan in pictures taken by Swedish photographer Torkel Korling, Mr. Packard rode his bike around Chicago, in search of its wild places. For a while he was convinced that Illinois' nickname was a misnomer. "I'd ask people what the words 'the Prairie State' meant and no one could tell me. I thought that was funny. Maybe it was a myth if no one knew what it was."

Eventually he found remnants of prairie lands just outside the Chicago area. One afternoon he wandered down the path in a nature preserve, using Korling's book as a guide, trying to identify the wildflowers he found. By then he'd done enough reading to learn that the American prairie was becoming endangered. Humans were illegally dumping waste on the grasslands. Precious swaths of land were being lost to development. Poachers illegally hunted birds and other creatures. And woody species were gaining a toehold on turf where they didn't belong.

You see, the American prairie had evolved a perfect way of restoring equilibrium every time a woody tree species grew too rampant. During lightning storms, trees, which had not adapted to fire, died off, and tough prairie grasses grew back abundantly. Thousands of years ago, Native Americans who lived in the area hunted the deer, elk, and bison that lived on the prairies and oak savannas.

These animals ate grass to survive, and the Native Americans hunted the animals in return. When the Indians realized that the grasslands always grew back, they occasionally set fire to the prairie lands. They called the burning prairie "red buffalo," because of the way it looked and sounded as it whipped across the dry plains in the dark of night. The trees burned and died, the grass grew, and the bison and other animals thrived. In this way, the Native Americans could ensure their way of life.

"Humans have been involved in the prairie as far back as 12,000 years ago," says Mr. Packard. "We played a role in how the prairie looked. We're the reason it was the way it was."

These so-called controlled burns, so necessary to American Indian culture, disappeared as Europeans moved into the prairie states. The early colonists plowed the grassland to plant corn and soybeans. The roots of old native plants made rich soil for crops. In fact, the Midwest used to be called "the empire of locked roots" because those native plants had such strong roots. Settlers invented a steel plow to break it up. They couldn't take the chance that all their hard work might go up in smoke. Seeing fire as dangerous, they fought natural prairie fires wherever they flared up. As a result, the prairie was dying out.

Sitting in these diminishing fields, Mr. Packard believed that if humans intervened, they could save the prairie. Some scientists had experimented with burning entire fields, then sowing native seed, and watching as long-dormant or rare species came back.

Mr. Packard knew he was contemplating a huge task. He knew that many people would think his efforts were wasted. "Who cares about grass, really?" they would say. "Why does it matter if one species of grass lives or dies?" Mr. Packard thought about this question long and hard and came up with a good answer.

"That's like saying, 'Who needs museums? Comic books and newspapers have pictures, don't they?' Well, the prairie is an ancient ecosystem. It's a snapshot of what this part of the world looked like thousands of years ago. What's left is one one-hundredth of one percent of what was once here. The tropical rainforest has a better chance of survival than the prairie! If the prairie goes, we lose ancient, beautiful, precious things like butterflies and birds and snakes and frogs that used to live here. No one has even counted all the fungi and nematodes that would be lost. Now, true, the world won't come to an end if they're gone, but we have to live with the fact that more ancient things have been destroyed."

When prairies disappear, he adds, we all lose. "The species that take root in a dying prairie are things like buckthorn, which is a shrub that grows tall and shades out the grass," he says. "What's left is nothing but dirt. Without that dense network of roots to hold it in place, the soil erodes dramatically. When it rains, ravines get bigger and bigger, our waterways fill up with runoff dirt, and we get more severe flooding."

The major prairie grasses were shaped by evolution to be warm-season plants. That means that they grow best when the weather is hot, absorbing carbon dioxide at a time in the year when we most need pure air. Exotic species such as buckthorn are cold-season plants; they thrive and grow in cooler months, preferring to lay dormant during hazy, hot, and humid seasons.

That's why restoring the prairie is so important to Chicago. If citizens want a city that feels cool on summer days, a city with cleaner air, a city that is less vulnerable to dangerous floods, they must save the prairie. On a purely selfish level, the prairie can save their lives.

Shortly after that revelation in the fields, Mr. Packard and other volunteers spent weekends chopping down and burning piles of nonnative plants and trees, then reseeding the land with plants that make a healthy prairie. It was exhausting work that required expertise. They needed to know exactly which plant was which. They needed to collect the native seeds intelligently. Someone had to test the seeds to learn how they grew and pass that knowledge along to the volunteers.

Many nights Mr. Packard returned home with aching bones, his shins and hands scraped and cut. The fields looked worse: The ground was bare where they had cut away the brush. Mr. Packard and the other volunteers learned to be patient. "We'd only work in an acre or a half-acre at a time," he says, "so you could see what you were doing. And over time, little by little, you could see the results over the years. Every year, the prairie looked better and better."

Mr. Packard eventually became a leader in his field. He organized Chicago Wilderness, a coalition comprising more than 170 different organizations devoted to preserving Chicago's wild places. Today those lands are a model for the rest of the nation. Nearly 11 percent of Cook County, the county that contains Chicago, consists of preserved places.

On a recent walk, Mr. Packard, now 61, showed a visitor the result of his work. "What I'm walking through is a restored savanna. We started working on this site 25 years ago," he says. "This is a kind of natural community that was thought to be extinct. These plants wouldn't be here without the work that's been done. The birds that are singing, they wouldn't be living here, raising their babies here, if the habitat hadn't been restored. All the grass was planted. Some wild orchids are thriving here, a number of them endangered species, because people did the work here to put them back.

"I compare restoration to building a cathedral," he continued. "There are so many types of skills involved, and it takes generations. The pace of restoration is picking up. We used to tackle areas that were an acre or two. Now we're working on a thousand and making plans for 10,000-acre pieces."

People volunteer, Mr. Packard says, for a reason that is more rational than emotional. "It's the same thinking that inspires people to become scientists or painters or to be kind to children. They want to do something good and do something beautiful. The acres we saved were like foundlings left on our doorstep, needy but showing great promise."

Questions

1. What is a prairie?

____

____

2. Why does the author of this article say the prairie can save lives?

____

____

3. How can American prairie lands help cool neighboring cities and reduce air pollution?

____

____

4. How can prairies lessen the affect of strong rains?

____

____

5. What happens after a prairie burns?

____

____

6. Why did Native Americans encourage prairie burns?

____

____

7. How did the coming of European settlers change the prairie?

____

____

8. Why are scientists and others burning prairie lands today?

____

____

9. List three possible reasons Mr. Packard gives for saving native plants.

____

____

10. Which of his reasons makes the most sense to you? Why?

____

____

Name ________________________ Date ________________

HANDOUT 11.1: Classmate Diversity Scavenger Hunt

Read the questions below. You have 15 minutes to find as many classmates as you can to sign their name next to any statement below that applies to them. Have fun!

1. Knows the state tree

____

____

2. Knows the state bird

____

____

3. Has been white-water rafting

____

____

4. Has been camping overnight

____

____

5. Has been to a national park

____

____

6. Has held a snake

____

____

7. Was born in another country

8. Can speak another language

9. Can name a plant native to your area

10. Has been to an ocean, a mountain, and a desert

11. Has ever held a part-time job

12. Can play a musical instrument

Name ______________________________ Date ____________________

HANDOUT 11.2: Getting to Know Your Own Backyard

1. What kinds of plants do you see growing? On what surfaces are they growing? If you know their common names, write them in the space provided. If you know their Latin names or can look them up, fill those in as well. Sketch or photograph the plants and bring them into class to share with others.

Plant	Common Name	Latin Name	Surfaces in Which They Grow

2. What kinds of animals do you see using these plants? How are they using them?

Animal	Plants	Use

3. Which of the plants you found are considered native plants in your area? (You may need to use a local nature guide or refer to a local nature website to answer this question.)

Plant	Common Name	Latin Name

Name ______________________________ Date ____________________

HANDOUT 11.3: Overfished and Contaminated

Fish are among the healthiest foods humans can eat. But which ones should you eat? That's hard to answer, because two major issues complicate our choices.

Issue 1: Many fish species are contaminated with dangerous chemicals such as PCBs or elements such as mercury. When choosing fish in restaurants and supermarkets, then, consumers need to be mindful of which fish species are safe to choose and which should be avoided.

Issue 2: Many species are being overfished—that is, too many of these fish are being harvested for food, leaving too few in a particular area to reproduce in large numbers.

How can you keep these two important concerns in mind and make the choice that is healthy and environmentally sound? Creating a Venn diagram, an intersecting circle diagram, is a helpful place to start.

Instructions:

1. On another sheet of paper, draw one big circle and label the left side "SAFE TO EAT." Draw an intersecting circle and label the right side "PLENTIFUL."
2. Study the lists of fish that follow. Write the names of the safe fish in the correct circle. Write the names of the plentiful fish in the correct circle.
3. Do some fish appear in both circles? Write those in the center of the Venn diagram, where the circles intersect.

CONTAMINATION LEVEL OF POPULAR FISH

Most contaminated

Avoid eating the following fish:

- Grouper
- Mackerel
- Marlin
- Orange roughy
- Shark
- Swordfish
- Tilefish

Contaminated

Don't eat the following fish more than three times monthly:

- Bass (saltwater)
- Bluefish
- Croaker
- Halibut
- Lobster (American/Maine)
- Sea trout
- Tuna (canned, white albacore; fresh bluefin; ahi)

Less Contaminated

Don't eat more than six servings of these fish monthly.

- Carp
- Cod
- Crab (blue, Dungeness, snow)
- Herring
- Mahi Mahi
- Monkfish
- Perch (freshwater)
- Skate
- Snapper
- Tuna (canned, chunk light; fresh Pacific albacore)

Least contaminated

These fish are safe to eat—*bon appetit!*

- Anchovies
- Butterfish
- Calamari (squid)
- Catfish (farmed)
- Caviar (farmed)
- Clams (farmed)
- Crab (king)
- Crawfish/crayfish
- Flounder
- Haddock
- Hake
- Herring
- Lobster (spiny/rock)
- Oysters (farmed)
- Perch (ocean)
- Pollock (wild-caught in Alaska)
- Salmon (wild-caught)
- Sardines
- Scallops
- Shad
- Shrimp, pink (from Oregon)
- Sole
- Sturgeon (farmed)
- Tilapia (farmed)
- Trout (freshwater)
- Whitefish

SUSTAINABLY FISHED MARINE CREATURES

Plentiful Fish

These fish are caught in environmentally sensitive ways or are raised under well-managed farm conditions.

- Catfish (farmed)
- Caviar (farmed)
- Char, Arctic (farmed)
- Clams (farmed)
- Crab (Dungeness; imitation; snow; stone)
- Halibut (Pacific)
- Lobster, spiny
- Mussels (farmed)
- Oysters (farmed)
- Pollock (wild-caught in Alaska)
- Salmon (wild-caught in Alaska)
- Sardines
- Shrimp, pink (from Oregon)
- Striped Bass (wild and farmed)

Sturgeon (farmed)
Tilapia (farmed)
Trout, rainbow (farmed)
Tuna (albacore, bigeye, skipjack, yellowfin caught on lines)

Not as Plentiful

These fish are OK to eat if the others are not available. There are some concerns about their populations and the way they are harvested or farmed:

Clams (wild-caught)
Cod, Pacific (trawl-caught)
Crab (blue, king)
Flounder/sole (caught in U.S. or Canadian Pacific waters)
Lobster (American/Maine)
Mahi mahi/dolphinfish
Oysters (wild-caught)
Scallops (bay, sea)
Shrimp (Caught in U.S. and Canadian Atlantic, U.S. Gulf of Mexico)
Squid
Sturgeon, white (wild-caught in U.S.)
Tilapia (farmed in Central America)
Swordfish (from U.S.)
Tuna (albacore, bigeye, yellowfin caught on a "longline"; canned)

AVOID

These fish are either overfished or farmed in ways that pollute the environment or hurt other creatures:

Caviar (wild-caught)
Chilean sea bass
Cod, Atlantic
Crab, king (from Russia)
Flounder/sole (from the U.S. Atlantic Ocean)
Groupers
Halibut, Atlantic
Monkfish
Orange roughy
Rockfish (trawl-caught)
Salmon (farmed)
Sharks
Shrimp (imported)
Snapper, red/vermilion (from United States)
Sole (from Atlantic)
Sturgeon (imported and wild-caught)
Swordfish (imported)
Tilapia (farmed in China or Taiwan)
Tuna, bluefin

Note: These lists may change according to the organizations that compile them, as a result of changes in farming or catching procedures.

UNIT 12

URBAN AGRICULTURE AND COMMUNITY GARDENS

TOPIC BACKGROUND

Where does the food your students eat come from? Ask them, and they'll most likely say, "From a supermarket!" Sadly, kids today are so removed from Earth and its processes that many do not automatically think of food as originating in America's soil.

The American food supply is probably the most reliable in the world. Even in the coldest winter months, Americans who live in snowbound locations have access to fruits and vegetables from warmer climes. But tremendous environmental trade-offs are made to achieve this. The varieties of crops planted are more often chosen for their ability to stay fresh over long distances, not necessarily for their flavor. For example, many apples are picked in Washington state and shipped all over the nation. More flavorful apples may be available, but supermarkets carry the "cultivars" that are least perishable. To transport these foods cross-country, vast amounts of gasoline are consumed by trains, planes, and trucks—which further pollute the air we breathe. Imagine the vast distances tropical commodities such as coffee, chocolate, and bananas must travel to reach our shores! (Your students might be very surprised to learn that the United States grows coffee only in Hawaii and grows no commercial chocolate or bananas.)

In days gone by, people who lived in cold climates preserved vegetables and fruits for winter in cans and jars. This is a practice that has fallen by the wayside, and we do not advocate this except as an interesting and rewarding hobby. But we advise you to use this unit to teach your students about the value of "thinking globally and eating locally." For several months during the year, it is still possible to buy local produce. (Some clever farmers grow great hardy produce—such as kale, spinach, and carrots—during winter months too!) If you and your students' families begin buying from these people, you will pump much-needed money into your local economies, help your neighbors make a living, and reduce energy consumption and air pollution caused by long-distance transportation.

In this unit, we focus on two types of growers: farmers who are growing crops in the downtowns of major cities and amateur gardeners who grow their own foods and flowers in community gardens. Both practices contribute to a city's goals of sustainability. Open ground and plants capture water and reduce the sweltering asphalt heat of cities and suburbs. Free fresh vegetables are always welcome in an urban environment, reducing the need for food to be trucked in from outside the city and consuming precious fuel. Your students will see numerous stories of professional and amateur gardening throughout the *Edens* films. In particular, Philadelphia has encouraged such agriculture, turning over vacant lots to community groups, farmers, and others after razing dangerous abandoned buildings. In Los Angeles, an inspiring mentoring program for young Latina women uses gardening as a way to empower these young women, to awaken their self-esteem, to show them that they can coax precious food from the soil by their own efforts. Also in this unit, we introduce the concept of Community Sponsored Agriculture (CSA) as a means for buying fresh local produce and supporting local farmers. In the CSA model, consumers subscribe to a particular farmer's produce by paying a fee at the beginning of the growing season.

Throughout the season, as the crops become available, subscribers are entitled to a bag or box of groceries as a share of the crop. This is a wonderful system, bringing fresh vegetables to city and suburban dwellers who do not grow their own food, but it does have its drawbacks. If weather is bad that season, crops can be lost, and the farmer may have little to distribute to the paying customers. In a sense, subscribers share the farmer's risk with him or her. The reality of lost crops is one major reason, we should add, why America has shifted to such an industrialized system of food production. Because our produce comes from all over the country and world, we are no longer at the mercy of single crop failures or local weather events. Our food supply can be said to be "safer" because of this. But in this unit, we want students to explore alternatives.

Purpose

In this unit, students learn about the origins of the food they eat and discover how some people are farming in major cities and maintaining their own community gardens. They learn that these gardens are good for the general health of these cities because they provide open space, cool off the city in summer, and capture rainwater.

GROUP ICEBREAKERS

1. Ask students where they think most of the food eaten in the United States is grown. What about during the winter? (The longest U.S. growing seasons are in California and Florida, so much produce comes from these two states. During winter months, much produce comes from South America, which enjoys the summer season when we are in winter.)
2. Ask how many students have ever visited a farm or seen one from the window of a car, bus, or plane? (Accept all answers.)
3. Would a farm feel out of place in their neighborhood? Why or why not?
4. Can they think of some benefits of buying their fruits and vegetables from a local farmer? (Buying from a local family farmer allows you to support a small business in your own community. Produce that is grown locally travels a smaller distance to get to you and conserves resources such as gasoline.)
5. While you're at it, you might ask how often students eat fruits and vegetables. This could lead to another good discussion about nutrition and a possible activity in which students use a log to keep track of their eating habits.
6. Ask students how they would feel about picking the food their family eats every week. Does it sound like fun, or utter drudgery, to visit a farm in the region one day a week to help a farmer pick the food that their family will eat? (Accept all answers but have students explain why or why not. By the way, this is a description of a CSA farming operation, or Consumer Supported Agriculture.)
7. Tell them that some schools have actually started their own farms on school property and raised food that students have prepared for school lunches. Do they think this is a good idea? Why or why not? (Accept all answers provided they explain their reasoning.) Is this something they would want to participate in if it were ever possible at their school? (Accept all answers but make sure you know your school's position on this possibility, in case students are intrigued by the idea.)
8. Have they ever heard of or seen a community garden? What goes on there? (A community garden is a relatively small plot of ground that shared and farmed by city dwellers to raise vegetables, flowers, fruits, and other items. City dwellers team up to work the land because people in cities often don't have their own land on which to plant a garden.)

ACTIVITY 12.1
Growing All Around

Materials

different colored pens
notebook
library
Internet research materials
phone book
Handout 12.1

Primary Subject Areas and Skills

science, social studies, critical thinking

Purpose

Students use library and Internet resources to complete an activity sheet about local agriculture. The activity is designed to get students thinking about what they eat, where it comes from, and what role local agriculture does—or does not—play in their daily life.

Every day, meals are prepared and eaten, and rarely a thought is given to how the foods got from the animal or plant to the plate. The food ingredients used to prepare these meals often come from far away. They may have been grown long ago and then packaged for storage and transport.

Today, smaller farms and urban farming initiatives are growing in popularity, providing consumers with more choices about what they put in their bodies and how they decide to support the business of agriculture.

The term *urban agriculture* refers to working farms that are within, or very near, urban centers. A Community Supported Agriculture group, or CSA, is a service that offers individual consumers the opportunity to buy a stake in the harvest of local farmers. For a set fee, individuals can "buy in" to a particular region's offerings and receive fresh produce all year. In doing so, they assume some of the same risk as the farmer does. Community gardens are another growing movement and have long been a way for neighbors to come together—brightening up a block, spending some time together, and growing foods that they can share.

1. Ask students if they think about where their food comes from. Are they curious? Do they think the foods' origin has an effect on the quality?
2. Discuss the differences and similarities between urban agriculture, CSAs, and community gardens.
3. Copy and distribute Handout 12.1 to students. Explain to them that they should first see how many questions they can answer on their own, without doing any research or asking for help.
4. After students have worked through the questions once, tell them to swap papers with another student. Together, the students can discuss those questions for which they did not know the answers. They can then use research materials, the Internet, or a phone book to find the rest of the answers.
5. Discuss the answers to all the questions as a class. Take a survey of the class: How many students can answer most of the questions without doing any extra research?

ACTIVITY 12.2

Zoning in on Hardiness

Materials

pen or pencil
encyclopedia or maps of North America
Handout 12.2 (color downloads will be available online at www.edenslostandfound.org)

Primary Subject Areas and Skills

science, social studies, math, critical thinking

Purpose

Students analyze a USDA Hardiness Zone map to learn about the different growing zones in the United States and how plants are affected by the temperatures in different areas.

One of the first ways that students can become more engaged with local agriculture is to become familiar with the growing season and crops in their area. A wonderful and simple exercise that fosters a deeper understanding of the agricultural offerings of the United States is to study and discuss a map of the growing or "hardiness" zones of the United States. The map, though complex, provides the basis for discussions about local, regional, and national growing trends.

1. Discuss growing seasons with your students. Have any of them ever been members of the 4H club, or do they have parents or family members who garden? (Answers will vary.) If so, do they participate as well?
2. Ask students if they have heard of, or are at all familiar with, the hardiness zones of the United States. If so, how did they learn of them? (If they have heard of them, it may be because they or someone in their family is a gardener.) Discuss the cycle of growth throughout the year: planting, growing, harvesting, pruning, and so on.
3. Ask your students to name three vegetables that can be picked in cold weather months. (Kale, spinach, or carrots. Accept others provided they can verify their claims.)
4. Distribute Handout 12.2 to students. (The black-and-white version is tended for reference only. We recommend that you download and print the color version, available free online at www.edenslostandfound.org.) Explain that the temperatures on the map refer to the lowest temperatures in these areas over a long period of time. A plant that cannot survive these temperatures cannot be grown this area.) Instruct students to study the map carefully for trends as well as specifics and then answer the questions.
5. After students have completed the map activity, discuss the answers. How does the climate in your area affect the kinds of crops that grow? (The hardiness zone map determines which crops grow in an area and which do not. A good example is the orange, which does not grow in the majority of U.S. states because the weather is simply too cold.)

ACTIVITY 12.3

Food on the Move

Materials

pen
notebook or sketchbook
Handout 12.3

Primary Subject Areas and Skills

science, social studies, math, writing, critical thinking

Purpose

Students read a pictograph to learn how far certain foods have traveled to reach a central U.S. market.

How far did your food travel before it landed on your table? This activity encourages students to consider this question, perhaps for the first time.

1. *Day 1:* Ask students how many miles they think the average bunch of peas or broccoli travels before it lands on a consumer's table. Ask them to write the number down.
2. Next, tell students to choose an item of food from their kitchen—a bag of produce, a can of tuna, or the like—and look on the packaging to find out where it was packaged, grown, or distributed. Tell them to bring the information to class the following day.
3. *Day 2:* Start the class with a discussion of the information the students found on the packaging of their food. Make a list on the board of some of the locations they found.
4. Distribute Handout 12.3. Explain to students that they will be looking at information about how far, on average, some produce travels to get to the Chicago Terminal Market. Instruct students to study the charts and answer the questions.
5. After students have completed the questions, discuss their reaction to the information. Are they surprised how much gasoline must be used to bring food to their supermarket? Bothered? Unaffected? (Answers will vary. Accept all, provided students can articulate their feelings clearly.)

Detective Activity

Remember that information students found out by looking on the packaging of some of their foods at home? For a long-term extra credit project, have students see how much more they can discover about where their food comes from. Once they've found the distributor, can they find the companies that supply the ingredients that go into that food? Where are the ingredients grown? Where are they processed? Have students create a flowchart or map that shows all the steps on the journey from the various food sources to processing and packaging plants, distributors, grocery stores, and, finally, home. As part of a current events discussion, have students research newspaper and magazine clippings or websites and report on the origin of food that was involved in a particular *E. coli* outbreak. How far had the foods involved traveled?

Local Food Activity

Ask students to write an opinion piece for the school or local paper about the availability of local produce and foods in your area. What would they like to see change about their local food supply? Emphasize that students should make as many environmental connections as possible to strengthen their argument.

Extensions

1. Besides freshness, ask students to name at least three (try for many more) benefits of buying produce and other foods that are produced locally. (Accept all reasonable answers. Encourage them to think about things like packaging, fuel costs, CO_2 emissions, storage, supporting local families and farms, and the like.)
2. There are some groups who feel that foods should be more completely labeled with information about where they came from. Ask students if they would be surprised if one package of their favorite hamburger meat contained contributions from several different animals. Would they want to know that? Do they think they have a right to know? Why or why not? (Accept all reasonable answers. Most people would want to know as much as possibly about the origins of food that they consume.)

ACTIVITY 12.4

Your Own Garden

Materials

pen
notebook or sketchbook
resources listed in unit directory

Primary Subject Areas and Skills
science, social studies, math, art, critical thinking

Purpose
Students do research to decide the feasibility of building and tending their own school garden.

Many of the people profiled in *Edens Lost & Found* either had gardens in their backyard or planted flowers and vegetables on a small plot of land in a community garden in their neighborhood. Community gardens allow people who have little or no garden space at home to garden in a group-run location. Everyone gets her or his own plot to plant whatever they like. Many schools (such as the Edible Schoolyard project listed in the "Resources" section, page 184) allow students to plant gardens on school grounds. The students raise food for family use, the school lunch program, charitable organizations, or to sell to local restaurants. Planning a garden—even going so far as to start one of your own—is a great way to get kids in touch with the cycle of growth, the importance of growing and buying local foods, and the basic science required to successfully nurture plant life.

In this activity, you will encourage students to think about designing and tending their own garden. The activity is designed to be theoretical—an actual garden does not need to come to fruition. At the very least, you may be able to plant a container garden and grow a few spring plants in your classroom. Another possibility is to grow fruits such as blueberries, raspberries, and blackberries, which are among the costliest to buy in supermarkets. Bushes of all three of these plants are sold relatively cheaply at home center stores. You may wish to look into planting a school fruit garden and tending it over several seasons. Another alternative is to contact local arboretums or nurseries and investigate the possibility of students volunteering their time there.

Note: We encourage you to confer first with your principal or supervisors to gauge your school's willingness to permit an onsite garden run by students and teachers. If this is not possible, don't give up. A local nature center or garden group may be able to furnish your class a plot in a community garden. Know your options before presenting the idea to your students.

1. Ask students if anyone has ever had a garden of his or her own. What did they grow? How much work was involved? If they had to do it again, how would they do it differently? (Accept all anecdotes.)
2. Review some of the gardening operations viewed in the course of the *Edens* program. Some people use gardens to earn money. Some do it as a hobby and to supplement the food they feed their families. For others, such as senior citizens, gardening is form or relaxation, exercise, or therapy. If your class were to start a garden, what would it be for, fun and good nutrition or for profit? (To keep your project simple, we strongly suggest that the former should be the goal.)
3. What other decisions are needed before starting a garden? As students call out their ideas, write them on the board. (Ideas might include: Deciding on the dimensions in feet and inches of the garden plot; what to grow; which varieties prosper best in the region; what is the typical growing season in the area; what is the area's hardiness zone; how much needs to be planted to feed the class or sell to local restaurants; how the garden will look.)
4. Divide students into small groups and ask each group to research each of these "business" decisions. Let them use a variety of research tools, including the Internet. Ask them to compile a "business plan" for their garden and to support their written work with drawings, designs, maps, clippings from seed catalogs, and the like.
5. Arrange to have each group report on their findings. If the class is serious about this idea and the possibility exists to make it a reality, you may wish to have them polish their work and re-present it to school or local leaders.
6. If the sale of food is a possibility, you might challenge the students to visit local independent restaurants—franchises or chains may be locked into group buying agreements—to see if they would be interested in purchasing your class's products. If so, students need to generate a sample financial plan as well: how much they expect to sell their products for, how much it will cost to grow crops, and so on.
7. If the students are simply expecting to use the food for their own or for classroom use, ask them to generate menus based on the food they will be growing. Have students share recipes, hold a local food festival, or make dishes at home to share with the group.
8. *Note:* Before embarking on a serious gardening effort, be sure to inquire if students have any allergies to food, flowers, bees, and the like.

EXTENSIONS

Pick, Pick, Pick

Take a field trip to a local farm where you can pick your own fruits or vegetables.

Go to the Source

Research and visit a CSA, farm, or farmer's market in your area. Invite one of the farmers to come talk to your class about the rewards and challenges of local agriculture.

The Bee's Knees

Do you have a beekeeper in your area? Find out from your local Department of Agriculture or contact the National Honey Board at (303)776-2337 to find out. Then visit her or his place of business, see how the honey is extracted, and learn more about the role these valuable creatures play in pollination.

Gardens in the News

Find out who the gardening or home editor is at your local paper. Invite that person to speak to your class about gardening and produce issues in the news.

RESOURCES

- **Garden Web**
  www.gardenweb.com/
  This site provides a series of bulletin boards where hobbyist gardeners trade ideas and advice.

- **Chefs Collaborative**
  www.chefscollaborative.org/
  Here can be found information on a network of people across the country who promote sustainable cuisine and support local, seasonal, and artisanal cooking. Find a chef near you and have him or her visit your classroom; or take a trip to their restaurant.

- **Food routes**
  www.foodroutes.org/
  This site explores the benefits of buying locally and offers information on farms, farmer's markets, cooperatives, and CSAs across the United States. This is a great resource for several of our activities.

- **Food safety**
  www.consumersunion.org/food.html
  Consumers Union's food safety efforts seek to focus public attention on food safety risks and regulatory deficiencies that can result in harm to the public.

You can take action on food safety by visiting our sister site, **The Edible Schoolyard**

- www.edibleschoolyard.org/homepage.html
  The Edible Schoolyard, in collaboration with Martin Luther King Junior Middle School, provides urban public school students with a one-acre organic garden and a kitchen classroom. Using food systems as a unifying concept, students learn how to grow, harvest, and prepare nutritious seasonal produce.

- Center for Eco Literacy
  www.ecoliteracy.org/publications/getting-started.html
  From the folks at the Center for Eco Literacy, at this site you can find free guides for creating school gardens. This is a valuable and free resource!

Reading Handout Unit 12:
PHILADELPHIA EATS WHAT PHILADELPHIA GROWS

How hard are you willing to work to succeed at your dreams? In this reading, you'll learn about a woman who started a very unusual farm on an abandoned factory site in downtown Philadelphia. If you are able to watch the Philadelphia movie in the *Edens Lost & Found* DVD set, you can see Mary Seton Corboy talk about her farm. After you read the story, answer the questions.

Philadelphia is home to thousands of people who would rather plant a tomato plant or a handful of carrot seeds than a flower or a tree. Most do this as a hobby. For others, it's a living and a tough one at that. Farming is hard enough in the countryside, where you have open space, fresh air, and the friendship of other farmers. Imagine trying to grow vegetables in the soil of downtown Philadelphia, where you stand out like a sore green thumb and everyone thinks you're nuts.

That's Mary Seton Corboy's life. On St. Patrick's Day, 1997, she started Greensgrow Farm on a big plot of land in the Philadelphia neighborhood of Kensington that was once the site of a galvanized steel plant. Land such as this pops up often in big cities. Such plots are called "brownfields," because these lands are so contaminated, or believed to be contaminated, that they cannot be easily used. Ordinarily, you would not grow crops in soil this polluted. But Ms. Corboy did not want to run an ordinary farm. She wanted to grow crops hydroponically. That means she would grow her crops not in dirt but in tanks or troughs of water placed above the surface of the soil. The roots of the plants would suck up nutrients without needing soil. It has been a very difficult way to make a living. Ask Ms. Corboy if her farm is an Eden, lost or found, and she says, "It remains to be seen if I'm Eve."

"I didn't come up with the idea of farming on a brownfield," she says. "I knew that the land wasn't going to be everything that I hoped it would be, so I spent time developing a hydroponic system that I thought would work in the city. Why grow crops in a city? Well, I live in the city, and I didn't want to live in a rural area. I wanted to live in a city, and I wanted to pursue a green business, producing food for restaurants. At one point I had been a chef, and so I knew that restaurants, even in the middle of the Pennsylvania growing season, weren't getting Pennsylvania produce. I wanted to give restaurants in Philadelphia an alternative."

Many people today are interested in buying and eating locally grown produce. "Buying food from local farmers has a huge impact on the region," explains Judy Wicks, who owns a Philadelphia restaurant called White Dog Café. "First of all, it supports the small family farms that have been going out of business at a frightening rate and selling out to land developers. When you buy from a small farmer, you are helping him or her make a living, saving farmland from developers, and keeping the family on the farm."

When Ms. Corboy first started Greensgrow, she thought that the people in the neighborhood would think that what she was doing was weird. She didn't count on becoming very attached to her new community.

"When I started," she says, "I just wanted to grow lettuce and sell it to fancy restaurants downtown. When you work outdoors like I do, you see the neighborhood pass by you every single day. A lot of people still walk here. They walk to the store, they walk to visit each other, they sit on their stoops, and they hang out of their windows. So you hear this whole life that's going on around you. Quite honestly, it wasn't a life that I was familiar with."

Over time, Ms. Corboy stopped growing just for restaurants and started growing food to sell to her neighbors. Every time someone bought something, she told them about the importance of eating local produce, about why organic food—food grown without harmful pesticides—was important to their diets, and lastly, why she had chosen such a difficult way of making a living.

"I would like to think that we're an important part of the neighborhood," she says. "Our goal is to both educate people about food and to provide access to the highest quality food for them. The same is true for the nursery that we run in the springtime. I think it's important that people, especially in this kind of cement environment, have access to beautiful living things. So we pride ourselves on the fact that we sell high-quality locally grown plants."

Ms. Corboy now has a great deal of affection for her adopted city. "I say it's a city that loves you back. I'm not a native Philadelphian, I moved here. Most of my friends are people who moved here. I think that we as a group are much more optimistic about Philadelphia than people who grew up here. Philadelphia has all of the resources to be a really great city. It has still the power to come back. If you can play some role, however tiny, in making it be all that it can be, then you've done your job."

Some days, though, Ms. Corboy is not sure if she'll be able to live up to that promise. "If I had known how hard this life was going to be," she says. "I wouldn't have done it. It's interesting: People say I'm a pioneer. They say that now. At first I was just a nut case. I don't think that I was really cut out to be a pioneer. After a couple of years in which things didn't always go exactly the way I wanted, I could have given up. Now it's pure obstinacy that keeps me here. I'm determined I'm going to beat this thing. I don't care what it takes."

Questions

1. What is a brownfield?

____

____

2. What does it mean to grow crops hydroponically?

____

____

3. Do you think it is easy to grow food like this? Why do you think Ms. Corboy wanted to do this?

____

____

4. What reasons does Ms. Wicks give for buying and eating local produce?

____

____

5. How does the way that Ms. Corboy thinks about her neighborhood change over time? Why do you think she changed?

____

____

6. Do you think it is important for a farmer to educate her or his customers about the food they are buying? Why or why not?

____

____

7. Ms. Corboy says if you can play a role in your city, you have done your job. What role is she playing in her neighborhood?

____

____

8. Do you think farms like Ms. Corboy's should be encouraged or discouraged in cities? Why?

____

____

Name ______________________________ Date ____________________

HANDOUT 12.1: Growing All Around

See how many of the following questions you can answer without doing any research. Then swap papers with another student. Together try to fill in as many blanks as you can, using resources in your classroom, library, or computer lab.

1. Are there any CSAs (Community Supported Agriculture groups) within 50 miles of where you live? What are they and where?

   ____

2. If something is organic, is it always locally grown?

   ____

3. If something is locally grown, is it always organic?

   ____

4. Name three agricultural crops grown in your region or state.

   ____

5. How long is your growing season?

   ____

6. Name two farms within 50 miles of where you live.

   ____

7. What is the most important crop to your state's economy?

   ____

8. What fruits grow in your region in the summer?

   ____

9. Is there a community gardening group in your town? Where is it?

   ____

10. Are there farmer's markets in your town? Where are they located, and when do they take place? What do they sell?

    ____

Name ______________________________ Date ____________________

HANDOUT 12.2: Zoning in on Hardiness and Growing Seasons

Study the map below showing all the hardiness zones of the United States. The temperatures shown in the left column are the lowest temperatures recorded over a period of time. Chances are high that if a plant cannot survive outdoors when temperatures reach these recorded lows, then farmers and gardeners should not expect to raise them except as short-term plants. They will not survive the winter without warmth and shelter.

Be sure to look at the key and pay attention to the different trends that you see. Then answer the questions.

Zoning in on Hardiness and Growing Seasons

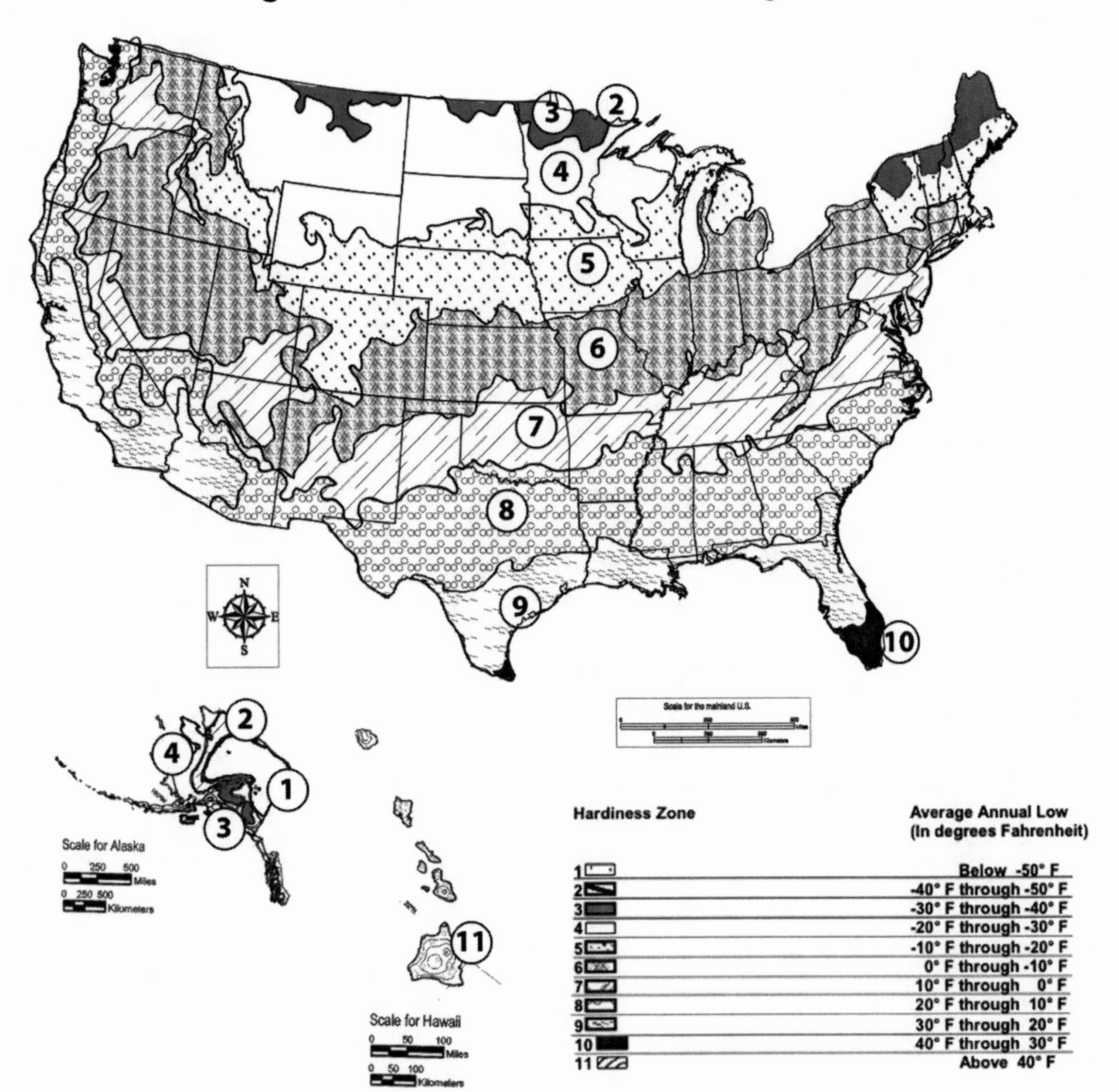

1. In which zone is your town located?

____

2. Which zone is considered the warmest?

____

    a. What is the average annual minimum temperature in that zone? ____________

    b. Name a town located in that zone.
    (Use a map if you need to.) ______________________________

3. Which zone is considered the coldest?

____

    a. What is the average annual minimum temperature in that zone?____________

    b. Name a town in that zone. ______________________________

5. Name a region you think would have a short growing season, based on this map. What do you think might grow there?

____

6. Name a city located in zone 5.

____

7. Why do you think some of the "colder" zones are located to the south of some "warmer" zones? What might affect their average annual minimum temperature?

____

____

8. Calculate the average annual low temperature range of zone 10, in degrees Celsius.

____

Name ______________________________ Date ____________________

HANDOUT 12.3: Food on the Move

Does your food look tired? If so, it's probably because it had to travel so far to get to your plate. How far? Look at the following charts to get an idea. Then answer the questions.

Food on the Move

Average distance by truck to Chicago Terminal Market (Continental U.S. only)		# States supplying this item	% total from Mexico
Grapes	2,143 miles	1	7
Broccoli	2,095 miles	3	3
Asparagus	1,671 miles	5	37
Apples	1,555 miles	8	0
Sweet Corn	813 miles	16	7
Squash	781 miles	12	43
Pumpkins	233 miles	5	0

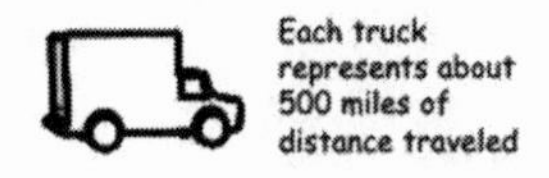

Each truck represents about 500 miles of distance traveled

Source: Leopold Center for Sustainable Agriculture

Weighted average source distance* (WASD) estimations for produce arriving by truck at the Chicago Terminal Market, 1998

Fresh produce type	Continental U.S. only* (miles)	Number of states supplying	% Total originating from Mexico
Apples	1,555	8	0
Asparagus	1,671	5	37
Beans	766	11	10
Blueberries	675	6	0
Broccoli	2,095	3	3
Cabbage	754	17	<1
Carrots	1,774	6	3
Cauliflower	2,118	3	2
Celery	1,788	4	3
Sweet Corn	813	16	7
Cucumbers	731	17	36
Eggplant	861	8	36
Grapes (table)	2,143	1	7
Greens	889	11	2
Lettuce (iceberg)	2,040	7	0
Lettuce (Romaine)	2,055	6	0
Mushrooms	381	4	0
Onions (dry)	1,675	15	10
Peaches	1,674	8	2
Pears	1,997	4	0
Peas (green)	2,102	1	30
Peppers (bell)	1,261	12	27
Potatoes (table)	1,239	14	0
Pumpkins	233	5	0
Spinach	2,086	6	<1
Squash	781	12	43
Strawberries	1,944	2	15
Sweet Potatoes	1,093	4	0
Tomatoes	1,369	18	34
Watermelons	791	14	2

* The weighted average source distance is a single distance figure that combines information on distances from production source to consumption or purchase endpoint. For these calculations, USDA Agricultural Marketing Service arrival data for 1998 were used to identify production origin (state or country). Distances from production origin to Chicago were estimated by using a city located in the center of each state as the production origin and then calculating a one-way road distance to Chicago using the Internet site Mapquest (mapquest.com). Estimations do not include distance from the Chicago Terminal Market to point of retail sale.

Source: Leopold Center for Sustainable Agriculture, Rich Pirog and Tim Van Pelt, 2002.

1. Which food travels the most miles? How far is that? How many states supply that food?

____

2. What percentage of asparagus comes from Mexico?

____

3. Which produce is supplied by the most number of states?

____

4. Which produce is supplied by the least number of states?

____

5. What do you think the relationship is between how many states supply a food and its price?

____

____

6. What do you think the relationship is between how far a food travels and its price?

____

____

7. If each gallon of gas burned results in 19 pounds of CO_2 emitted, calculate how much CO_2 one average trip for spinach will generate. Assume the truck will get approximately 7 miles per gallon. Show your work.

____

____

8. What is your favorite vegetable? How far does it travel? How many states supply it? How much CO_2 is generated by transporting it? Could you grow your favorite vegetable at your house?

____

____

UNIT 13

URBAN FORESTRY

TOPIC BACKGROUND

The term *urban forestry* refers to the care and management of planted or naturally occurring trees in urban settings. It's part of a new national movement, supported by the U.S. Forest Service, to plant more trees and more effectively protect existing trees in urban areas. The word *urban* does not refer solely to cities but also to large metropolitan areas that are made of cities, suburbs, and exurbs. The more people who are living in a given geographical area, the greater the chance that energy use will be high in this area. In summer months, the center city will be hotter than outlying areas. And flooding will be a problem throughout the metropolis because so much concrete and asphalt has replaced the excellent drainage provided by permeable soil.

These are costly dilemmas that all Americans communities are facing. As these issues become persistent, there has been a growing movement to return to natural methods of dealing with excess heat, excess rainwater, and the demand for greater energy. In recent decades, people who sought to protect natural forestland have been derided. But now we are seeing that trees can perform vital functions that cost us billions of dollars each year to do through human engineering. They capture rainwater, cool buildings, and clean air. That's why we say trees are part of our "urban infrastructure." Rather than spend tax dollars to artificially correct problems, politicians and urban planners are realizing that bulking up a region's trees and protecting its existing forestland are excellent ways to nip these problems in the bud.

Your students are likely to be thrown by the use of the word *forestry*. Help them to understand that a forest is not just a remote stand of trees in the wilderness. A million trees standing on street corners throughout a vast city also constitute a forest. Share these facts with them: One hundred million additional mature trees planted in American cities would save $2 billion per year in energy costs. If you place trees in your backyard in a spot where they can block prevailing winter winds, you can lower your family's heating bills 10 to 20 percent. Shade trees planted to the east and west of a home cut cooling costs 15 to 25 percent. Street trees shade concrete and help cool entire neighborhoods. Flowering, fruit, and nut trees feed humans, insects, and wildlife. Trees keep land from eroding, rivers from flooding, and landslides from occurring. Besides the scientific benefits, studies show trees have an enormous effect on the psychological appeal of a city.

Purpose

By the end of the unit, students will have a solid understanding of how urban forestry can help cities cope with high energy bills and increasingly extreme weather events. You will be able to solidify these concepts further if you can show footage from the *Edens Lost & Found* DVD set or video resource library. All four films in the *Edens Lost & Found* series deal with trees and their protective qualities.

GROUP ICEBREAKERS

1. Ask students to define the word *forest.* Does a forest have to be in the wilderness? (No.) Can a forest be right where you live? (Yes! Many forests are part of local parks.) Help them to see that scientists regard any vast group of trees in a specific geographic region as a forest.
2. What does the term *forestry* mean? (Students may equate it only with lumbering activities. Help them to see that forestry is primarily about stewardship—human involvement and management of a natural resource.)
3. Off the tops of their heads, can your students list some important things trees do for us humans? (They should be able to refer to what they've learned in previous units. If they get stuck, remind them of the problems: heat, flood, energy use, and so on.)
4. Considering all the environmental and economic

benefits of trees, why do students think some cities do *not* have a tree-planting program? (Answers might include lack of education, poor planning, lack of funds, and the like.)

5. Based on students' impression of their own city, town, or neighborhood, should more trees be planted here or not? (Answers will vary, but we would argue that most communities would benefit from the planting of more trees.) How could they verify their impression? (They could devise a simple audit project and have classmates analyze how many trees are located on their home street, their neighborhood, or their walk home from school.)
6. Ask students what they would say to convince the average person to plant more trees in their area. (They could stress the emotional as well as scientific and economic benefits. Trees give us shade, are beautiful to look at, reduce our need for air conditioning, keep our neighborhoods cool, and capture rainwater.) What could a city government do to encourage its citizens to plant more trees? (Local plant-a-tree days would be fun and educational.) How could your class help spread the word or do some of the tree-planting work? (Students could help by writing, designing, and distributing educational materials and then helping with the physical planting.)

ACTIVITY 13.1

Did You Know . . . ? Fun Facts about Urban Forestry

Materials

pen or pencil; Handout 13.1

Primary Subject Areas and Skills

science, social studies, critical thinking

Purpose

Students complete a warm-up quiz designed to assess how much they already know about urban forestry.

The number of ways trees can positively influence a community and its environment is constantly on the rise, with new research and findings coming to light each year. It is truly amazing that the simple act of planting a tree can affect so many different aspects of modern life and living.

This activity is designed to give students just a small idea of the sweeping effects that planting and maintaining trees can have. You may wish to begin class with a discussion about your own town and region.

The quiz on Handout 13.1 is designed to be an introduction to the topic of urban forestry. However, if you choose to do so, you can also use this quiz as an actual assessment tool toward the end of the unit.

1. Ask students to think about your town and region. Are there a lot of trees? Do students think there should be more? Have they noticed if the number of trees has increased or decreased during their lifetime? Which neighborhoods or other populated areas seem to have the most trees? (Answers will vary. Accept all reasonable answers.)
2. Explain to students that they will be taking a quiz to make them more aware about the role that trees can play in urban life.
3. Copy Handout 13.1 and distribute the quiz to students. Explain that they should work alone and complete the quiz without using any books or other research materials.
4. After students have completed the quiz, call out the answers and have student self-correct. (*Note:* If you opt to use this as an assessment tool, simply have students turn their papers in to be graded.)
5. After the answers have been given, talk about the results and discuss each question. Which facts were the most surprising to students and why?

Extension

What kind of trees are the most dominant in your area? Ask your local Parks Department for information about the amount and type of tree coverage that has existed in your area over the years. You may also contact your county soil and water conservation district for copies of aerial photographs. How has the area changed? Ask students to analyze the information and write a report, including graphing, that shows these changes. Can they think of any possible explanations as to why this has occurred? (In most cases, the level of trees or open farmland will have dropped as local towns and cities have grown. Trees and farmland have been eliminated to make room for new homes.)

ACTIVITY 13.2

Getting to Know Your Trees

Materials

pen and pencil
notebook and/or sketchbook
fabric tape measures (one for each group of students)
Handout 13.2

Primary Subject Areas and Skills

science, social studies, art, critical thinking

Purpose

Students use a simple math formula to determine the age of some common trees in their neighborhood.

We walk and drive past trees every day, sometimes looking up but rarely stopping to think about that tree's history and role in our community. Encouraging students to stop and examine the natural history that surrounds them each and every day is a wonderful tool in engaging them in further environmental study.

This activity is designed to make students more aware of the trees in their community, while strengthening science and math skills. The activity can be done on school grounds during a school hours. However, for a more extensive experience, an afternoon in a park or downtown area may provide a greater variety of both trees and settings.

1. Explain to students that they will be calculating the ages of a variety of trees on their school grounds or other designated area.
2. Divide students into small groups of three or four students. In each group, one of the students will record the group's findings while the others work to measure the tree and calculate the age. These roles should rotate as students move on to other trees so that everyone gets the opportunity to participate.
3. Take students outside on school grounds during class time or choose an afternoon and arrange for the class to be at a designated place in town or a nearby park.
4. Students should bring paper and pencil and a sketchbook, if they'd like. Teachers should provide one fabric tape measure for each group.
5. Distribute Handout 13.2, "How Old Is That Tree?" to students. Instruct each group to calculate the ages of at least five different trees. Students should show their work. (*Note:* If they cannot find specimens of the nine trees shown on the list, have them calculate a range of possible ages for their trees, using the smallest growth rate [2 inches per year] and the largest [7 inches per year]).
6. If students choose to, or perhaps for extra credit, have them sketch the trees they have measured. Draw the leaves. Are they sure they know what kind of tree it is? If not, have them collect enough information about the trees they are examining so that they can use their sketches and notes to determine the species of tree.
7. After students have completed the activity, ask them to share their results. How could students have measured the trees without the tape measure? (Encourage discussion of estimation, think about hand spans, string, and so on.)
8. How old was the oldest tree they found? What do students think life was like at that time? How do they think the tree's surroundings have changed since it first sprouted?

Extensions

1. Ask students to calculate the ages of trees near their home.
2. Instruct students to pick one of the trees that they measured or to choose an old tree at a particular location near their home or in town. Writing in first person, as the tree, instruct students to tell their life story. What have you seen? How has your life changed? What were the good ol' days like? Encourage students to incorporate as much of your area's history as possible.

ACTIVITY 13.3

Trees and You: A Numbers Game

Materials

pen and pencil
calculator (if necessary and at teacher's discretion)
Handout 13.3

Primary Subject Areas and Skills

science, math, social studies, critical thinking

Purpose

Students read and interpret a double line graph and a pie chart to quantify the beneficial role that trees can play in their community.

Being faced with the benefits that planting trees can have on the environment of any town or city is an inspiration. Such a simple act—the planting of a tree—can have curative, positive effects on generations of inhabitants. And the economic benefits are impressive as well, especially when the amount of money that can be saved on personal utility bills is taken into account.

This activity is designed to give students even more "hard" evidence about the beneficial role that trees cans play in their communities and towns, by engaging students in the statistics and numbers behind some of the studies.

Once they have analyzed the data below and answered the questions, have a discussion in class about what students feel they can or should do with this information. Discuss the possibility of making a presentation to you local government or community groups. Organize a tree-planting task force. Encourage students to take this information and use it to make informed decisions about actions they can take to be more invested members of their schools, communities, and town.

1. Distribute Handout 13.3 to students.
2. Explain to students that they will be reading and analyzing data that relate to the role of trees in the environment and in a consumer's everyday life.
3. After students have completed the exercise, discuss the answers. Ask students whether the information here will affect actions they take in their daily life. Will they mention this to their parents? Does it inspire them to action? Why or why not? (Answers will vary, but accept all reasonable ones and try to encourage them to share the information with others.)

ACTIVITY 13.4
At the Podium

Materials

Research materials, available in your library or on the Internet; copy of the Judges' Evaluation Sheet from Unit 1, page 14.

Primary Subject Areas and Skills

science, social studies, reading comprehension, writing, research, critical thinking, public speaking

Purpose

Students participate in a pro–con debate on the wisdom of planting more trees in new housing developments and as part of routine town improvements.

The statement to be debated is as follows:

> *Minimum tree-planting guidelines should be established for all new developments and town improvements.*

Point

Yes. The benefits of a growing urban forest to all residents—on personal, environmental and economic levels—make tree planting a priority and a civic responsibility in every community.

Counterpoint

No. If businesses are to thrive and grow, and individuals are to be given a fair chance to build new homes and expand their businesses, they should be able to do so without having to adhere to even stricter standards, which will only raise the costs of already costly enterprises and discourage economic development.

1. Divide students into two groups of at least four students each. Depending on class size, be sure to leave a group of at least five (preferably an odd number) of students to act as judges. Assign each group to Point or Counterpoint as described above.
2. Instruct students to research and report on their point of view. They should devise statements in support of their position, and be prepared for rebuttal from the opposing team.
3. For the debate itself:
   a. Judges should be concerned with organization, evidentiary support, and overall presentation, as well as politeness and poise.
   b. Debaters should clarify who will speak for the group. (Debaters may take turns, if that is decided beforehand).
   c. The affirmative or Point team will speak first for three to five minutes to present their argument. The Counterpoint team will then have three to five minutes to present their argument.
   d. Each team will be given three to five minutes to respond to the other team's argument.
4. Have judges rate the arguments of each group and present their findings to the classroom.
5. Discuss among yourselves the emotions and

ideas that came up during the debate. Did anyone change his or her mind about the topic at hand?

EXTENSIONS

Just Do It!

Clearly, one of the easiest things you can do outside your classroom to drive home the importance of this unit is to plant some trees! Your school grounds, town center, or a vacant lot will work. Ask others at school to get involved or even invite other schools in your area. There's no need to wait for Earth Day or Arbor Day—although those are great events for a tree planting. You can also investigate state and local forestry resources, as there is a very good chance that you can get trees for free.

Ask a Pro

Invite a representative from the National Forest Service or a certified arborist to visit your class and talk about the appropriate trees to plant in your type of urban environment.

RESOURCES

Nominate a Tree to Be a Champion

National Register of Big Trees has documented the largest known specimens of every tree in the United States. The largest tree of its species in the country is the National Champion. You can nominate your own tree by going to the site or even send in a "big tree postcard" featuring a big tree from your community. Kid-friendly information on the site tells you how to determine the height, trunk circumference, and crown spread of the trees as well.

- www.americanforests.org/resources/bigtrees/

U.S. Forest Service's Community and Urban Forestry

United States Forest Service's Community and Urban Forestry is a great federal government resource. Copies of education materials and pamphlets are available here.

- www.fs.fed.us/ne/syracuse/
- www.urbanforestrysouth.org/

TreePeople

Andy Lipkis and the folks at TreePeople in Los Angeles are trying to make the world a better place, one tree at a time.

- www.treepeople.org/

American Forests and Katrina Releaf

American Forests, the same organization that organizes the National Register of Big Trees, provides extensive information about tree planning efforts on their site, including the Katrina Releaf project that is helping to plant trees in areas damaged by Hurricane Katrina.

- www.americanforests.org/

Good links and activities from the U.S. Forest Service

The site for the USDA Forest Service Pacific Southwest Research Station has fantastic research and a number of studies for future class activities and research. It also has a wonderful collection of links.

- www.fs.fed.us/psw/programs/cufr/

Tree Link

Tree Link invites you to "log in and branch out." This organization seeks to improve urban and community forests. Great tools on this site, including tree flash cards, are a wonderful teaching tool.

- www.treelink.org/

The National Arbor Day Foundation

The National Arbor Day Foundation provides all the information you need about planting and caring for trees. You can also research trees in your specific area. This is a great resource for schools, communities, and individuals who want to increase the green in their area.

- www.arborday.org/

Reading Handout Unit 13: THE MAGIC OF TREES

How important are trees to the livelihood of cities and towns? Are they just something pretty to look at, or do they perform an important public service? In this reading, you'll learn why some communities are valuing their trees more and more. The role of trees is a major theme throughout all the films in the *Edens Lost & Found* DVD set. After you read the story, answer the questions.

The phrase *urban forestry* is used to describe the care and management of natural and planted trees in cities, suburbs, and towns. Traditionally Americans have thought little of the role trees play in urban settings. Trees were regarded as pretty things that are sometimes useful as providers of shade on a hot day. But that is changing as our cities grow and millions of people spill over into suburbs.

Now we are beginning to appreciate that trees perform important jobs. For example, tree roots capture water, thousands of gallons at a time. The more trees a city can plant, the more water it will have stored underground. To feed themselves, trees absorb sunlight and carbon dioxide (CO_2), the very ingredients that can make large sprawling urban areas difficult places to live. CO_2 is the chief ingredient of climate change. Extra heat makes cities hot, uncomfortable, sweaty, and sometimes dangerous places to live. Trees can save us from all that. That's why all the cities in the *Edens Lost & Found* series practice large-scale tree planting.

America is waking up to the notion that cities thrive best when they have a "green infrastructure." What is infrastructure? Simply put, it is the basic necessities needed to run a community, such as roads, bridges, mass transit, schools, and so on. No one would find it surprising to hear a mayor say, "We need to invest in our roads and bridges," or "We need to invest in our schools," or "We need to invest in our older and younger citizens." All these statements sound perfectly reasonable because everyone understands that a city depends upon people, schools, roads, and bridges to survive.

Well, a green infrastructure is now being recognized as equally necessary and important. If you have lots of trees in your city, your city feels cooler on hot summer days and nights. The trees shade the buildings, so people don't need to run their air conditioners so high, thus saving money on energy bills. In Philadelphia, citizens learned that when they planted trees in their neighborhoods, three things happened: The prices of homes went up. Crime went down. And more people wanted to live in those neighborhoods.

That's not all. In recent years, researchers have discovered other benefits of trees in cities. For example, they have discovered that people would rather shop in downtown areas where trees are present. If given a choice, people are more likely to stop and browse downtown areas that seem wooded, shady, and inviting. For some reason, people prefer to eat and shop on attractive tree-lined avenues. A city that understands this interesting human quirk can help create attractive downtown settings and serve the public good at the same time.

Another cool finding: Drivers tend to speed when driving their cars down treeless streets. They tend to slow down when driving down tree-lined streets. There's an interesting theory that might explain why this is happening. On a treeless street, drivers have few landmarks to gauge how fast they are going. But a driver who zips past eight big trees in a minute quickly senses that he's going too fast. It may also be that drivers are instinctively more cautious when their visibility is reduced. Or perhaps tree-lined neighborhoods appear friendlier to the eye. Maybe drivers slow down because they want to drink in some of that atmosphere.

These are just a taste of the findings by psychologists working at the University of Washington and the University of Illinois. Taken alone, these are simply quaint findings that point to interesting features of human personality. But coupled with everything else we know about trees in the urban setting, there's not a city manager alive who should resist encouraging more green space in cities. "It's not just about the trees," says Susan Mockenhaupt, an expert in urban forestry who works with the U.S. Forest Service. "If it's just about trees we will lose the battle in American cities."

Recently, some urban planners and thinkers have tried to put a dollar value on the trees in their cities. Look, they say, everyone knows that if a city lost a bridge in a flood, it would cost money to rebuild it. Well, trees are valuable too. In fact, the older a tree is, the more valuable it is, because it is capable of capturing more water when it rains, removing more carbon dioxide from the atmosphere and giving more shade. In some cities, such as Seattle, when new buildings are built, a tree expert inspects the building site and carefully assesses the value of the existing trees. The expert, or arborist, then assigns a value to a tree and hangs a sign on it. At one Seattle work site, the cost of each tree is written on a sign attached to a wire fence around the tree: "Bigleaf Maple, Acer Macrophylum, appraised value: $42,365," "London Plane, Platanus X Acerifolia, appraised value: $35,666," "Douglas Fir, Pseudotsuga Menziesii, appraised value: $19,248," "Lawson Cypress, appraised value: $24,417."

Each fence carries this stern warning to everyone working on the site:

Tree Protection Fence

No trespassing on critical root zone of this tree without direct approval of owner's representative. Work within the critical root zone shall result in a fine of $1,500 or the appraised landscape value, whichever is greater.

The most expensive tree on this work site is valued at $71,000. If a worker digs too close to the roots of this tree, and the tree dies, that worker or the company he or she works for may be forced to pay that amount.

Some people think this system is wrong and outrageous. If you bought a tree at a tree farm, they say, it would only cost a few hundred dollars. Why are you forcing someone to pay thousands to replace one tree? Arborists and others see it differently. A tree from a tree farm is only one to three years old and very small. An old tree is worth more because it has grown to a mature age and can do more "work" for a city than a little tree can.

Questions

1. What does *urban forestry* mean?

____

____

2. How can trees help a city on hot summer days?

___

___

3. How can trees help save a city from too much rain and damaging floods?

___

___

4. How can trees help humans reduce the danger of climate change?

___

___

5. What is a green infrastructure? It is something you build?

___

___

6. What three things changed in some Philadelphia neighborhoods that planted trees?

___

___

7. Why do you think people prefer to shop in downtown areas where there are trees?

___

___

8. The author suggests a few reasons why people tend to slow down when driving down tree-lined streets. Can you suggest a few ideas of your own?

___

___

9. Do you think it is fair to fine people who harm older trees on work sites? If so, how much should they be fined? If not, what should be their penalty?

___

___

Name ______________________________ **Date** ______________

HANDOUT 13.1: Eco Facts Urban Forestry Quiz

Answer the questions below to the best of your abilities. Circle the correct answer or fill in the blank.

1. On a tree-lined street, which of the following is the average person most likely to do?

   a. Drive faster

   b. Drive slower

   c. Drive the same speed

   d. Stop

2. Name two ways in which trees reduce the amount of CO_2 in the atmosphere.

   1. ______________________________

   2. ______________________________

3. In the last 30 years, the natural forests in major metropolitan areas have declined by an average of 30%.

   True

   False

4. Which of the following have trees *not* been proven to do?

   a. Increase property values

   b. Reduce crime

   c. Increase population

   d. Reduce noise

5. According to studies, hospital patients with a view of trees have faster recovery times following surgery.

   True

   False

6. On average, how much CO_2 does a mature urban tree reduce from the atmosphere each year?

   a. 15 pounds

   b. 50 pounds

   c. 115 pounds

   d. 200 pounds

7. Name two ways in which trees reduce runoff and help prevent erosion.

   1. ______________________________

   2. ______________________________

8. How many gallons of water can a 100-foot-tall oak tree capture and hold?

   a. 570

   b. 5,700

   c. 57,000

   d. 570,000

Name ______________________________ Date ____________________

HANDOUT 13.2: How Old Is That Tree?

Using the information below, calculate the ages of at least five different trees.
Sketch the tree and its leaves. Write down any additional observations you may have.

1. Using the tape measure, measure the distance around the tree, or its **circumference.**
2. Find the **diameter** of the tree by dividing the **circumference** by 3.14. Round to the nearest tenth of an inch.
3. Think about it: What math formula is question 2 based on? Write it here. ____________
4. Now, multiply the diameter by the growth rate of the species of the tree. This will give you the approximate age of your tree. *Note:* If you cannot find any of the trees listed, calculate a range of possible ages for your tree using the smallest (2) and largest (7) growth rates.

Tree	Growth Rate per year in inches*
American elm	4
Birch (average)	4.25
Dogwood	7
Hickory	6.5
Oak (average)	4
Poplar	2
Maple (average)	4
Walnut	5
Wild Cherry	3

Example:
You find an oak tree with a circumference of 60 inches.
Dividing by 3.14, you find the diameter to be 19.1 inches.
Multiplying 19.1 inches by the average growth rate of an oak gives you 76.4.
The approximate age of the oak tree is 76.4 years.

**The growth rates listed here (per year in inches) are based on averages for various species within the same family. Your trees may be older or younger than the estimate calculated with this method.*

ONE LAST THING . . .

List two other species of trees in your area that are not listed here. Using the library, Internet, or local nursery, find out what their growth rates are. Then find an example of each of the species and estimate their ages.

Name ______________________________ Date ____________________

HANDOUT 13.3: Trees and You: A Numbers Game

Study the information and questions below. Then answer the questions.

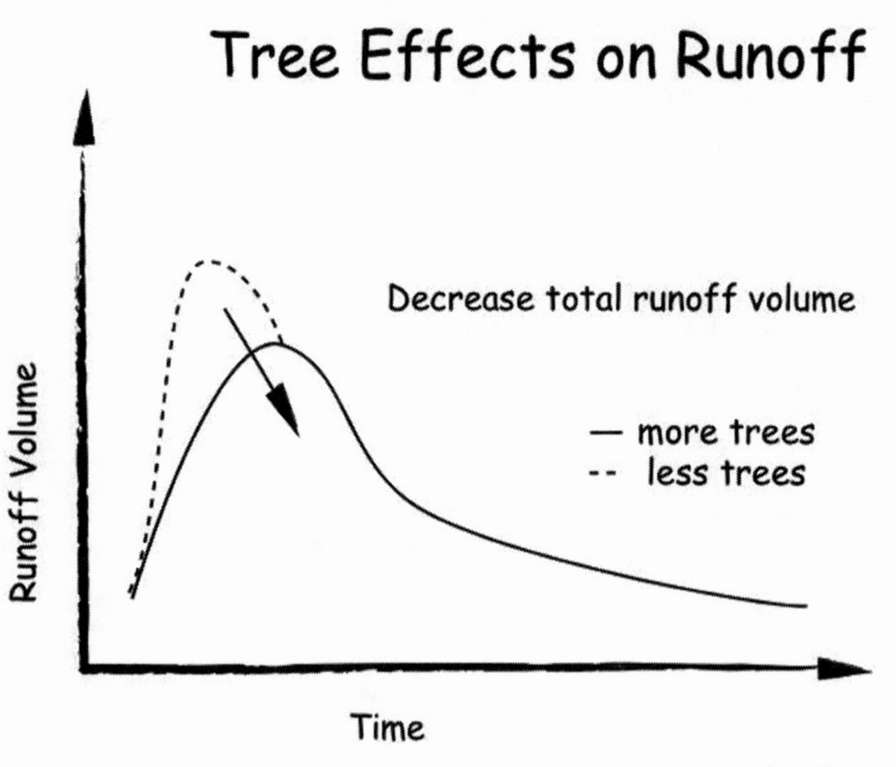

1. Look at the graph above. What are the two variables being plotted on this graph?

   ____

   Can you describe the effect that trees have on runoff as described by this graph?

   ____

   ____

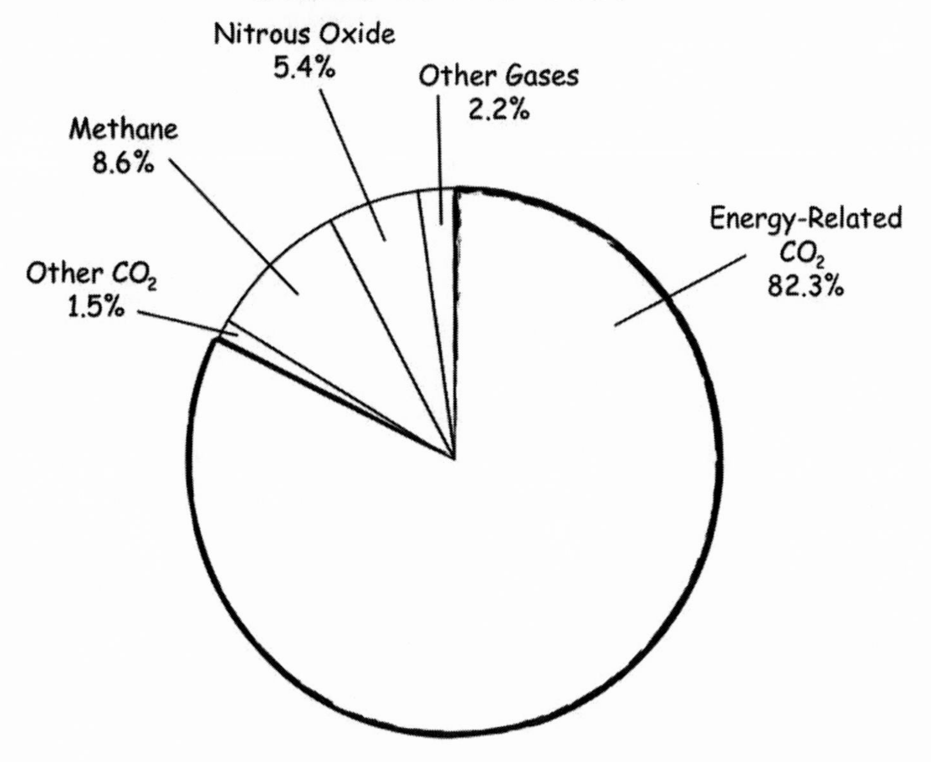

Source: Energy Information Administration 2006

2. Methane makes up what percentage of greenhouse gases? Can you name a source of methane? Use a book to research this, if you need to.

____

____

3. Which is the main greenhouse gas contributing to climate change? What percentage of greenhouse gases does it represent? Can you name three sources of this gas? Again, use a book to research this, if you need to.

____

____

4. Explain what "energy-related" means in this graph. Be specific.

____

____

5. An 80-foot beech tree removes the amount of carbon dioxide produced by two single-family dwellings daily. A specific housing area has 200 homes.

    a. How many beech trees should be planted to help offset the amount of CO_2 those homes generate?

    b. How might that affect the utility costs of the neighborhood?

    c. List as many other benefits as you can that may result from planting these trees.

___

___

6. Trees planted in a spot where they can block prevailing winter winds can lower a family's heating bills 10 to 20 percent. One family's heating bills average $150/month during December, January, and February.

    a. What is the most the family could save each month?

    b. How much is that for all three months?

    c. At this rate, how much would they save over 10 years?

7. Shade trees planted to the east and west of a home can cut cooling costs 15 to 25 percent. One family's average cooling costs are $80/month during June, July, and August.

    a. What is the most the family could save each month?

    b. How much is that for all three months?

    c. At this rate, how much would they save over 10 years?

8. Reducing the need for heating and cooling costs reduces the need for energy and therefore power plants. Thinking along these lines, what are some of the long-term effects—on air, water, soil, ozone—that might result from an extensive tree planting efforts in inhabited areas? Be specific. Write your answer in paragraph form.

___

___

___

___

___

UNIT 14

URBAN PLANNING

TOPIC BACKGROUND

Urban planning is a branch of architecture and design that focuses on the division of the physical spaces where people live. Urban planners design cities. They are as interested in laying down the street design of a new city as they are in determining the best way to redesign an older city so automobile traffic and human activity flow better. In the *Edens Lost & Found* films, students will see that some cities benefited from intelligent urban design while others did not. The most famous example is that of Quaker William Penn, who decreed that Philadelphia would be laid out in a grid pattern with a number of verdant parks sprinkled throughout the grid of city blocks. No matter how much of the natural land the growing city developed, it always had to keep some land undeveloped. Calling Philadelphia his "Green Countrie Towne," Penn believed that open spaces allowed air to circulate and reduced the chance of disease. Much later, Philadelphians wisely set aside a vast tract of land as a buffer to protect their drinking water from contamination. Today this "buffer" is Fairmount Park, the largest city park in the United States.

Architect Daniel Burnham saw Chicago as a "Paris on the Prairie." He dreamed of building fabulous public monuments like the ones seen throughout Europe, particularly France. He envisioned wide boulevards, public fountains, and a glorious central city hall with a dome. The cornerstone of the "Burnham Plan," as it was called, was a ring of forest preserves encircling the city and green space along the city's lakefront. Visitors and native Chicagoans can thank Burnham every time they drive up beautiful Lakeshore Drive.

The moral of the story is that when urban planners carefully think about a growing city's welfare, they are bound to incorporate green spaces as part of the city's design. When they don't, the city is bound to grow up with some problems.

In the Los Angeles film, viewers will see that this great American city—famed for the motion picture industry—was designed for easy automobile access. Early city planners didn't think they had to make space for huge swaths of parks or wild nature because the dream of Los Angeles was being able to hop in your car and drive off to a pristine destination. As a result, 70-odd years later, Los Angeles has the lowest amount of park space per person of any major city in the United States. When dangerous floods occurred in the early 20th century, the Army Corps of Engineers encased the Los Angeles River with concrete as a way of ensuring that heavy rains would forever wash out to sea. Today, deprived of a naturally absorbent riverbed, the city routinely floods. Instead of capturing precious rain that could be used for drinking water, the city has been designed to drain all that water to the ocean.

Many other wonderful examples of urban planning exist throughout the United States, such as New York, Washington, D.C., and Boston. Incidentally, the "grid" plan so popular in the United States was not invented here. It dates back to 2600 B.C., to ancient cities in Pakistan and northern India. In 1573, when the Americas were being settled by Europeans, Philip II, King of Spain, sent very precise guidelines to the New World describing how streets should be laid out in a grid plan radiating outward from centrally located plazas.

Your students may find it hard to understand the idea of designing a city. To them, and indeed to many adults, cities and suburbs just exist and always have. Many people have never thought about the ideas and planning behind their communities. But the truth is that they are constantly being revised. When a building changes hands, when old roads are repaved, when bus routes are planned, someone is—or should be—carefully making plans about how it will all end up. Today, for instance, urban planners mindful of sustainability seek to boost the "density" of cities. That is, they want to increase the number of essential services—supermarkets, dry cleaners, banks, post offices, and the like—in a place so people don't have to travel far to get their needs met. If people can walk to their destination, they'll use cars less often, and less energy will be used to get them where they need to go.

Purpose

In this unit, students will learn the definition of urban planning and why it is important for the survival and development of current and future cities. By the end of the unit, they will understand the stakes behind urban planning and how a good plan will foster the goals of sustainability. Encourage them to develop a critical eye as they survey their own community for possible design improvement.

GROUP ICEBREAKERS

1. Ask students what they think it means to "plan" a city? What aspects of a city can be planned, and which do they think just "happen"? (Student may not realize that the city grids are planned. They may think that only parks or individual buildings are planned. Help them to understand that whole cities are laid out according to a what city leaders think is important to its citizens.)
2. Ask them if they've ever noticed traffic corners in their town or city that feel poorly designed. Maybe there are no stoplights or stop signs at a busy corner. Perhaps there are few or no crosswalks. Can they suggest ways to correct these problems? (Add stop signs, flashing lights, or regular three-color traffic lights. Some corners might even benefit from separate turn lanes or protected green turn signal.) Congratulate students on this idea and let them know that they have the makings of good city planners.
3. Why do they think city planning is a topic that adults take very seriously in city government? (City planning is taken seriously because it affects a large number of people. Once buildings are built, they cannot be easily taken down.)
4. Ask them to come up with a checklist of important ingredients for a healthy city. (The list might contain such things as shops, sidewalks, roads, parks, gardens, and the like.)
5. Where do they think people get their ideas for laying out cities? (*As a hint:* The long boulevards in Chicago and Washington, D.C., were inspired by famous cities in Europe, particularly Paris.)
6. Can they think of one attribute of their town that makes it stand out from other towns or cities? Is there a way to bring more attention to it? Could it be spruced up in some way? (Answers will vary. Accept all creative answers.)

ACTIVITY 14.1

Our Town: A Closer Look

Materials

pen or pencil
notebook or sketchbook
Handout 14.1

Primary Subject Areas and Skills

social studies, art, critical thinking

Purpose

Students perform a walking tour of a local neighborhood, collect data on visible attributes, and then brainstorm ways to improve that area as a prelude to more discussions and larger projects on urban planning.

It's remarkable how familiar a town or city can become after a given amount of time spent there. Streets blend into one another, block after block, and residents can become blind to the character and uniqueness of a place after seeing it day after day.

When engaging students in the topic of urban planning, it is therefore necessary to get them to look at their own surroundings first, and with fresh eyes. The intricacies of the plans designed by legends such as Olmsted and Penn will be lost on those who haven't sought to really examine what makes a town flow or what makes it click.

This activity is designed to help students take a fresh look at parts of their community, town, or city, and then break the parts down into its basic elements and determine what purpose each element serves.

1. Divide students into groups of three or four.
2. Explain to students that they will do a walking tour of a specific area of their town. They will record highly detailed information about the characteristics of the blocks they survey.
3. Have each group choose a five-block area—or, if that's not possible, a particular street, intersection, or the like—of their choice. Keep a list of these assigned areas, making sure that no areas are duplicated.
4. Students should arrange among themselves to meet at their designated area after school or on the weekend. Alternately, the teacher can arrange

for a class outing, if all groups are studying areas close to each other.

5. Students will walk their designated area and use Handout 14.1 to gather information about the characteristics of the area. As they record their findings, students should remain focused on the *kind* of establishment, building, or service they find, rather than the "name." For example: clothing store, bank, coffee shop, dry cleaners, library, police station, and the like. They should also note the presence or absence of sidewalks and bike lanes, trees, crosswalks, benches, and so on.
6. Have students bring their findings to class and share five suggestions for improving the area they studied with the rest of the class.

Extensions

1. Get in touch with your local public works department or mayor's office. These can be wonderful resources for maps and charts of your area. You may even be able to arrange a speaker for your class or class "tour" of various nearby locales that will make it easier to look at your surroundings with different eyes and glean valuable information from these professionals.
2. Ask students if they noticed any similarities among the suggestions made by the groups. Is your town missing sidewalks in many areas? Is there a real shortage of parks or libraries? Are there particular businesses students feel would liven up their town?

Encourage students to put their thoughts in a proposal to the local planning commission.

ACTIVITY 14.2

Penn's Plan: A Look at Old Philly

Materials

pen or pencil; Handout 14.2

Primary Subject Areas and Skills

science, social studies, math, critical thinking

Purpose

Students study a city plan created by William Penn for the city of Philadelphia, with an eye toward understanding the value of a simple grid pattern for urban organization.

William Penn is considered one of the great city planners in American history. The city he laid out between the banks of the Delaware and the Schuylkill is still considered a monumental achievement in urban planning.

Although the plan looks simple at first glance, a lot of thought went into laying out this grid system. The aim was to ease transportation and encourage commerce, give green space to residents, and allow access to the outside world via waterways.

This activity is designed to get students to look more closely at this "simple" plan and come up with some reasons of their own to explain some of the choices Penn made. They should also think about whether this type of urban planning would work today and if they see it playing a role in the cities of the future.

1. Ask students if any of them has ever been to Philadelphia. If so, what were their impressions of the city? Was it easy to get around? Why, or why not?
2. Distribute Handout 14.2 to students. Explain to students that they will be looking at a copy of William Penn's original plan for Philadelphia and answering questions about it. Tell students they should think creatively about what they're looking at and to try to "get inside" Penn's mindset, rather than look for the "right" answer.
3. After students have completed the worksheet, discuss their answers as a class. What would they have done differently and why?

Extension

Discuss the following or assign as an essay assignment for students: How is Penn's plan similar to towns that exist now? (Many cities today are laid out on a grid pattern.) How is it different? (Not all cities mandate parks as integral part of their plan.) How do students think Penn would fare as a city planner today? What has changed about the way we live now that might make it more difficult to put a plan like his into action? (People tend not to have their own farms in cities, and cars and public transportation complicate all city plans.) How can Penn's plan be incorporated into future sustainable cities, and why would it work? (The value of parks and open spaces cannot be overstated. If we could learn only this from Penn's vision, cities would be better, healthier, more sustainable places to live.)

ACTIVITY 14.3

You're the Planner

Materials

pen or pencil, grid paper
ruler
library or Internet research materials
various town and city maps for reference (see Resources, page 211)
colored pens or pencils
materials for a diorama (if applicable)
Handout 14.3

Primary Subject Areas and Skills

science, social studies, math, art, critical thinking

Purpose

Students participate in a long-term project that is designed to put them in the planner's shoes and give them an opportunity to map out a town of their own, focusing on issues and characteristics that are important to them.

After surveying their own town, looking at Penn's plan for Philadelphia, and reading about Olmsted and Burnham, students should have a clearer idea about what goes into planning a city. Now it's their turn.

Distribute Handout 14.3 to students. Ideally, students should work individually, but they can also work in pairs. Each student should use grid paper, a straight-edge ruler, and colored pens and pencils to create a map of a town of his or her own creation. Streets and municipal buildings should be listed, as well as roadways, housing areas, schools, and open spaces. Each map should have a clearly defined legend and scale. A selection of maps detailing downtown areas should be available for reference.

Extensions

1. For extra credit, or as a separate project, have students create a diorama of their plan. After students have completed their projects, have a show-and-tell in class. Each student should be prepared to answer questions about the choices she or he made. After everyone has shared his or her creation, discuss the following: Were there any particular features that were popular with the majority of students?
2. Consider displaying student results in your classroom or in the school library. Invite local planning commissioners and other municipal leaders and public works professionals to come and critique the designs and participate in a planning Q&A. You could even have them vote on the best plan and award a ribbon to the winning student.

ACTIVITY 14.4

Frederick Law Olmsted's Vision

Materials

Internet or library research materials

Primary Subject Areas and Skills

science, social studies, writing, research, critical thinking

Purpose

As part of discussion about the life of Frederick Law Olmsted, the father of landscape architecture, students brainstorm, then choose and write about a memorable experience in nature, using research to flesh out their anecdotes.

Frederick Law Olmsted (1822–1903) was a famous landscape architect. His primary goal was to make nature accessible to those who were probably not going to venture out on their own into the wild. He designed parks that were extensive but manageable, a place where even city dwellers would be comfortable. His projects brought an aspect of nature into the cities and towns. Some people believe "true" nature is not controlled by humans and would prefer to keep as much wild, untamed land untouched by humans as possible. This activity asks students to think about and write how they think and feel about nature and its role in their lives.

1. As a class, discuss the idea of nature as a part of towns and cities. What role does it play in communities and towns? (Answer will vary; accept all reasonable ones.)
2. Ask students what their experiences with nature have been: summer camp, parks on the weekend, camping or hiking, visiting national parks. What effect did these experiences—or lack thereof—have on their perception of nature and its role in their lives? Remind students that experiences with nature can range from walking to school or having a picnic or even dreaming of going to their favorite national park. Nature is always around you, no matter where you are.
3. Ask each student to pick one memory or opinion

that he or she has about nature—be it a good experience or a bad one, recent or long ago—and write about it.

4. Students should adequately research their subject, no matter what it is, and look for articles, books, or studies that work with their topic of choice. For example: Someone who has a pleasant memory of Yellowstone should be able to discuss the history of that park. Alternately, if a student claims to have had no important experience with nature to speak of, she or he can research studies that have been done about the effects of nature on self-esteem and creativity. Or that student can research and write about a natural place he or she has always wanted to visit but never has. What is it about the place that appeals to them?
5. Encourage students to be creative, as well as thorough. They may wish to present their feelings in a play format, for example, or as a piece of journalism.
6. Once the essays have been completed, they should be shared with the class. Copies of each essay can be distributed to all students. If possible, several minutes of each class period could be allotted for individual students to present their subject and share their experience.

Extension

Consider publishing your profiles in a class booklet or online. This is also an ideal activity to incorporate into a classroom blog.

EXTENSIONS

Old Local Maps

Visit your municipal center or a local library to inspect old maps of your city or town. Have students compare old and new maps to see what changes they can find between then and now.

Go to the Source

Invite a city planner from your local city hall or from a local university to address the class on their profession. If he or she can discuss aspects of sustainability, so much the better.

Emerald Necklace

Many cities dreamed of creating an "emerald necklace" of forested lands around their city. Challenge your students to see how existing parks or woods around their town might be linked to form an unbroken necklace of greenery. What would have to change to make this dream a reality?

Green-Lined Waterways

Rivers, streams, and other small waterways make excellent spots for designated parks. Have students trace a section of waterway in your town or city and have them design a park along this stretch of water. What will they need to do to keep the park green?

RESOURCES

Google Earth

A fascinating free software program from Google allows you to inspect satellite images of any city in the world: http://earth.google.com/

A Layout of Ancient Rome

A clickable, color-coded map of ancient Rome lets you see how the each of the different types of buildings fits into the layout of the old city: www.geocities.com/Athens/Forum/1274/ancient-city-by-type.html

William Penn's Life

An encyclopedia entry on the life of William Penn, Quaker and Philadelphia's founder, can be found at: http://en.wikipedia.org/wiki/William_Penn

Free Design Program

A free software program that allows your students to design on-screen models of buildings, cities, and other 3-D shapes can be found at: www.sketchup.com

Grid Plan of Major Cities

An encyclopedia entry on the history and evolution of city grid plans, with numerous links to city examples and maps, can be found at: http://en.wikipedia.org/wiki/Grid_plan

Fairmount Park

The official website to Philadelphia's famous Fairmount Park provides a history of the park's origins and its function as a watershed: www.fairmountpark.org/

How Washington, D.C., Came to Be

Two thorough online articles about the designing and building of Washington, D.C., the nation's capital, can be reached at http://xroads.virginia.edu/~CAP/CITYBEAUTIFUL/dchome.html and www.ncpc.gov/about/histplann/histplann.html

Chicago's Designer

Learn the story of the famed architect who designed downtown Chicago at http://en.wikipedia.org/wiki/Daniel_Burnham

Los Angeles's Master Plan

The story and images of the ill-fated 1930 city plan for Los Angeles that never came to be can be found at two websites, www.calnaturalhistory.com/books/pages/8995/8995.report.html and www.arlisna.org/artdoc/vol20/iss1/02.pdf

Reading Handout Unit 14:
WILLIAM PENN'S "GREEN COUNTRIE TOWNE"

What is the value of good city planning? Does it really matter if a city is laid out intelligently, with ample space for man, nature and industry? William Penn thought so. In this reading, you'll learn about the founder of Philadelphia and his early city planning ideas. The Philadelphia film in the *Edens Lost & Found* DVD set provides further information and examples. After you read the story, answer the questions.

William Penn came to the New World in 1682 to build a new kind of city. In his hands he clutched a charter from King Charles II of England, granting Penn a massive parcel of land west of the Delaware River. Penn dreamed of a place where people could live free, regardless of their religious beliefs. Back home in London, he and his Quaker brethren had been harshly persecuted for their religion. Penn had caused trouble for himself and his staunch Anglican family ever since he began spouting his ideas in his 20s. His father's position as an admiral in the Royal Navy may have protected the young man until now, but family connections went only so far. To young Penn, the King's offer must have seemed like a godsend, proof that his convictions were finally bearing fruit. The King no doubt saw it another way: Finally, London would be rid of this troublemaker once and for all.

Once in America, Penn embarked on his "holy experiment" in a place called Pennsylvania, or "Penn's Woods." The city he founded, Philadelphia—"the city of brotherly love"—was the first in America to guarantee all citizens equal rights under the law, regardless of race, gender, or religion. As governor, he proclaimed his lands open to free press, free enterprise, trial by jury, education for both men and women, and religious tolerance. Penn's progressive vision included even the local Native Americans, whom he won over without weapons and to whom he insisted on paying a fair price for their land.

Today, Penn's words seem like common sense to our ears because he summed up what America was all about, although almost 100 years would have to pass before his ideas would be echoed in the Declaration of Independence, written and signed in the city he founded.

Penn was ahead of his time, not only in his notion of liberty but in his environmental concern and city planning as well. He laid out his city in a simple grid pattern, each block resembling a square. Penn remembered that wooden homes back in England burned easily, so he mandated that houses in his city be built of brick. He stipulated that owners build their homes in the center of their lot, to allow room for gardens on each side. He would never forget the squalid slums of London and worked hard to avoid replicating them in his new paradise. If you had enough green space, he believed, you could prevent the disease that ran rampant in big European cities. And so he laid out grand parks in his city—the first public ones in North America—and encouraged all homeowners to tend their gardens. He called Philadelphia his "Green Countrie Towne," and called for one acre of trees to be left untouched for every five acres that were cut down.

From that moment on, Philadelphia became the intellectual and horticultural center of the New World. Philadelphia's John Bartram, the first horticulturist in America, built the first botanic gardens and sent exotic New World species back to England. For a time even Ben Franklin financed some of Bartram's expeditions. Still later, when President Thomas Jefferson sent Lewis and Clark on their western journey, the pair departed from Philadelphia and sent all their specimens back for study at Philadelphia's Academy of Natural Sciences. In 1819, the city began work on a series of beautiful classical buildings at the edge of the Schuylkill River that would become the Fairmount Waterworks. Later in the century, as sections of the river became a dumping ground for industry, the city realized that the only way it could ensure the purity of the drinking water was to set aside a massive plot of land as a watershed. This 4,180-acre holding, just up the river from the Waterworks, became the largest landscaped park in the United States. The park—which today numbers 9,100 acres—was a commitment to the environment early in the city's history.

All this shows that Philadelphia had a long legacy of incorporating nature into the fabric of city life. But why does that matter?

Penn's city went on to become the great intellectual center of our then-young nation: the seat of America's Enlightenment; the city of the Liberty Bell, the Declaration of Independence, and the birthplace of American freedom. It was also a major manufacturing hub. After World War II, however, middle-class families fled American cities in droves, believing that urban areas had become too big, too noisy, too expensive, or too crime-ridden and unsafe.

In 1950, Philadelphia's population peaked at 2.1 million. Today, it stands under 1.6 million. More than 500,000 people just picked up and left and haven't returned. The outlying suburbs grew while the city shrank. In Philadelphia, the loss of those people, their incomes, and their potential tax revenues was devastating. Poverty, crime, and drugs soon filled the void.

Philadelphia has struggled in the last 50 years to stop the slide and turn itself around. Despite the city's problems, thousands of people moving back into its downtown area believe in its future. With its solid and historic architecture, its excellent location along the Northeast corridor, an extensive public transportation system, and bustling cultural scene, the city is poised for a rebirth.

Environmentally, Philadelphia is already far ahead of other cities. Because the city incorporated a huge park and many little parks into its design when it was still young, citizens there now enjoy green spaces and wilderness right in the heart of town. And the big park system continues to serve as a giant filter that purifies the city's water.

These are wonderful efforts, but a massive challenge remains. Philadelphia must remake its old inner-city neighborhoods, which have been hit hardest by the downturn in fortunes in recent decades. Positive signs are emerging. Builders are building new houses, new small businesses are popping up, and vacant land is being turned into attractive spaces with grass and trees surrounded by wooden fences. In many neighborhoods, residents are working together to make community gardens and reclaim derelict parks.

For guidance and inspiration, they are looking to the past. Philadelphia's salvation appears to lie in the dreams of the idealistic Quaker who first breathed life into it. Today's heroes and visionaries—today's William Penns—have hit upon a winning formula:

To clean the city, to dream the city, you must green it.

Questions

1. Why was William Penn persecuted in England?

2. What was special about his new colony, Pennsylvania?

3. What rights did he grant to most people who lived there?

4. What ideas of Penn's later found their way into the Declaration of Independence?

5. Why did he suggest that houses in Philadelphia be built of brick?

6. Why did he urge citizens to build their homes in the middle of their square lots?

7. Why do you think he urged citizens to leave one acre of wild land for every five they cut down?

8. Why was Fairmount Park ahead of its time? What special function did it provide for the city of Philadelphia?

9. The author seems to think modern-day urban planners in Philadelphia owe a great deal to William Penn. Why might that be? Do you agree?

____

____

10. Modern historians point out that Penn owned and traded slaves. He believed that masters should treat their slaves with kindness. (In his day, and after, Pennsylvania was home to many Quakers who were among the earliest antislavery proponents in America.) Do you think this information about Penn changes the way history should view his accomplishments?

____

____

Name ______________________________ Date ______________

Names in Group ______________________________________

HANDOUT 14.1: Your Town: A Closer Look

1. What area are you surveying? (Be specific: intersection of, and so on.)

____

2. Walk up and down the streets you've been assigned and list all of the establishments you see according to the purpose they serve (bank, dry cleaner, police station, and so on).

____

____

3. Now list all other characteristics you notice (parks, trees, sidewalks, and the like).

____

____

4. Is there anything in particular that you notice a lot of?

____

____

5. Is there anything in particular that feels missing or that there should be more of?

____

____

6. Draw a sketch of the area you studied on a separate piece of paper.

7. List five suggestions for improving the area you studied.

____

____

____

____

____

Name ______________________________________ Date ______________________

HANDOUT 14.2: Penn's Plan: Old Philly

Study this map of William Penn's plan for the city of Philadelphia. Then answer the questions.

Penn's Plan: Old Philly

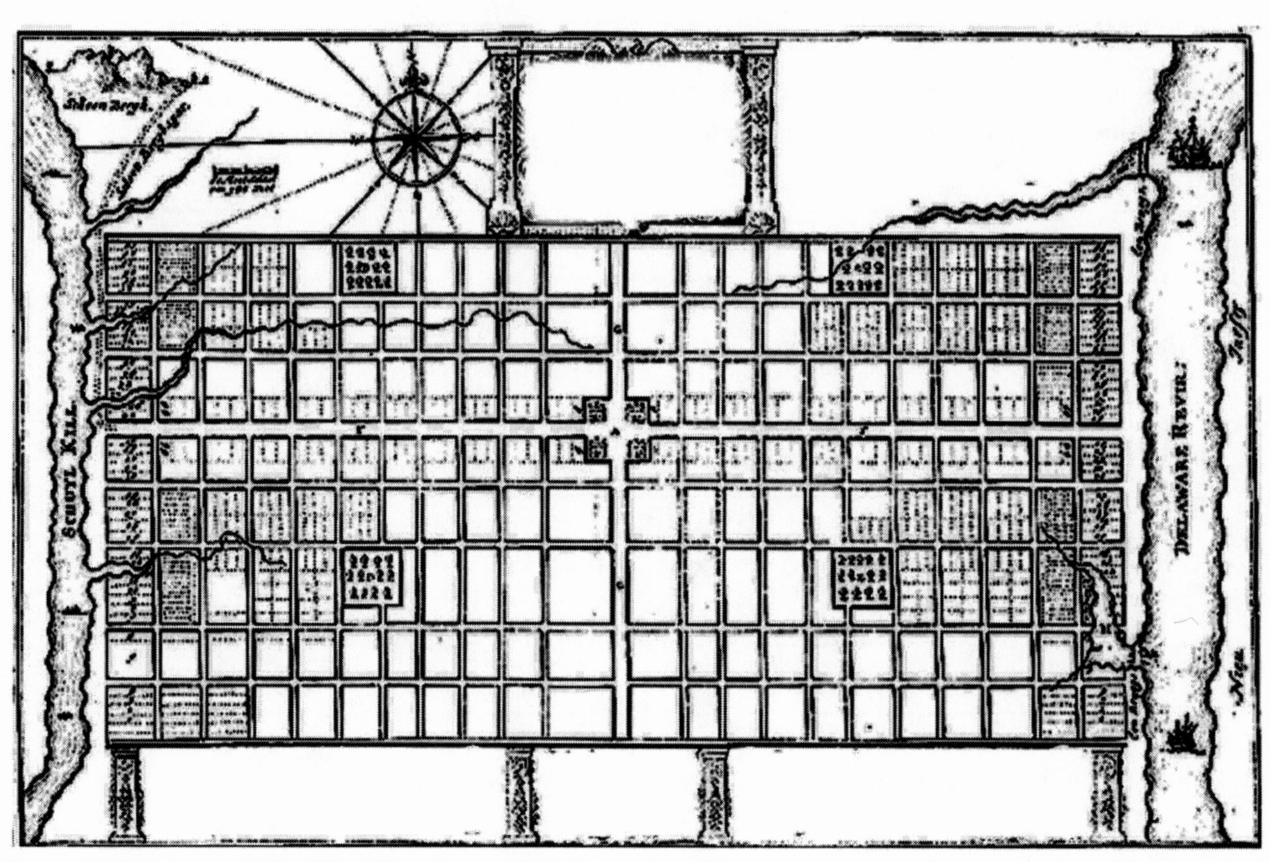

Source: Adapted from image from The Library Company of Philadelphia.

1. How many main sections does the city appear to be divided into?

____

2. What do you think goes at the center of this map?

____

3. What two objects shown on the map were not made by humans and forced Penn to build his city around them? Where are they located? Why do you think this spot was chosen for a city?

____

____

4. Why is it helpful to have the streets laid out in small blocks, the way they are?

____

____

5. Are all the blocks the same size? Why do you think this is? What do you think was planned for the larger blocks? For the smaller blocks?

____

____

6. What aspects of today's towns and cities still adhere to some of Penn's city plans?

____

____

7. Why might a plan like Penn's not be workable in a modern city?

____

____

HANDOUT 14.3: Checklist

You and your teammates, if any, will design your own plan for a town or city of your own creation. Your design must contain all of the following items:

❒ Give your map a title—the name of your newly created city.

❒ Be sure to include a legend that explains what different colors, lines, and so on, represent.

❒ Include a scale.

❒ Include three police stations, three fire stations, two libraries, six schools, public spaces, parks, greenways, and administrative buildings (city hall, and the like).

❒ Include services such as restaurants, homes, and businesses. Indicate where these important things will be laid out.

❒ Decide and mark: Are streets one-way, two-way? All the same size? Paved? Dirt?

❒ What about pedestrians and cyclists? Is there public transportation? Where do people live? Where do they work?

UNIT 15

POPULATION GROWTH AND INTEGRATED RESOURCE MANAGEMENT

TOPIC BACKGROUND

Everyone knows Earth's population is growing. Every year more and more people are born. At this writing, 6.5 billion people live on planet Earth. The world's two largest nations are China (1.3 billion) and India (1 billion). The United States is third on the list, with only 298 million people. Those numbers are deceiving, because although the United States has only 5 percent of the world's population, Americans consume 25 percent of the world's resources. This is why we Americans have a special obligation to live more lightly upon Earth. *If everyone lived as Americans do, we would need four more planets to support human life.*

The more people we have on the planet, the more land, food, water, and other resources we consume. Population Connection, an organization devoted to educating people about population issues, lists the following sobering statistics, which you may wish to share with your students:

> *The world's population has grown faster since 1950 than the previous four million years. As humans stress the planet more and more, we harm the environment. Eighty percent of the rainforests have been cleared or harmed in some way by human expansion. Each year, 27,000 species of animals die off. Right now, 505 million people live in water-troubled areas; by 2025, 48 percent of Earth's humans will be living in places where there are water problems. (Some American cities already face water shortages.) Fresh water is relatively scarce to begin with. Of all the water on Earth—more than 70 percent of the planet's surface—just three percent is fresh water, but most of that is trapped in the icy poles. Less than one percent is fresh water, suitable for drinking and readily available to us. Another 40 percent of all water that is stored underground is polluted and very difficult and costly to purify.*

What does all this mean? It means all humans must use our resources more intelligently. One way of doing this—integrated resource management (IRM)—is explored in greater detail in this unit. Simply put, IRM is a way of carefully studying the whole system of resources available to a community and figuring out a way to use them with the least amount of waste. This practice is slowly catching on across the nation, as larger American cities and suburbs are realizing that they can no longer afford to spend millions or billions of dollars every year cleaning water that has become contaminated or lost through wasteful practices. For example: Because communities can no longer afford the high cost of hauling grass clippings and other yard waste to dumps, many U.S. states have banned this so-called "green waste" from public landfills and insist that towns and cities compost it.

Admittedly, this is a sobering theme to set before young people, who have done little to cause these problems but must nevertheless inherit them. To teach these lessons in a more sensitive manner, we provide activities that focus on actions we can do to help the problem. The issue of population growth is implied through much of the *Edens Lost & Found* films, but it is most apparent in the Los Angeles film. Urban sprawl is one of the major issues in the American west; each year new suburbs consume more wild land than our population growth seems to warrant. In the Los Angeles film, a dedicated group of citizens employed integrated resource management to solve, in one fell swoop, a host of different problems, from water contamination and loss to the lack of employment opportunities. Remarkably, they did this by modeling their communities on wild forests—the finest example we have of ecosystems that perpetuate themselves forever without needless waste.

Purpose

Students will learn and understand the important issues surrounding Earth's growing population and be able to explain why integrated resource management can be an effective tool to check unsustainable practices. With the activities in this unit, they will be prepared to become effective spokespeople for change in their communities.

GROUP ICEBREAKERS

1. Ask: "How many people do you think live on the planet?" (6.5 billion.) "Of these, how many are Americans?" (298 million.) Tell them that this figure is about 5 percent of the world's population.
2. Here's a good way to imagine this figure: If an ordinary 12-inch ruler represents all the people on the planet, Americans would take up approximately the last 1/2 inch of the ruler. The rest, about 11 1/2 inches, would be all the other people on the planet. You might pass a ruler around the classroom to have students visualize this.
3. Next, ask: "How much of the world's resources do you think Americans consume?" (25 percent.) Now, if the ruler stands for all the world's resources, this figure is 1/4 of the ruler, or 3 inches. That means Americans consume 3 inches of the ruler every year, leaving the remaining 9 inches to be consumed by the rest of the world. You may also wish to demonstrate this by dividing a large, soft cookie into 12 equal pie-shaped pieces. First show the students how much the United States would consume (three of the 12 pieces). Then take just one of those pieces and divide it in half. Explain to students that *this* one tiny piece represents our portion of the world's population.
4. On the board, draw a picture of Earth as best you can, then ask: "If everyone on Earth lived the way Americans do, how many Earths do you think we'd need?" (Five Earths.) Don't tell them the answer just yet. Allow them to guess. Draw one more Earth and say, "How many of you think it's one more Earth?" Draw another and say: "How many think it's two more?" And so on.
5. If they think this scenario is too far fetched, tell them that two of the world's fastest-growing economies are China and India, which have the largest populations on the planet. (1.3 billion and 1 billion people, respectively.) Say: "If more and more people in China and India are able to afford cars, do you think that will increase or decrease climate change?" (Increase.)
6. Have them suggest ways all humans could work together to reduce consumption of the world's natural resources. Say: "Is there anything we can do right here?" (Accept a variety of answers, such as recycle, reuse, and reduce goods consumed, drive more efficient vehicles.)
7. Introduce the concept of integrated resource management and follow up with the unit reading, which has diagrams to illustrate this potentially confusing concept.

ACTIVITY 15.1

Pop Goes the Population

Materials

pen or pencil
Handout 15.1
Reproducible 15.1 for overhead

Primary Subject Areas and Skills

science, social studies, math, critical thinking

Purpose

Students read a table and multiline graph to learn about the world's population and how it will continue to rise.

One of the hardest things to gauge is just how quickly the population is growing and what effects that might have on our society and the way we live our lives. When discussing population, it is important to discuss the amount of resources that are being used *per capita* by various countries.

This activity is designed to help students think about the number of people in the world and how that number continues to rise at an exponential rate, despite global conditions and lack of resources.

1. Ask students, in general, what they feel about the world's population. Is it growing too quickly? (Answers will vary. Be prepared that they will have no idea how fast the population is growing.) Do they think there will be enough resources when they are age 70 or older? (Answers will vary. Most will guess that there will not be enough resources when they are age 70 or older.)
2. Copy Handout 15.1 and distribute to students. Instruct them to study the graphs and answer the questions.

3. After completing the activity, have students self-correct as you read the answers aloud.
4. Ask students if they were surprised by any of the information.
5. What did students think about the U.S. population as compared to other countries? (Answers will vary. Most will point out that though the United States is in the top 3 in population worldwide, it still has markedly fewer people than China and India.) Remind them of the stats they learned at the beginning of the unit, which noted that Americans consume 25 percent of Earth's resources. Do they think the United States consumes more or less, per capita, than other countries? (More.) How do they feel about this? (Answers will vary.)
6. Next, display Reproducible 15.1 on overhead, which shows predictions for world population growth until 2050. Explain to students that the numbers on the graph are expressed in millions. So, for example, 1,600 million is 1,600,000,000, which is 1.6 billion.
7. Ask students the following questions.

Questions for Reproducible 15.1

1. How old will you be in 2050? How many people are predicted to be living in the United States then? (Answers will vary for the first question. By 2050 the U.S. population is predicted to be 400 million.) How many more people is that than the number living in the United States today? (About 100 million more people.)
2. Do all countries grow at the same rate? (No.) How do you know? (The steeper the line on the graph, the faster the growth rate.)
3. Which countries show signs their population might be leveling off? (Indonesia.)
4. How many more people are projected to be living in India than in the United States in 2050? (1,200 million, or 1.2 billion more.)
5. Is it predicted that any countries will show a decline in population? (Yes.) If so, which ones? (China.) What might be the reason for that decline? (China has taken serious steps to decrease its population.)

ACTIVITY 15.2 •
Growing, Growing . . . Gone?

Materials

pen or pencil; Handout 15.2

Primary Subject Areas and Skills

science, social studies, math, critical thinking

Purpose

Students use a simple graph and calculate proportions to illustrate the carrying capacity of a group of organisms.

Carrying capacity is a term used to describe how many living things an ecosystem can support, given its resources, without negative effects. In general, if a group of living things is existing *below* carrying capacity, the population will increase. If a group of living things is existing *above* carrying capacity, the population will decrease. The ecosystem tries to maintain equilibrium, and if that is thrown off, it can damage the organisms living there, as well as the environment itself. Students will likely be generally acquainted with the idea of overpopulation in the animal world, especially if they live in areas where, for example, there are too many deer.

There are a number of examples in the animal world that illustrate this, and carrying capacity is a hot topic among economists, scientists, and government leaders. Some people believe we are already at or beyond carrying capacity, while others do not. Can technology provide answers to resource management for an exponentially growing population? Or are we headed, as many believe, to a future of severe water and food shortages, in addition to an energy crisis?

1. To jump-start class discussion, write the following on a board or overhead projector:

$$\text{Human Condition} = \frac{\text{Available Resources}}{\text{Population}}$$

2. Ask your students to consider what they think this equation means and to write at least three sentences describing their thoughts.
3. Ask several students to share their thoughts and encourage other class members to comment.
4. Distribute Handout 15.2. Tell students to complete it on their own.
5. Go over the answers in class. Did the answers affect how they think about the equation above?

(Answers will vary. In general, they should begin to see that as population rises, the more resources must be shared, leading to a more uncomfortable state for us all.) Do they think the equation is an accurate reflection of what happens when populations grow out of control? Why, or why not? (Answers will vary. But here are the two predominant positions on this issue: Yes, it is accurate; the more people we have on Earth, the worse off we will all be. No, it is not accurate; humans are clever and will begin to solve the population problem when the situation gets intolerable. We will conserve resources or discover new ones.)

Extension

For a more advanced discussion, bring the ideas of "limiting factors" and "carrying capacity" into the analysis of the Boomington graph. Point out that certain limiting factors—such as lack of food, space, clean water—probably affected the fish population in Boomington. What limiting factors could affect the human population on Earth? (Frankly, humans may well face the same problems: lack of food, little space, lack of clean water, depleted resources.)

ACTIVITY 15.3

Greening Your School

Materials

pen
notebook or sketchbook
resources listed in unit directory
Reproducible 15.3A–E for overhead

Primary Subject Areas and Skills

science, social studies, math, art, critical thinking

Purpose

Students will study some "before" and "after" photos of a school whose grounds were redesigned in line with the principles of integrated resource management with the goal of conserving rainwater, eliminating toxic runoff, reducing the urban heat island effect, and providing shade for students and the indoor school areas. Student will then design a retrofit for their own school grounds. By the end of the activity, they should be able to explain their design decisions in light of integrated resource management and be able to answer the question: What problems does the redesign solve?

The Reading in this unit focuses on the city of Los Angeles and its plan to solve many of its citywide problems by planting more trees, installing cisterns, and designing landscaping to capture as much rainwater as possible. When a group or agency makes decisions in a unified way that helps them manage vastly different resources for long-term stability or sustainability, the approach is called *integrated resource management.* In this case, the resources being managed are water, air, soil, plants, and energy. As illustrated in the Reading, trees, cisterns and creative landscaping can solve a number of problems. Many of these initiatives are taking place on the grounds of California public schools. School properties are often large. With a little creativity, they can be turned into small parks that give back to the community as a whole.

The activity we outline is designed to be theoretical—a real retrofit never has to happen—but it is always worth knowing your options before presenting the idea to your students. At the very least, you may be able to plant a wall of trees in a spot perfect for shading certain classrooms on hot days.

Note: We encourage you to confer first with your principal or supervisors to gauge your school's willingness to permit, at the very least, an "official" presentation by your students of their ideas to the school administration, PTA, or Board of Education. It's entirely possible that your school district has been thinking of ways to save money on monthly expense such as heating, cooling, water, and the like. The ideas outlined here may actually help them accomplish those goals.

1. Use the Reading earlier in this unit to review the definition of *integrated resource management.*
2. Mention that California's new approach toward sustainability is being implemented in public schools. Begin by showing Reproducibles 15A and 15B. (*Note:* Color versions of all of these reproducibles are available on the curriculum website.) Ask: "What changes do you notice in the 'before' and 'after' photos of these three schools?" At first glance, student will assume that they simply added some grass, fields, and trees. Ask: "What are the cons of having asphalt in warm climates with water issues?" (Asphalt retains heat, contributing to the heat island effect and generally warming up the entire city or town. Rainwater runs right off asphalt, creating toxic runoff, washing into public water systems, or being washed out to sea.) Ask: "What are the pros of having green surfaces and trees?" (Grass,

fields, and trees absorb heat and help shade and cool the school grounds, ultimately cooling the city or town. They also absorb rain, reducing toxic runoff and contamination of public water supplies.)

3. Next, before showing Reproducible 15.3C, say: "Another important feature of this new landscaping is something you can't see—because it's underground." Show Reproducible 15.3C. Work through the cistern installation diagram with them and help them understand how cisterns work and their benefits. (See Reproducible 15.3C for benefits.) You might ask if they can think of any cons of having more grass, fields, and trees. (See Reproducible 15.3C ,"trade-offs." In general, greenery requires more maintenance than asphalt. Depending on your perspective, this may be a pro or con. It adds jobs to the community, but it also adds costs because maintenance workers' salaries need to be paid. Refer to the Unit 15 Reading and remind them that Los Angeles was interested in creating jobs. Also, leaders felt that the pros outweighed the cons. Mr. Lipkis calculated that the amount of water saved in a year in one Los Angeles neighborhood alone would be $200 million!)
4. Next, show Reproducibles 15.3D and 15.3E, depicting "before" and "after" images for the same school campus. Here, they can see the complete change in an entire school campus. Work slowly through Reproducible 15.3D to note the changes. (In general, asphalt has either been replaced with fields or "softened" with trees and shrubs. Notice that the faculty parking lot has been redesigned with trees on the edges, and a small grove of trees can be used as an outdoor classroom. They installed a cistern under the large field. Between fields and buildings they installed swales, indicated in the photo by another name—*riparian edge*. This particular school had a garden before and after.)
5. Ask: "How does all this work reflect the goals of integrated resource management?" (The schools were redesigned to conserve not just one resource but several at a time, among them air, water, soil, energy, and the community's financial resources. Because water is being saved, they don't have to spend as much to buy it. Because runoff is reduced, they don't have to spend money to clean up polluted streams, rivers, and reservoirs. Jobs are created.)
6. You may choose to end your lesson here. If you have additional time, challenge students to come up with a redesign of their own campus that reflects the goals of integrated resource management. They will need a map of the current school grounds, and some art supplies to create a sketch similar to Reproducible 15.3E. Be sure they carefully list the pros and cons of their concepts and clearly state what special needs your school has. You may wish to have them present their ideas to local officials or your school administration.

ACTIVITY 15.4
The Population Debate

Materials

pen or pencil
notebook or sketchbook
Handouts 15.4A–B

Primary Subject Areas and Skills

social studies, science, research, critical thinking, public speaking

Purpose

Students participate in a mock summit on population issues, role-playing as delegates from various different nations.

This activity is designed to put students in the shoes of individuals from countries other than their own to try to view the population debate from as many perspectives as possible. This is also a wonderful way to encourage debate about policy and the role that government plays in the lives of its citizens. It also illustrates that when it comes to the environment, population, and preserving resources, there are no easy answers.

1. Explain to students that they are going to participate in their own population summit as members of different countries.
2. Students will work with a partner (chosen or assigned), and groups must have no more than three people. All members of each group must complete Handouts 15.4A and 15.4B.
3. Assign each pair or group of students a country to represent at the Human Population Conference. For simplicity's sake, choose from the list of nations listed on in Handout 15.1.

4. First, instruct students to use Handout 15.4A to research their country. Next, each nation will complete Handout 15.4B, using it to solidify their positions on each mandate. At the end of their research, they will present their findings to the class.
5. Students should include at least one visual aid or chart in their presentation.
6. Remind students that they are to answer questions and comment on the mandates *as members of their assigned country.* They must be prepared to answer questions about their opinions and priorities based on what they know about the country they are representing—not what they, themselves, think.
7. At the end of their presentation, each group will present a list of recommendations they propose their country abide by to address population concerns.

Extension

After the discussion, ask students to write an essay about what they learned and what they, personally, think about the population crisis. How do they think so many different countries with so many different needs and priorities can come together and face these issues?

EXTENSIONS

Stand Up and Be Counted!

Go to town archives and research the history and trends of your town's, county's, and state's population. Is it on the rise? Declining? Graph your results and analyze the data based on economic growth, emigration and immigration, cost of living, and so on.

Count on Your Feathered Friends

The Audubon Society's Christmas Bird Count is a bird census that takes place from December 14 to January 5 each year. Volunteers follow specified routes through a designated 15-mile-diameter circle. They count every single bird they see or hear all day. Become a part of it as a class or community. More information can be found at: www.audubon.org/bird/cbc/index.html

Integrating Your Resources

Challenge students to design their own model of integrated resource management using the Venn diagrams shown in the Unit Reading. Can they come up with their own Venn diagrams to illustrate a potential solution of a problem currently plaguing their school or town?

RESOURCES

Interactive Graph of World Population

Wondering how the population has changed in your state or in, say, Portugal? Use this handy interactive graph to track population changes since the U.S. Industrial Revolution: www.usatoday.com/news/graphics/300million_popchart/flash.htm

GeoHive

GeoHive is one-stop shopping for information about population and resource reserves and consumption. All of the material is free to use: www.xist.org/charts/population1.aspx

Population Connection

The people at Population Connection (formerly known as Zero Population Growth) have a variety of tools and resources that help educate young people about various actions needed to stabilize world population at a level that can be sustained by Earth's resources. Wonderful fact sheets are available at: www.populationconnection.org/

U.N. Population Fund

The U.N. Population Fund focuses on the relationship between population and development: www.unfpa.org/

Population Statistics

"The State of World Population 2005," as discussed by the U.N. Population Fund, can be found at this site, which includes news features, charts, photos, studies, and more focusing on the state and future of the world's 6 billion inhabitants: www.unfpa.org/swp/swpmain.htm

U.S. Census

The U.S. Census Bureau's population clocks feature up-to-date tracking of birth and death rates, as well as population, in the United States and around the world: www.census.gov/main/www/popclock.html

Reading Handout Unit 15: WATER IN, WATER OUT

Did you ever hear the expression, "The left hand doesn't know what the right hand is doing"? Often in big cities, government agencies may enact policies that directly contradict the policies of one or more other agencies. In this reading, you will see how Andy Lipkis, founder of an organization called TreePeople, convinced the city of Los Angeles to save water by employing the principles of *integrated resource management.* You can learn more about Mr. Lipkis's work in the Los Angeles film in the *Edens Lost & Found* DVD set. After you read the story, answer the questions.

A forest is sustainable because each of the processes that exist in the forest work together harmoniously. Leaves fall, rot, and become food for the trees that shed the leaves. Each time it rains, water is captured by the thick, spongy layer of rotting leaves and by the roots of trees. The spongy layer of rotting leaves, or *humus,* is perfectly suited to the needs of seeds and nuts that fall to the ground. In an unsustainable system, natural processes don't work well together. Sadly, this is what has happened in Los Angeles and in many other cities.

Andy Lipkis is a man who has worked to save California's trees since he was a teenager. Recently, he has become interested in the way his city manages its water. Los Angeles is an ever-growing city whose demand for water is greater than its supply. Los Angeles could meet half of its water needs if it could somehow capture and retain its rainfall each year. But since two-thirds of the city is paved, concrete and asphalt cover a living, functioning ecosystem. Each year, 85 percent of that water slips out of the city's grasp, washed out to sea and never seen again. "Because we don't save that water," Mr. Lipkis says, "we have to buy water that falls as rain in Salt Lake City. It's the height of absurdity."

The more Mr. Lipkis studied the problem, the more he saw waste. For example, half the water used in Los Angeles is for irrigation—watering plants and lawns—not drinking, cleaning, or showering. Forty percent of the garbage the city sends to the dump is green waste—grass clippings, tree branches, leaves—that could easily be mulched and spread on plants. Green waste has the potential to mitigate, prevent, or lessen flood conditions. In the forest, as we saw, green waste isn't waste at all. It is the essential factory of nutrients, which acts as a giant sponge lying at the floor of the woods, absorbing and purifying water. But if you remove from your property everything you cut, rainwater rushes off your property. If people mulched all that waste instead, their city would save all that money driving big trucks around neighborhoods picking up clippings and hauling them back to a landfill.

TreePeople's twentieth anniversary occurred close to a pivotal event in Los Angeles's history—the Rodney King trial and subsequent riots. Sociologists looked at the problems leading up to the riots and concluded that the city needed about 50,000 more jobs for urban youth. The city was at a loss as to how to create those jobs. The most intelligent estimates said the city would need $500 million to make those jobs a reality. It was a whopping figure, seemingly too big to ever become a reality.

Mr. Lipkis wanted to help but didn't know how. After the riots, he read in the papers that the Army Corps of Engineers was planning to raise the concrete walls of the Los Angeles River yet again to prevent flooding.

The high walls formed a basin that collected rainwater and sent it out to sea. "This is insane," he thought. "We should be taking those walls down." When Mr. Lipkis read further, he saw how much the city was about to spend on those walls: $500 million!

Lights went off in his head. If the city could come up with a more sustainable solution to its water problem, it would have all the money it needed to help employ its citizens. "We're hurting people because we're spending that money on water," he thought to himself.

On a visit to Australia, Mr. Lipkis saw how citizens used cisterns, or underground tanks, to capture runoff on that parched continent. Cisterns are hardly a new concept. The Greeks and Romans used them more than 2,000 years ago. They carved chambers into solid rock and built their homes or gardens above them. When they needed water, they pumped it out of the cistern. Modern cisterns can be enormous, capable of holding enough rainwater to fill a swimming pool. Back home, Mr. Lipkis started designing his own cisterns.

He sat in on Los Angeles county water meetings. At the time, the county government had one agency responsible for flood control, and another responsible for buying water from other cities to slake Los Angeles's thirst. The flood folks were in charge of getting rid of rainwater as fast as possible. Neither agency had ever thought about *capturing* rainwater. The idea of installing millions of cisterns all over the city or building swales to absorb the water into the ground seemed strange and expensive. It was cheaper to keep doing what had always been done.

If you install cisterns, Mr. Lipkis told them, they will be expensive, but you will have water *forever*. If you keep doing what you've always been doing, you'll *always* need to buy water—at the cost of billions of dollars a year.

Mr. Lipkis wanted to show everyone that trees, swales, and cisterns would be cheaper in the long run. He formed a team of experts—scientists, economists, and others—who designed a "natural" redesign for the city of Los Angeles. As a test subject, his group picked a small, single-family house in South Central Los Angeles.

Mr. Lipkis's crew descended on an 80-year-old house in the neighborhood and installed a "dry well" at the end of the owner's driveway. Rainwater rolls down the driveway to a small grate and falls into a chamber to be filtered and cleaned, then released into the ground. Drain spouts, which normally expel rain into the street, were repositioned to send water into the owner's newly landscaped garden. The edges of the front and back yard were raised to form a bowl shape, as in Fig. A.

This method of landscaping is called a swale. It is designed in such a way that when it rained, water would seep into the ground instead of running away. In the backyard, TreePeople installed two 1,800-gallon plastic cisterns against the side fence, collecting water through a drain spout that runs down from the roof. The set-up is shown in Fig. B.

The yard still looks very much like the way it did before Mr. Lipkis's crew arrived, except that the flower and shrub beds have been piled high with chopped-up green waste.

When the house was ready, Mr. Lipkis called in the press and invited the public works officials to watch a demonstration. Workers lugged fire hoses to the roof of their test house. Officials in suits stood by, watching from the safety of their umbrellas. Someone gave the word, and water gushed from the hoses, spewing a gentle, simulated rainfall over the house. Two tons of water came down, but none of it ended up in the

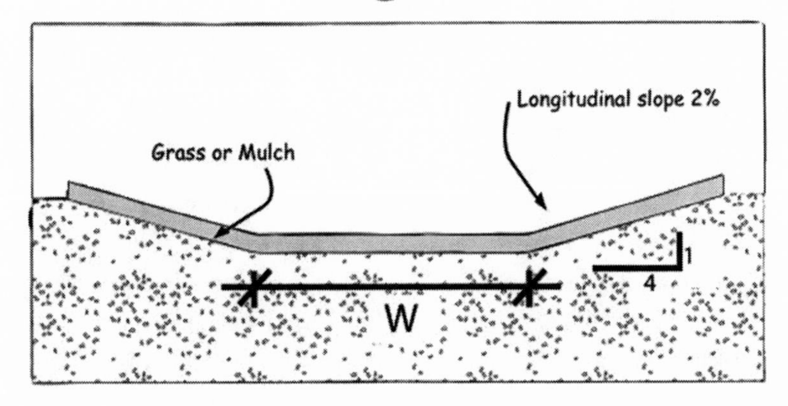

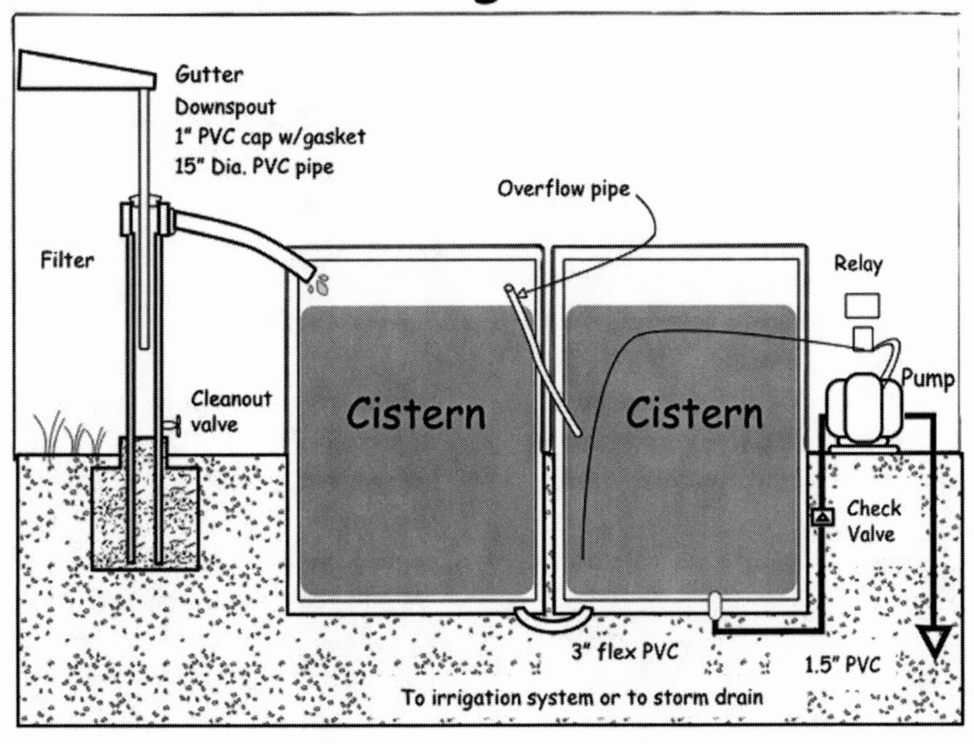

street. Every drop trickled into the cisterns or the ground.

When the hoses were shut off, Mr. Lipkis invited everyone to the backyard, where he turned on the water sprinkler, which was hooked up to the cistern. It started spraying away.

Newspaper and TV reporters loved it. "How much will this cost?" a reporter asked.

Mr. Lipkis told him that it could cost as much as $10,000 to retrofit a house. But it could be done more cheaply if the city worked hard to make the price affordable.

Four days later, Mr. Lipkis got a call from a man who worked on the county's water board. "I'm sorry," he told Mr. Lipkis. "We didn't understand your ideas. We think you've cracked the problem. It's got to be done. We have to do it throughout the county, and we have to start *today*."

Sitting at his desk at TreePeople's office, Mr. Lipkis finally felt vindicated. He'd been waiting for this phone call for six years.

The man told Mr. Lipkis that he could try out his ideas in Sun Valley, a Los Angeles neighborhood that was famous for floods. Mr. Lipkis's ideas weren't cheap. Installing underground cisterns in Sun Valley could cost as much as $300 million—six times more than the county's original project to install new storm drains to wash rainwater out to sea. No one wants to spend six times more than they must. But doing it Mr. Lipkis's way would put $200 million worth of water in Los Angeles's underground water supply in the first year. Mulching green waste and spreading it around people's gardens instead of hauling it to the dump would save the county $30 million alone!

Not long ago, Sun Valley broke ground on the first cistern project, a big infiltration system that will be installed under a soccer field.

What is it about Mr. Lipkis's idea that offers such hope?

Mr. Lipkis's plan addressed each of the major air and water quality problems facing Los Angeles. On the next page are two diagrams that Mr. Lipkis shows to explain the problems Los Angeles faces. In Fig. C, you can see six major problems.

In Fig. D, you see that the circles are now linked. Instead of each problem standing alone, each *solution* leads to another. In other words, each link in the chain is *integrated* with the next, a perfect model of something called *integrated resource management*.

Here's how the integrated approach works.

- The problem of green waste? *Solved:* Residents and schools would mulch their waste on site and feed their flowerbeds.
- The problem of too much or little water? *Solved:* Hundreds of thousands of gallons would be captured and stored during storms or else fed into the ground to recharge reservoirs or underground water supplies. When needed, the water could be pumped out of the ground or reservoirs.
- The problem of polluted water? *Solved:* Driveways and parking lots would be designed to reduce toxic runoff and improve water quality.
- The problem of high energy demands? *Solved:* Trees and grass would shade homes and schools, reducing the demand for air conditioning.
- The problem of poor air quality? *Solved:* The growing urban forest would filter out pollutants and CO_2.

Even the employment problem had been addressed. A project of this size—2,700 acres encompassing 8,000 homes—would require the employment of tons of people, from trained engineers and technicians to landscapers and laborers. In the years to come, trees would need to be tended, cisterns maintained, filtering systems installed or repaired, grass mown, and green waste mulched. An endless supply of good, honest work was awaiting hundreds of people in Sun Valley alone.

Today, thanks to Mr. Lipkis's urging, Los Angeles County's water and flood control administrations are doing a better job of talking to each other and *integrating* their policies.

LOS ANGELES TODAY

DIS-INTEGRATED APPROACH WASTES RESOURCES, DUPLICATES EFFORTS AND EMBRACES UNSUSTAINABLE SOLUTIONS.

Fig. C

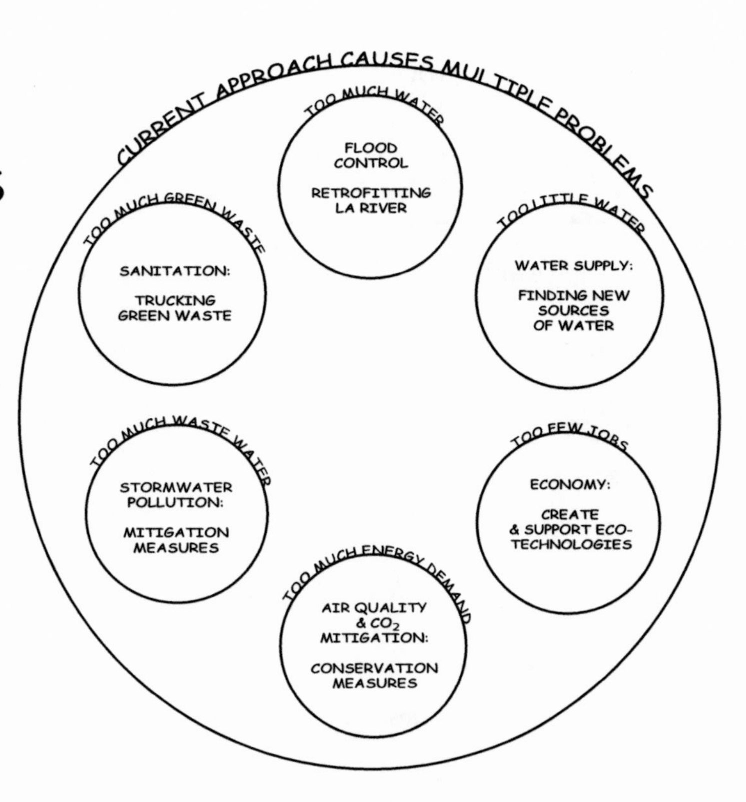

LOS ANGELES POTENTIAL

INTEGRATED APPROACH ALSO CREATES JOBS AND SUPPORTS EMERGING ECO-TECHNOLOGIES.

Fig. D

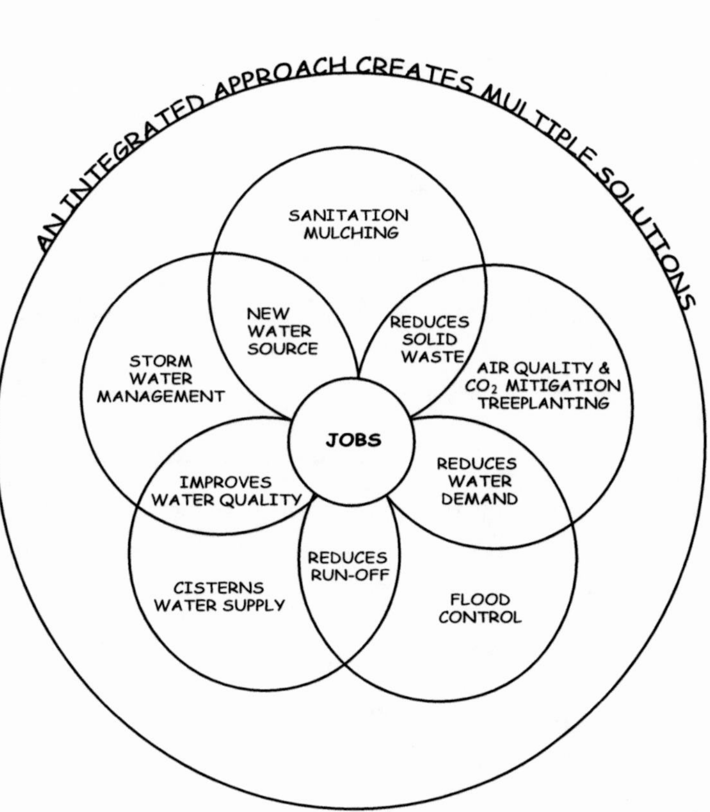

Source: Tree People

Questions

1. Why was the city's approach to water unsustainable?

2. What percentage of the rain that falls in Los Angeles washed out to sea every year?

3. Mr. Lipkis's ideas come from having observed the way forests and trees capture and retain water. Why are forests considered sustainable?

4. What are cisterns and swales, and how do they work?

5. Cisterns are expensive, so why did Mr. Lipkis think they were worth installing?

6. Can you describe what Mr. Lipkis's team did to the test house?

7. Why is it important that none of the water went into the street?

8. What does it mean to *integrate* something?

9. Can you define *integrated resource management*? How is Mr. Lipkis's idea of using cisterns, swales, and trees a good example of integrated resource management?

____

____

10. Mr. Lipkis says that people were hurting because the city spends too much on buying water. Can you explain his reasoning?

____

____

11. Why is green waste important to Los Angeles's water problem?

____

____

12. What role does green waste play in a forest?

____

____

13. Why is it wasteful to throw green waste in a dump?

____

____

14. How is the story "Water In, Water Out" an example of "the left hand not knowing what the right hand is doing"?

____

____

Name ______________________ Date ______________

Handout 15.1: Pop Goes the Population

Study the chart below that shows countries with the 25 largest populations in the world. Then answer the questions.

The 25 Largest (by population) Countries in the World

Rank	Country	Area (sq. km)	Population	
			July 2005	July 2006
1.	China	9,596,960	1,306,313,812	1,313,973,713
2.	India	3,287,590	1,080,264,388	1,095,351,995
3.	United States of America	9,631,418	295,734,134	298,444,215
4.	Indonesia	1,919,440	241,973,879	245,452,739
5.	Brazil	8,511,965	186,112,794	188,078,227
6.	Pakistan	803,940	162,419,946	165,803,560
7.	Bangladesh	144,000	144,319,628	147,365,352
8.	Russia	17,075,200	143,420,309	142,893,540
9.	Nigeria	923,768	128,771,988	131,859,731
10.	Japan	377,835	127,417,244	127,463,611
11.	Mexico	1,972,550	106,202,903	107,449,525
12.	Philippines	300,000	87,857,473	89,468,677
13.	Vietnam	329,560	83,535,576	84,402,966
14.	Germany	357,021	82,431,390	82,422,299
15.	Egypt	1,001,450	77,505,756	78,887,007
16.	Ethiopia	1,127,127	73,053,286	74,777,981
17.	Turkey	780,580	69,660,559	70,413,958
18.	Iran	1,648,000	68,017,860	68,688,433
19.	Thailand	514,000	65,444,371	64,631,595
20.	Congo, Democratic Republic of the	2,345,410	60,085,804	62,660,551
21.	France	547,030	60,656,178	60,876,136
22.	United Kingdom	244,820	60,441,457	60,609,153
23.	Italy	301,230	58,103,033	58,133,509
24.	South Korea	98,480	48,422,644	48,846,823
25.	Myanmar	678,500	42,909,464	47,382,633

©GeoHive

1. How many more people live in China than in the United States?

2. How many more people live in the United States than in the United Kingdom?

3. On which continent will you find the country with a population of 78,887,007?

4. How many of the countries here are in North America?

5. How many of the countries here are in Africa?

6. Which has a higher population: India, or the United States, France, Italy and Japan combined?

REPRODUCIBLE 15.1: Projected World Population Growth to 2050

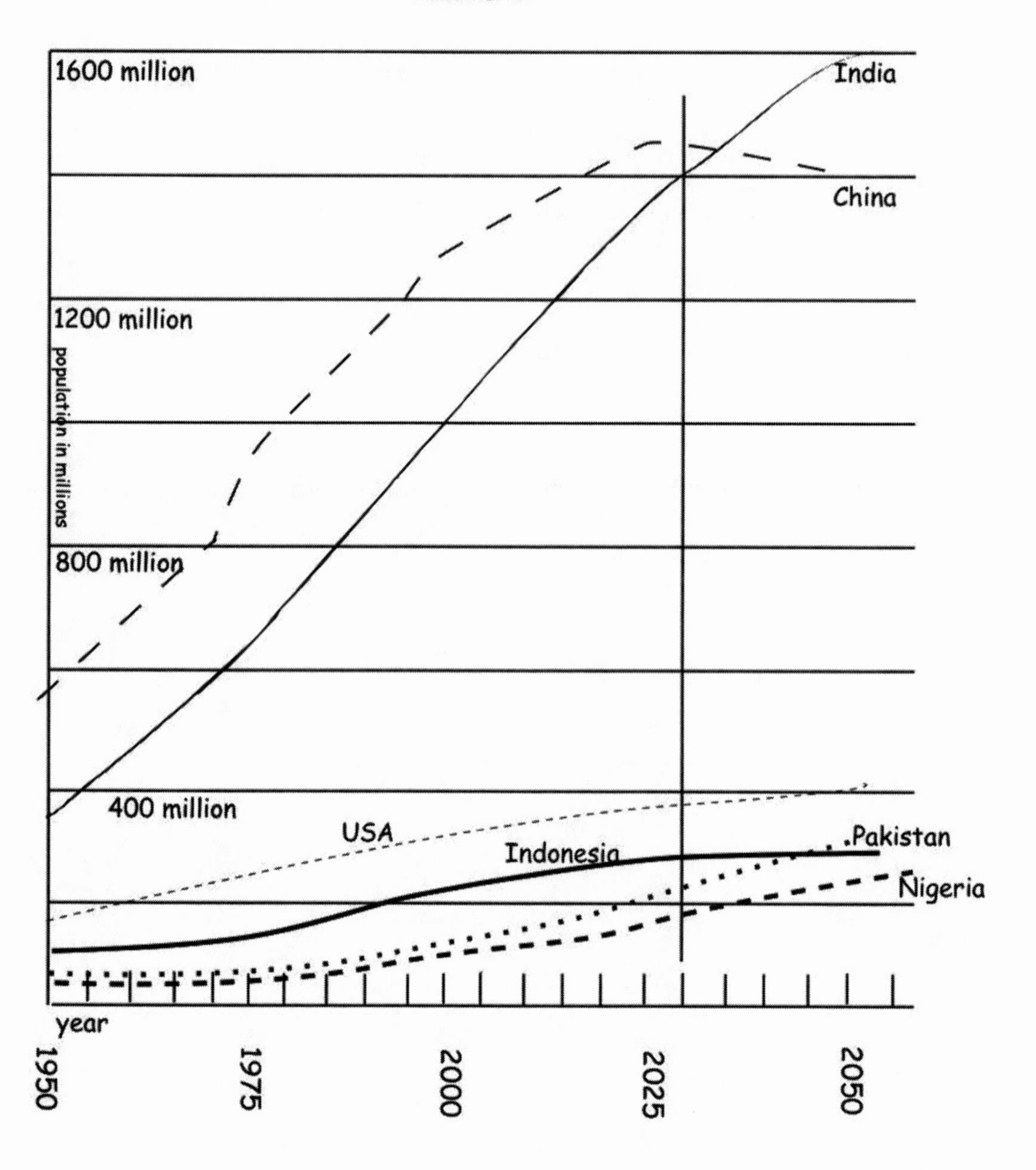

©GeoHive

Name ______________________________ Date ____________________

HANDOUT 15.2: Growing, Growing . . . Gone

Officials in Boomingtown, U.S.A., are concerned about their area's favorite fish sanctuary. They've been monitoring the population. Examine what they've found, graph the results, and then answer the questions.

YEAR	Number of BoomerFish
1996	15
1997	20
1998	30
1999	55
2000	60
2001	55
2002	50
2003	45
2004	50
2005	55

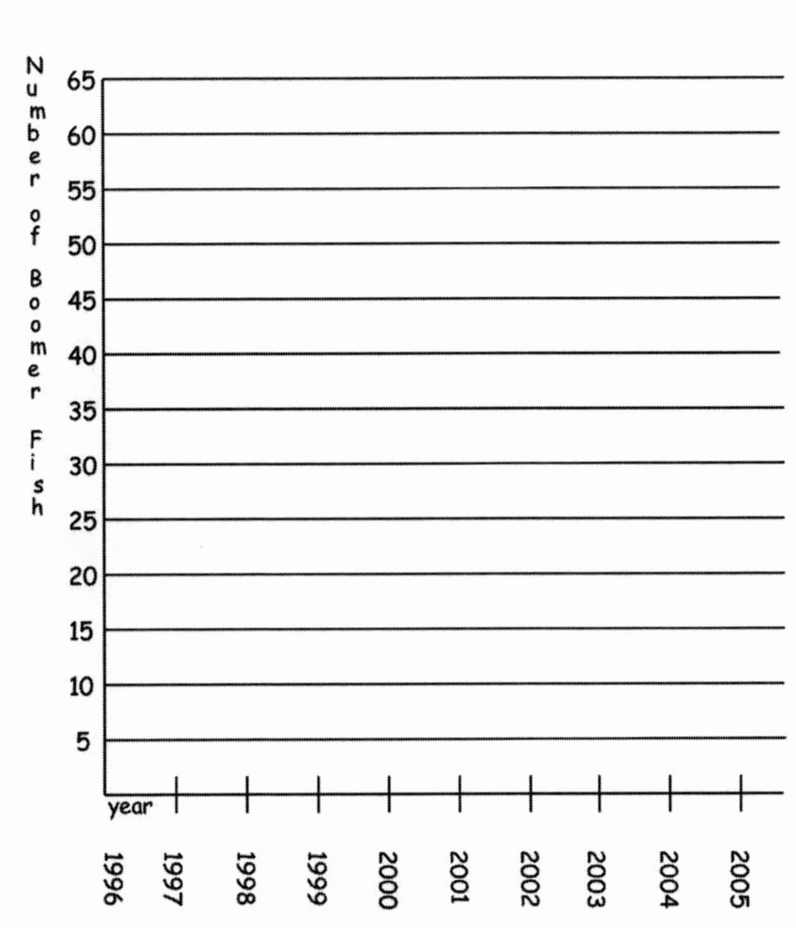

1. Label the years on the graph. Then graph the data in the chart.

2. How much did the population increase between 1997 and 2000?

____

3. Between which two years did the population increase the most?

____

4. Look at the graph as a whole: What happens after 2000?

____

5. What factors could be affecting the growth and decline of this population? (Assume there is no fishing.)

____

____

____

REPRODUCIBLE 15.3A

SCHOOL #1: "BEFORE"

SCHOOL #1: "AFTER"

Note: These photos were taken in slightly different positions, but they are of the same school.

REPRODUCIBLE 15.3B

SCHOOL #2: "BEFORE"

SCHOOL #2: "AFTER"

REPRODUCIBLE 15.3C: The Technology Behind School #2's Cistern

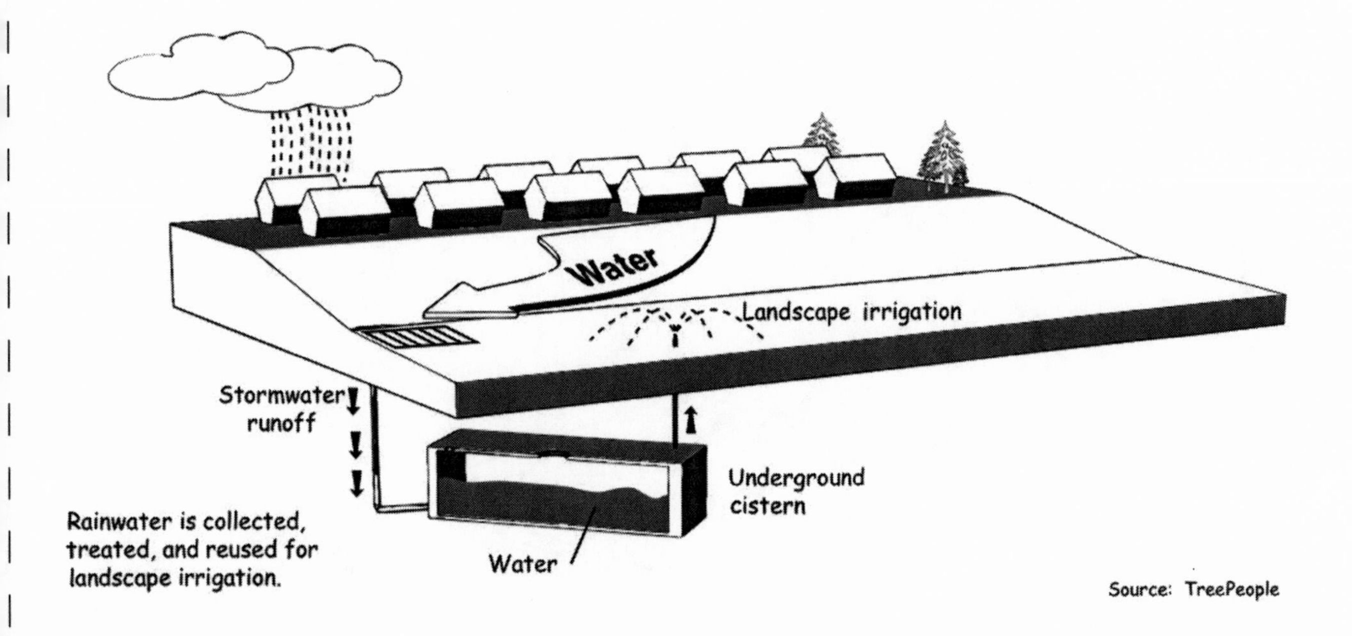

BENEFITS TO SCHOOL

- Collects rainwater (no waste)
- Filters out pollutants (no contamination)
- Water can be used again (saves money on water bills)

BENEFITS TO COMMUNITY

- Less water needed
- Water that doesn't go into cistern recharges underground water reserves and can be pumped out by the community when needed

ADDITIONAL BENEFITS TO SCHOOL AND COMMUNITY

- Plants and grass reduce heat island effect
- Trees shade buildings, reducing energy costs
- Generates green waste that can be used to absorb, retain more water

TRADE-OFFS

- Grass, plants, trees require more maintenance
- Provides jobs

REPRODUCIBLE 15.3D

SCHOOL #3: BEFORE

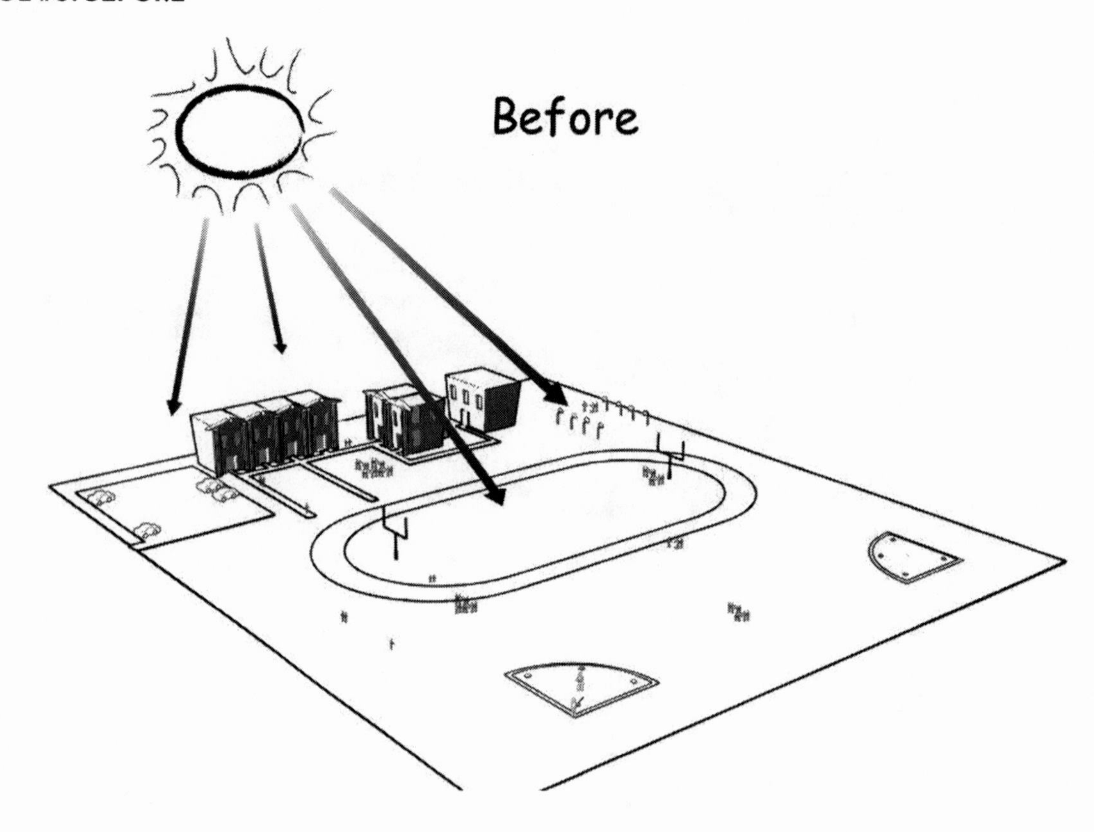

SCHOOL #3: AFTER

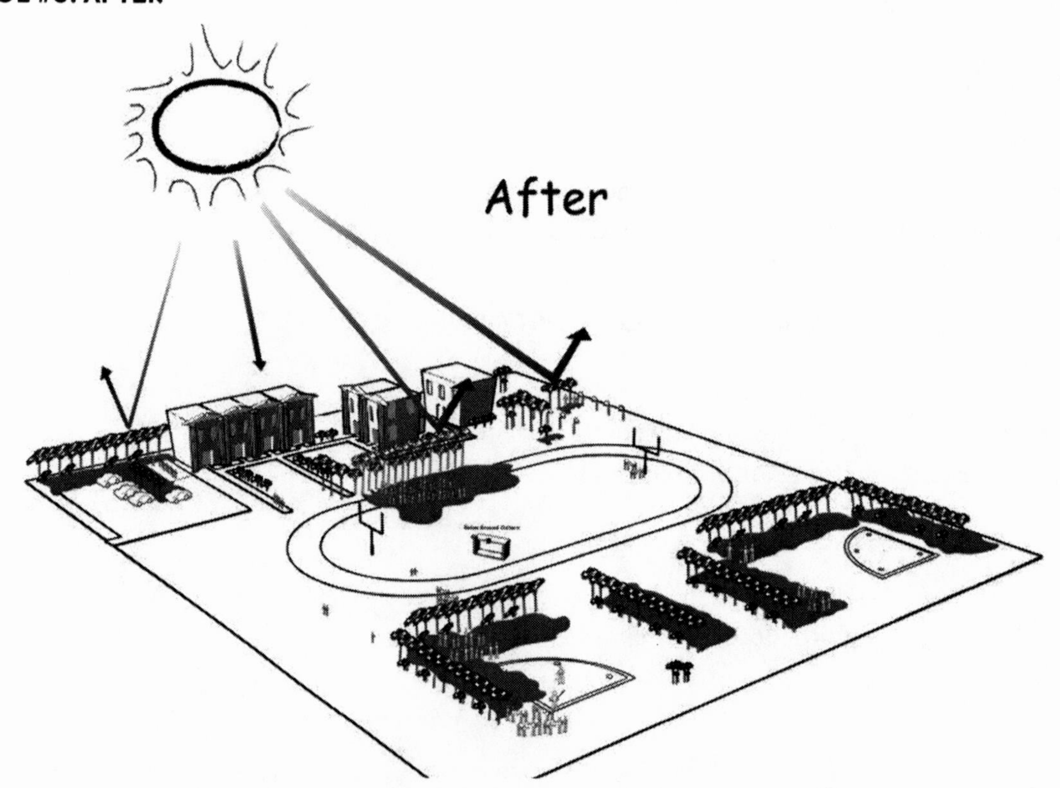

REPRODUCIBLE 15.3E

ARTIST'S VIEW OF SCHOOL #3'S REDESIGN

Scott High School

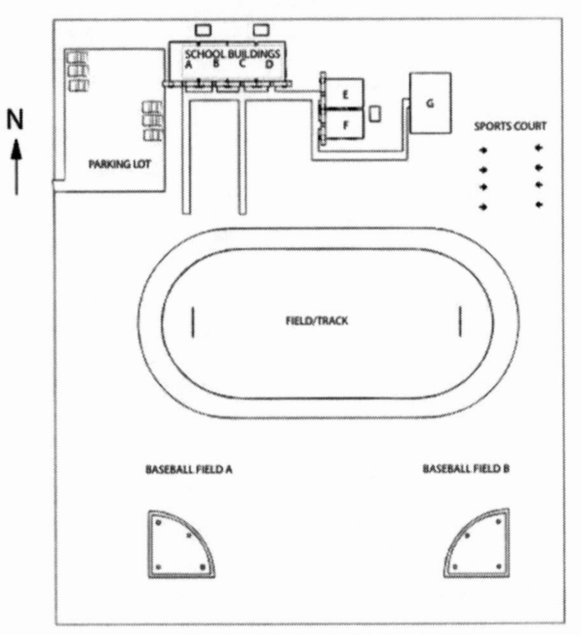

NO GROUND COVER

Scott High School

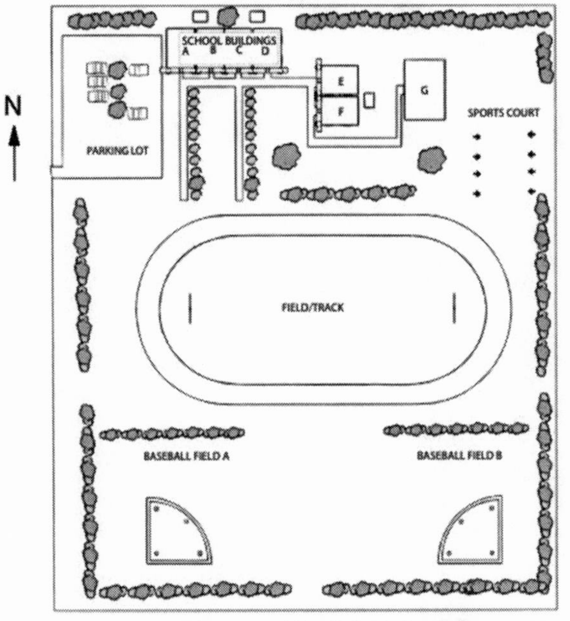

WITH GROUND COVER

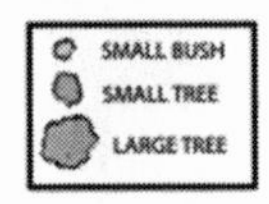

Name ____________________ Date ____________

HANDOUT 15.4A: Human Population Summit

KNOW YOUR NATION

Your Country Is:

Your Fellow Delegates Are: (write the names of everyone on your team)

With the other members of your country, research the following information for your presentation:

1. What is the crude birth rate and crude death rate of your nation?

 ____

2. What is the infant mortality rate?

 ____

3. In general, what are the conditions in which your citizens live?

 ____

4. What is the level of education and percentage of education for each gender?

 ____

5. How are women viewed?

 ____

6. What are the religions of your nation?

 ____

7. Are there cultural beliefs that encourage a large or a small family? What are these beliefs?

 ____

8. Are there currently laws or incentives in place that limit reproduction?

____

8. What is the literacy rate of your nation? What amount of education is mandatory?

____

9. What is the health care system like in your nation?

____

10. What is your nation's method of waste disposal?

____

11. What is your nation's rate of consumption of natural resources?

____

12. Is your population rising or falling?

____

Name ______________________________ Date ____________________

Team Members ____

Country you represent ____

HANDOUT 15.4B: Human Population Mandates

THE FUTURE IN YOUR HANDS

The United Nations has summoned you, Earth's leaders, to debate the population issues. The United Nations has decided that Earth does have a carrying capacity.

In 1994, the United Nations proclaimed we must reduce the following:

- Human population (total fertility rates)
- Consumption of Earth's resources

To meet these two goals, we have developed the following five mandates. The purpose of this summit is to create a Human Population Treaty to be enforced by all nations of the world.

GOAL ONE: REDUCE HUMAN POPULATION BY REDUCING TOTAL FERTILITY RATE

Mandate 1:
All nations are to practice "replacement fertility," meaning that all couples will be discouraged from having more than two children.

a) **If couples have more than two children:** They will be fined 5 percent of their gross monthly income per child over the allotted two children. The fine is to be paid to the United Nations and used to help enforce the treaty.

b) **If couples have only two children (or fewer):** They will receive housing, clothing, and food allowances. These benefits could be provided through tax incentives in developed nations. Most likely rations will be provided in developing countries.

1. Would your country support Mandate 1a and 1b (Yes or No)? Why would or wouldn't your country support this mandate? (You may give philosophical or political reasons.)

2. Edit the mandate so that your country would agree with it. You will need to present this during the summit. Use additional paper if necessary.

3. What do you, personally, think about the mandate?

Mandate 2:
All nations shall be required to provide family planning clinics and birth control to all citizens.

1. Would your country support this mandate (Yes or No)? Why would or wouldn't your country support this mandate? (You may give philosophical or political reasons.)

2. Edit the mandate so that your country would agree with it. You will need to present this during the summit. Use additional paper if necessary.

3. What do you, personally, think about the mandate?

Mandate 3:
All nations will be required to educate their women with a minimum eighth-grade education.

1. Would your country support this mandate (Yes or No)? Why would or wouldn't your

country support this mandate? (You may give philosophical or political reasons.)

2. Edit the mandate so that your country would agree with it. You will need to present this during the summit. Use additional paper if necessary.

3. What do you, personally, think about the mandate?

GOAL TWO: REDUCE CONSUMPTION OF EARTH'S RESOURCES

Mandate 4:
No nation will be allowed to use more than 10 percent of Earth's total availability of any one resource. All nations must be in full compliance by the year 2020. For each percentage point over 10 percent that the nation uses, they will be fined an amount equivalent to $10,000 U.S. dollars per day until the allotted 10 percent is achieved. Countries may pass this fine onto industry and citizens in the form of tax increases.

1. Would your country support this mandate (Yes or No)? Why would or wouldn't your country support this mandate? (You may give philosophical or political reasons.)

2. Edit the mandate so that your country would agree with it. You will need to present this during the summit. Use additional paper if necessary.

3. What do you, personally, think about the mandate?

Mandate 5:
After 2020, no country shall be allowed to mine or harvest any resource from another country. Each nation needs to become completely sustainable with the resources available with in its borders.

1. Would your country support this mandate (Yes or No)? Why would or wouldn't your country support this mandate? (You may give philosophical or political reasons.)

2. Edit the mandate so that your country would agree with it. You will need to present this during the summit. Use additional paper if necessary.

3. What do you, personally, think about the mandate?

UNIT 16

ENVIRONMENTAL JUSTICE

TOPIC BACKGROUND

Your students will be familiar with the word *justice* and may have even heard the phrase *social justice. Environmental justice* is yet another twist on the theme. As the U.S. Environmental Protection Agency defines it:

> *Environmental Justice is the fair treatment and meaningful involvement of all people regardless of race, color, national origin, or income with respect to the development, implementation, and enforcement of environmental laws, regulations, and policies. The EPA has this goal for all communities and persons across this Nation. It will be achieved when everyone enjoys the same degree of protection from environmental and health hazards and equal access to the decision-making process to have a healthy environment in which to live, learn, and work.*

Notice that this government definition stresses *protection* from health hazards. This concept of justice is based on the notion that all humans have to the right to breathe clean air, drink clean water, and live in neighborhoods that are not contaminated with toxins or polluted daily by illegal dumping or dangerous businesses that spew harmful chemicals into the local atmosphere. The federal government has an obligation to ensure these rights and to use laws or lawsuits to go to bat for its citizens.

As you and your students will see in this unit, very often the poorest citizens of our nation live in communities with the highest degree of contamination. Wealthier neighborhoods usually enjoy the healthiest conditions, but not always. Sometimes, a longtime local polluter will somehow escape notice because citizens just don't know of the danger. In the Unit 3 Reading, students learned about Michael Howard's fight to clean up his Chicago neighborhood, where vacant lots had become a dumping ground for toxic waste and local water was suspected of causing cancer and death. You may wish to use the same reading as preparation for this unit. Students will get practice accessing Internet data on this issue, as well as analyzing maps to see what environmental threats—if any—exist in their neighborhood.

Definitions vary, of course. Some people broaden the definition of *environmental justice,* saying that all humans have the right to experience wild nature, visit beaches, and so on. The Unit 16 Reading argues that Internet access is fast becoming an environmental issue. The Internet is an important tool for education and civic involvement, but 25 percent of Americans still do not have it.

Young people often possess a strong sense of justice, and they may enjoy ferreting out the truth about their neighborhoods, working like investigative reporters and spreading the word about what they discover. They may find it hard to believe that this information is so easily available yet hardly known by the general public. Congratulate them on taking the time to educate themselves about what's going on around them.

Purpose

By the end of the unit, students should be able to define *environmental justice* and know where their neighborhood stands on the pollution issue. As a group, they should be able to speak out for all citizens and take their message to the authorities and general public.

GROUP ICEBREAKERS

1. Ask students: "What does the word *justice* mean? Can you put it in your own words?"
2. Ask: "Who is responsible in our society for making sure that justice is carried out?" (Police, courts, lawyers—but also, each one of us has an important role to play. Each of us must stand up for what is right, notify police when we see, hear, or learn about something dangerous or wrong.)

3. Then ask: "What do you think the phrase *environmental justice* means?" Have students brainstorm types of injustice against the environment and how justice might be served. You might prompt them with suggestions if nothing comes up. (*Injustice:* A company dumping harmful chemicals or trash in a neighborhood. *Justice:* The company must clean up after itself.)
4. Remind them about the Michael Howard story in the Unit 3 Reading. Ask: "What environmentally dangerous situation did Mr. Howard and his neighbors face?" (Dumping in the vacant lots around them, lead in drinking water, no nearby nature for the children to experience.) If they do not recall the Reading or you have not yet used it, have them read it now.
5. Ask: "What did Mr. Howard and his friends do when their city didn't do anything to help them? (They took it upon themselves to raise money, buy water filters, seek donations, and clean up a whole city block, which they turned into a nature center.) If you have it, you may wish to play an excerpt of this story from the Chicago film in the *Edens Lost and Found* DVD series.
6. Ask: "What would you do if you found out someone or something was polluting your neighborhood. How would you help bring about justice?" (Answers will range from "calling the police" to "writing letters," "ask them to stop," and the like.)

ACTIVITY 16.1
What Is Environmental Justice?

Materials

pen or pencil
notebook
Handout 16.1

Primary Subject Areas and Skills

science, social studies, reading comprehension, writing, critical thinking

Purpose

Students read the EPA's definition of environmental justice and reflect on its meaning, using writing prompts, flowcharts, and diagrams to express what the concept means to them and what it may mean to members of their community.

There are basics in life that everyone wants: clean air, clean water, healthy food, safe environments for themselves, their children, and their loved ones. Though many individuals throughout the world feel concern for the future of the environment, another disturbing factor in this crisis is that many racial groups and individuals from economically challenged backgrounds are exposed to environmental toxins and pollutants at a much higher rate than their fellow citizens. Some people call this "environmental racism."

1. Write the phrases *environmental justice* and *environmental racism* on the board or an overhead projector. Ask students what they think the terms mean and what they refer to. Can they provide examples?
2. Review the EPA definition of *environmental justice* with students (see Handout 16.1).
3. Explain to students that the environmental justice movement is a growing one and that legislation such as President Clinton's 1994 Executive Order 12898, "Federal Actions to Address Environmental Justice in Minority Populations and Low-Income Populations," has been put on the books in an effort to increase awareness about the unfair concentration of pollutants in areas largely populated by minorities and the poor.
4. Copy and distribute Handout 16.1 to students. Instruct them to read the EPA definition carefully and then answer the questions. Have them write the essay described on the handout on a separate piece of paper.
5. You may wish to have students read their essays aloud in class to prompt a discussion or debate.

ACTIVITY 16.2
Environmental Justice on the Map

Materials

pen or pencil
notebook
Handout 16.2

Primary Subject Areas and Skills

science, social studies, math, critical thinking

Purpose

Students use a coded map to analyze neighborhoods that may or may not be affected by toxic waste sites.

This map is similar to one used by citizens and politicians to gauge the health of a neighborhood.

This activity is designed to put real facts behind theories and give students an opportunity to exercise their chart- and map-reading skills while learning more about an important topic. It works well as a precursor to Activity 16.3, in which students will do more extensive research of a similar kind.

1. Distribute Handout 16.2 to students.
2. Instruct students to answer the questions individually.
3. After they've answered the questions, have each student exchange papers with another student and compare their answers.
4. Discuss the activity. How did students feel about what they learned? (Many will be surprised to learn that (a) toxic waste is dumped where people live, (b) many of those people are minorities, and (c) this injustice can be so readily mapped.)
5. As a follow-up activity, use information on sites such as the EnviroMapper on the EPA site or HUD to take a closer look at communities that are important to your students. They may wish to check a favorite city or town or a region near a national park they've visited. Have them find out what kind of pollutants are the most prevalent there and which communities are the most affected by pollutants.

ACTIVITY 16.3

Environmental Justice Close to Home

Materials

pen or pencil
notebook or sketchbook
library or Internet research materials
Handout 16.3

Primary Subject Areas and Skills

science, social studies, math, critical thinking

Purpose

Students use a variety of research resources to gauge the level of environmental health of their neighborhood or region. They then use what they have found to judge if their area is environmentally just.

Environmental justice affects more communities that you might first imagine. Perhaps students in your class feel that theirs is a healthy environment, that your air quality is among the best in the nation, and that everyone—no matter their ethnicity or income—is exposed to the same level of toxins or environmental pollutants. But are their assumptions correct? The amount of information available to concerned citizens is rapidly increasing, putting valuable information in the hands of those if affects most. Use this information to take a closer look at your own town or region's health, primary polluters, and exposure to toxins. This is the first step toward creating a healthier and just life for all members of your community.

This activity is designed to develop students' research and critical thinking skills and, perhaps more importantly, to expand their perception of their own community and its environmental health. Primarily a research and evaluation exercise, this activity can be done in as short or as long an amount of time as you wish, depending on your class schedule. Research can be done using information provided in the Resource section of this unit (page 251) or using local sources.

1. Have an open discussion with students about your area and what they think about the quality of the local environment. Are there any areas in their town or region that they know of to be polluted? Are there any local industries or other activities that they know to be polluting their area?
2. Ask students how they think their community ranks when compared to other communities in the United States in the areas of air and water quality. Do they think their community is better or worse? Why? Get a class consensus and keep a record of it to refer back to after students have completed the research.
3. Divide students into groups or allow them to work individually. Instruct them to research your community on www.scorecard.org or by using information from your local office of the EPA, such as their "Where You Live" interactive environmental justice assessment tool at www.epa.gov/compliance/whereyoulive/ejtool.html
4. Distribute Handout 16.3 to students and instruct them to use the information to guide them through their research process and to answer the questions.
5. After students have completed their research, you may wish to have them prepare a presentation—individually or in groups—that may include charts of your area, graphs comparing pollutants, and so on.

6. Discuss the findings as a class. Were students surprised by the information they gathered? Does the information inspire them to take action? What kind of action would they be willing to take? (Answers will vary. In general, if they are alarmed by what they have learned, they may want to educate others and take the matter up with local leaders.)

Extensions

1. Prepare a report for your town council or other local government organization and encourage students to present what they've found.
2. Have students contact members of the local press to report on what they've found or ask them to write a guest editorial or letter to the editor.

EXTENSIONS

Bring in an Expert

Find out from your local Sierra Club or other environmental organization if someone can come speak to your class about environmental justice and what it means to your community.

RESOURCES

The EPA's site for Environmental Justice

This site features information on efforts across the country. You can check on what's happening in your region by clicking on their interactive map.

www.epa.gov/compliance/environmentaljustice/index.html

The Environmental Justice Geographic Assessment Tool

This interactive mapping tool from the EPA allows you to research areas and select your own views: add superfund sites, hazardous waste sites, air emissions producing facilities, and so on. You can overlay the map with demographics, such as income and race.

www.epa.gov/compliance/whereyoulive/ejtool.html

Scorecard

Who's polluting and where? This site allows you to search by ZIP code for the leading polluters in your county and see how your area compares to the rest of the United States. It also provides you with comparisons of exposures to environmental pollutants by ethnicity, income, and so on. This is a must-have for all community organizations and a great statistical tool.

www.scorecard.org

Environmental Justice Initiative from the Sierra Club

If you have an issue in your community that you would like to organize around, read the Sierra Club's information on "organizing guidelines" on the environmental justice page.

www.sierraclub.org/environmental_justice/

HUD

The HUD Homes and Communities site has a "map your community" application that allows you to enter your location and get a map of dumpsites, hazardous waste sites, and more.

http://egis.hud.gov/egis/

The TRI Explorer

This site tracks the Toxics Release Inventory for communities across the nation. Enter information on this EPA site to learn what kinds of toxins are being released in your area. This site is a great resource for map and chart activities, as well as valuable research to jump-start community action.

www.epa.gov/triexplorer/

The Right to Know

The Right to Know organization provides free access to a number of environmental information databases. You can use it to identify industries and how they pollute and then determine the people and communities affected.

www.rtknet.org

Green Action

This is a good site for learning more about how some other communities and schools have mobilized around environmental action issues that affect them.

www.greenaction.org/

Reading Handout Unit 16:
TOWARD A WIRELESS CHICAGO

The term *environmental justice* refers to the rights of all humans to enjoy a healthy, safe, and enriching environment. Does that include the right to a cutting-edge technology? In this reading, you'll learn about a Chicago neighborhood that recently received access to the Internet via wireless technology. You will learn more as you watch the Chicago film in the *Edens Lost & Found* DVD set. After you read the story, answer the questions.

A hallmark of a healthy community is the democratic distribution of information and opportunity. Some people might think it a stretch to consider information—in this case, information you get off the Internet—as a natural resource, but we think it is. No one questions the idea that all citizens should be able to enjoy clean air, clean water, open spaces, or a healthy environment to grow up in. These are all part of a concept called "environmental justice." These days, that justice must include access to Internet technology. Used correctly, the web can be a powerful tool for environmental improvement. Think about it: You can use the web to learn ways to decrease your energy usage. The web allows people to telecommute, thus reducing destruction of natural resources that would be consumed in the daily commute. When you pay your bills online, another envelope doesn't have to be delivered by trucks that use fossil fuels.

While the Internet might allow people to connect with a geographically broad community of people who share specific interests, it's more useful as a knowledge bank. Rich or poor, a person can use the Internet to research health information and educational opportunities and interact with local government. In this sense, access to the web has become an important tool for social change and an entrée to what's going on near you.

Yet many Americans do not have web access. Millions of young people cannot visit online libraries, read free books, explore colleges they might like to attend, research loans to help them pay for their education, learn more about the leadership of their cities, or even find out how to submit a complaint to the right government office. While it's certainly possible to do these things offline, if you are unplugged you are arguably missing out on opportunities.

Opportunities are what Chicago's Center for Neighborhood Technology (CNT) is all about. The nonprofit organization has devoted itself to coaxing the social and economic value out of cities. Recently, one of Chicago's major Internet providers announced it would not seek customers in lower-income neighborhoods because the company didn't see much of a market there. But CNT's project leaders already had a project in the works to bring free Internet services to at least 1,000 people in four different area neighborhoods.

The secret weapon in their plan is WiFi, the technology that allows one high-speed Internet connection to be shared by multiple wireless users. WiFi, an abbreviation for *wireless fidelity,* bypasses the labor-intensive process of installing wires throughout a neighborhood for high-speed Internet. The technology works the way walkie-talkies do: High-frequency radio signals radiate from a central transmitter in concentric circles and are picked up by other computers equipped with a computer card that receives and broadcasts back its own signal. The farther the signal travels from the original transmitter, the weaker the signal gets. But if extra transmitters, called repeater nodes, are mounted on houses or outside apartments in a neighborhood, the signal will bounce from node to node, growing stronger as it repeats. CNT had no trouble finding volunteers who let them mount nodes on their houses. How fast can people be up and running? Fast! In summer 2005, CNT held an old-fashioned "barn raising." They erected 100 nodes in a single weekend with help from community residents.

CNT received money in the form of a grant that allowed them to bring WiFi to as many as 250 people in each of four different Chicago neighborhoods: Pilsen, a largely Mexican American community on Chicago's near southwest side; North Lawndale, a primarily African American community on Chicago's Westside; Elgin, a working-class community about 40 miles northwest of Chicago that is becoming increasingly Latin American; and West Frankfort, a remote former coal mining community in downstate Illinois.

Residents who don't own and cannot afford a computer can receive a refurbished computer; most of these are donated by area colleges and local governments. CNT also asks for donations through their website and by writing to different companies, asking them to donate their old computers. They recently received another grant from the National Cristina Foundation, which pairs nonprofit organizations with computer donations. They bought a dozen used computers from Wayside Cross Ministries, a religious group, which were then fixed up by a youth technology program run by Korean American Community Services. Korean American kids learn how to fix computers, and others down the line get to use them. It was a win-win system all the way around. CNT expects to find homes for about 500 computers over the course of the project.

To get the best bang for their buck, they have given some computers to the Homan Square Community Center in North Lawndale. Residents are delighted to have this link to the community. One young woman used the web to apply to Loyola University online and now uses the web regularly to do coursework. Another woman had been sharing a dial-up connection at home with her mother and sister. Because all three women were taking college classes, it was virtually impossible to get anything done; the phone line was always tied up, and the computer connection was very slow. With the neighborhood's new high-speed connection, all three women can now surf the web without tying up the phone, and online coursework has become a breeze thanks to the higher connection speeds.

"Internet access has been shown to be one of the most important determiners of economic and educational success," says Bill Comisky, a CNT technical advisor. "So for a community like Homan Square that has been underserved by that kind of access, it's important that residents like this have that kind of access. The 'digital divide' used to be between people who have computers and Internet access, and people who *don't*. These days, the divide is between people who have high-speed, always-on Internet access and people who *don't*. That's what we're addressing here with the wireless community network."

Questions

1. What examples do the authors use to show that citizens can use the web to improve the environment or to impact their governments?

____

____

2. What is WiFi, and how does it work?

____

____

3. What's the difference between dial-up and high-speed Internet? Have you had any experience with either? Which is faster?

____

____

4. Do you think the repaired computer program is a good idea?

____

____

5. What is environmental justice? Why do the authors say they are stretching the definition?

____

____

6. What kinds of opportunities do the authors think people miss out on if they don't have Internet access?

____

____

7. Does the article make a convincing case that Internet access is an important part of environmental justice? Has high-speed Internet access become a "right" in modern America?

____

____

Name ______________________________ Date ____________________

HANDOUT 16.1: Understanding Environmental Justice

Read the EPA definition of Environmental Justice and then answer the questions on a separate piece of paper.

HOW DOES THE EPA DEFINE ENVIRONMENTAL JUSTICE?

Environmental justice is the fair treatment and meaningful involvement of all people regardless of race, color, national origin, culture, education, or income with respect to the development, implementation, and enforcement of environmental laws, regulations, and policies. Fair Treatment means that no group of people, including racial, ethnic, or socioeconomic groups, should bear a disproportionate share of the negative environmental consequences resulting from industrial, municipal, and commercial operations or the execution of federal, state, local, and tribal environmental programs and policies. *Meaningful involvement* means that: (1) potentially affected community residents have an appropriate opportunity to participate in decisions about a proposed activity that will affect their environment and/or health; (2) the public's contribution can influence the regulatory agency's decision; (3) the concerns of all participants involved will be considered in the decision-making process; and (4) the decision makers must seek out and facilitate the involvement of those potentially affected.

1. In your own words, and using the definition above, explain what *environmental justice* means to you.

2. Explain how—using a flowchart, diagram, or just your own words—a case of environmental injustice might arise. Think about the economical, geographical, and political possibilities.

3. "NIMBY"—Not In My Back Yard—is a phrase that is voiced by some citizens when dumpsites or power plants are looking for a new location. What does this phrase mean to you? How would you decide, fairly, whose backyard should be near chemical treatment plants and dumpsites?

4. Can you think of any examples of environmental justice issues in your community?

5. Write an opinion piece (or op-ed) for a newspaper, describing why your state needs an environmental justice action committee.

Name ______________________________ Date ____________________

HANDOUT 16.2: Environmental Justice on the Map

Study the map below and then answer the questions.

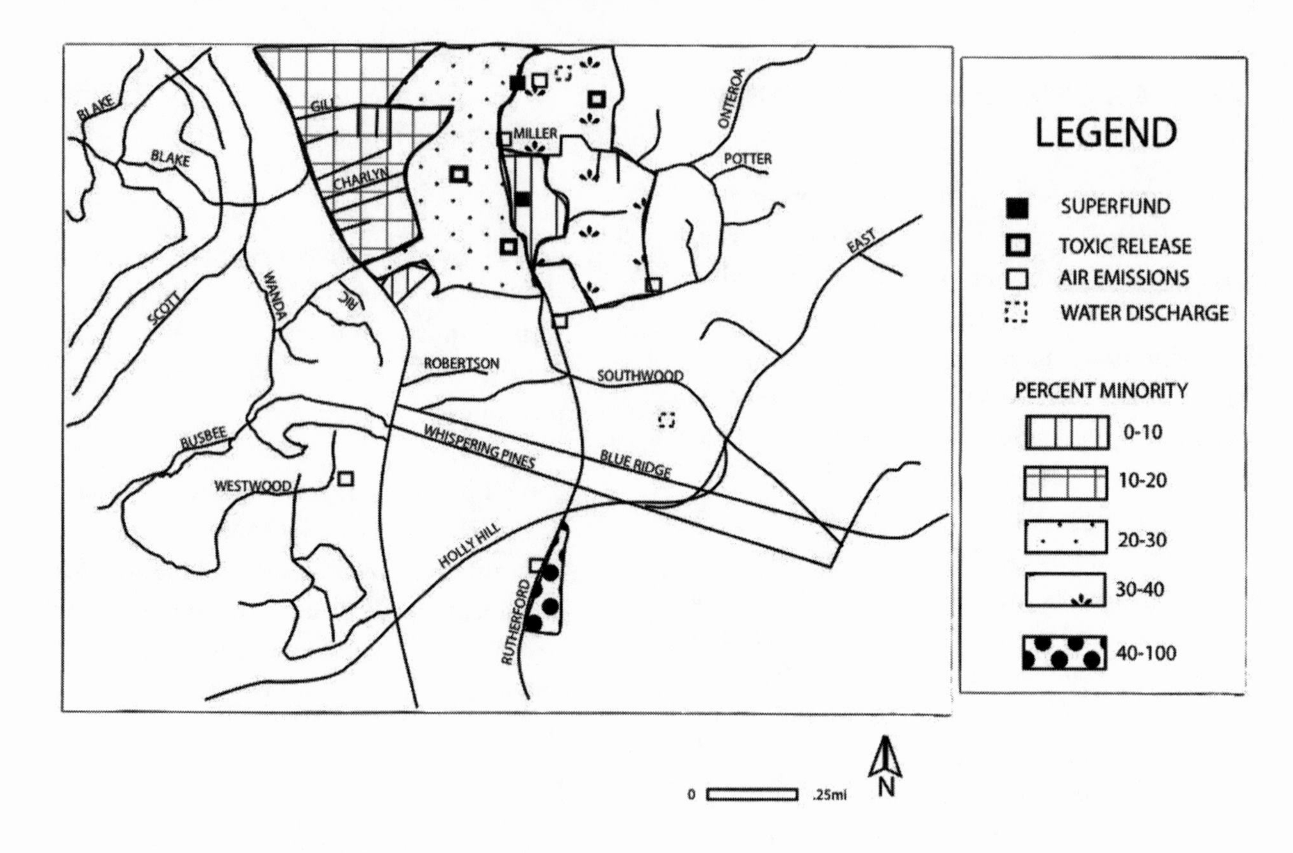

1. How many Superfund waste sites are shown on the map?

2. How many water dischargers are there? (A water discharger is a pipe a factory uses to emit waste water.)

3. What percentage of minority citizens live in Rutherford, below Holly Hill, where there is an air emission site?

4. How many of the Superfund or Toxic Release waste sites are in or border blocks with more than 30 percent minorities?

5. Write a short paragraph describing the trends you see on this map and your thoughts are about it. Use a separate piece of paper.

Name ______________________________ Date ____________________

HANDOUT 16.3: Environmental Justice

JUSTICE CLOSE TO HOME

Use the following questions to guide your research. Once you have gathered the necessary information, prepare a report and presentation for your class. You may use www.scorecard.org, your local office of the EPA, or the EPA's "Where You Live" tool at www.epa.gov/compliance/whereyoulive/ejtool.html to help you answer the following questions. You may also wish to refer to local environmental agencies or the public works department in your area for additional maps and resources.

1. How does your community compare to the national average in terms of air quality and water quality?

2. What are their reported primary pollutants, and what might be the possible effects of those pollutants?

3. What, if anything, is being done to prevent or lessen these pollutants?

4. Get a copy of a detailed regional map from your town hall or other city office. Find the location of the following: landfills, waste sites, power plants, water treatment facilities, chemical plants, and any of the polluting industries the students have identified. Mark the location of these on the map.

5. Using your local real estate board or the Internet, find out the property values of the houses and neighborhoods near these polluting facilities. Is the average property value of the homes near these facilities more or less than neighborhoods farther away?

6. Find out what you can about the communities most affected by the polluting factions in your area. Who lives in those areas? Is there a higher instance of asthma, cancer, or other disorder?

Extension

Do you think everyone should have the same access to clean water and air no matter how much money he or she makes? Why or why not?

UNIT 17

PUBLIC POLICY AND COMMUNITY ACTION

TOPIC BACKGROUND

If you were to ask your students who makes the laws we use to protect the places where we live, the air, the water, and parks, they would probably say the government, Congress, the president, and so on. But the truth is, the health of the environment is in the hands of us all. It is true that elected officials vote to enact laws and enforce them, but small nonprofit groups, larger coalitions of groups, big and small municipal agencies, and ordinary citizens can play an enormous role in the decisions to pass those laws.

Purpose

In this unit, we will look at some ways that citizens can have an impact on government and how even nonvoters, such as your students, can become involved in community action. This unit will give you ideas for how you can take the projects you've been working on all year and transfer them into community action.

To get started, work from the classroom out. You might remind students of a classroom activity that resulted in a letter to local officials or a presentation to school administrators. When they did this, they were acting as concerned citizens as part of a small group. Their action was not unlike the work of a "small community action group." If they wanted to improve their clout, they could team up with another small group, and another and another, to form a more powerful *coalition* that has similar goals at heart. (All the films in the *Edens Lost & Found* series highlight the work of large coalitions, such as Chicago Wilderness, which is made up of more than 100 smaller nonprofit groups.) When small action groups band together to form coalitions, they share their strengths. One group may be able to provide excellent media contacts. Another may be great at planning rallies and preparing signs for marchers to wave. Another group may excel at meeting with local business leaders or local politicians.

Most often, small action groups and the coalitions they form will be nonprofit. The terms *nonprofit* or *not-for-profit* are used interchangeably. The big difference between, say, a local chess club and a local church, synagogue, or scouting organization is that the last three are chartered organizations. They have a legal status and are recognized on the state or federal level. Environmental groups such as the Sierra Club are chartered groups. They are legal entities, much like corporations, except that they do not exist to earn a profit. They may earn money but only to pay their employees and run their organizations.

In our reading, your students will see how a chartered nonprofit organization and a large government body zero in on a similar problem and solve it in different ways. We provide you with a discussion activity that delves briefly into various types of environmental legislation, such as the Kyoto Protocol (a piece of international environmental legislation) and the Clean Air Act (national, or U.S.–based environmental legislation). Lastly, we provide you with a community action handbook that spells out exactly how you and your students can submit letters to the editors of local newspapers, write press releases, write op-ed articles for newspapers, plan and present rallies, hold press conferences, and so on. These activities are intended to be a good first step on the road to action in your community.

Note: Because students are minors, signed parental approval must be given before students participate in any public activities.

GROUP ICEBREAKERS

1. Ask students: "Can you name any environmental policies or laws?" If so, ask them to explain the policies as best they can. (Answers will vary. Remind them of the Clean Air and Water Acts mentioned in previous units.)
2. Ask: "What would happen if we didn't have laws protecting the environment?" (Answers

will vary. Some will say the environment will be spoiled by pollution; others will say that laws may not be necessary, that people will voluntarily choose to care for the environment.)

3. Then ask: "Should we be governed by laws that protect not only our local environment but the larger, global environment as well?" (Answers will vary. Accept all that are argued persuasively. Some students may feel uncomfortable imposing our laws on other nations, and vice versa.)
4. Ask: "Do you feel that wealthier countries should take more financial responsibility for protecting the environment than those countries with struggling economies?" (Answers will vary. Accept all reasonable responses.)
5. Ask: "What role do you think individuals should play in the development of policies that safeguard the environment and natural resources?" (Answers will vary. They will say that politicians make the laws and that they are elected by the people. Remind them that voters let politicians know what issues they care about, through opinion polls, rallies, petitions, and the like.)
6. Ask: "What do you think you would do if you disagreed with a law or public policy on the environment? What kind of action would you be willing to take?" (Answers will vary: write letters, complain by phone, send an e-mail, and the like.)

ACTIVITY 17.1
Understanding Public Policy

Materials

pen or pencil, notebook
library or Internet research materials
Handout 17.1

Primary Subject Areas and Skills

science, social studies, reading comprehension, writing, research, critical thinking, public speaking

Purpose

The aim of this activity is to require students to become more familiar with policies here in the United States and with the role that the United States plays in the global arena of environmental policy making and enforcement.

There have been some major advances regarding the enactment of public policy that aims to protect the environment over the years. Some of the most important and well-known policies, such as the Clean Water Act and Clean Air Act, are milestones in U.S. environmental history, even if they are considered too lenient and not adequately enforced by many citizens.

On a global scale, Agenda 21 from the U.N. Commission on Sustainable Development covers just about everything pertaining to sustainable development, from tourism and consumption patterns to hazardous waste and land acquisition.

The Kyoto Protocol to the U.N. Framework Convention on Climate Change concerns itself with the reduction of greenhouse gases. When this book went to press, the United States had yet to ratify the protocol and was therefore not bound by it. The lack of ratification by the United States—despite the fact that over 160 countries have signed on—has been a contentious issue both in the United States and abroad, since the United States is a leading producer of greenhouse gases.

1. Ask students what they know about the above-mentioned policies. Invite students to share their understanding of these policies. (Answer will vary.)
2. Divide students into at least four groups. Each group will prepare a well-researched presentation about one of the above-mentioned policies. If you want groups to be smaller, you may wish to divide the class into eight groups and assign two groups to each policy.
3. Distribute Handout 17.1. Students will need one handout for each policy they are researching. Explain to students that they will research and prepare a profile of their assigned policy. These should be well researched and written in report form and should include a detailed list or bibliography of their sources.
4. After the reports are finished, each group should present their policy profile to the class. Guidelines for what should be covered are included on the handout.
5. Take time after each presentation for a question-and-answer period.

ACTIVITY 17.2
Getting Organized: Making the Most of the Media

Materials

pen or pencil, notebook

library or Internet research materials
Handout 17.2

Primary Subject Areas and Skills
science, social studies, reading comprehension, writing, art, critical thinking, public speaking

Purpose
Community action means knowing how to bring your message to other citizens. Students familiarize themselves with the different aspects of organizing and publicizing events and how to make the most of dealing with the press.

A key component to acting together as a group—be it school, neighborhood coalition, or nonprofit—is successfully dealing with the media and helping them help you get the word out about your activities and goals.

Handout 17.2 is intended as an informational guide. Here are some ways to incorporate this information into your classroom:

- Have students write a sample press release for an event of your choosing—or theirs. The event can be real or fictitious.
- Have students pick a topic and write a letter to the editor.

ACTIVITY 17.3
Putting It All Together

Materials
pen or pencil
notebook
library or Internet research materials

Primary Subject Areas and Skills
science, social studies, writing, art, research, critical thinking

Purpose
Students use the tools they have assembled thus far in this curriculum to plan the logistics for a publicity campaign to bring attention to an important environmental issue.

Much of this curriculum has focused on bringing what students have learned into the community. This unit has provided practical tools that can help students continue to do that and to take their action to the next level by organizing and acting together.

You might remind students of a classroom activity that resulted in a letter to local officials or a presentation to school administrators. When they did this, they were acting as concerned citizens as part of a small group. Their action was not unlike the work of a "small community action group."

This activity is designed to compel students to bring together many of the tools they've learned to create a fully thought-out approach to promoting a special event or initiative.

This is a very important activity that takes time and effort. But students will see the value of their efforts if they take the time to do this to their best ability and choose an event or issue that means something to them.

1. Explain to students that they are going to create and plan a publicity campaign around an issue that is important to them or their community
2. Divide students into groups of three or four.
3. Student should pick a particular issue or event—recycling, an Earth Day demonstration or talk, the development of a local nature trail, or the like—that they will use as the basis of their campaign.
4. Instruct students that they, as a group, should prepare the following:
   a. An event publicizing their issue: This could be a protest, a rally, road race, marathon, tree planting, recycling effort, stunt, and so on.
   b. An organizing meeting that will give further structure to the publicizing of this event: Whom will they invite? Who will be members of their coalition?
   c. A press release that publicizes this event: Write it. Send it out.
   d. A contact list of local media that they will contact about this event: How did they choose them? Are they the appropriate contacts?
   e. A press conference describing this event: Confirm and follow up with media professionals and give them story ideas.
   f. An op-ed or letter to the editor that voices an opinion about this event.
   g. A follow-up plan for further action and coalition building based on the information that was gathered during this initiative.
   h. A chart, graph, or other visual that details the individuals, groups, or other supporters you engaged during this publicity campaign.

Extension

The best way to drive home the importance of this activity is to have students put it into action. They should not just "act as if" they are taking the steps but actually take them. Have them send the press release out, call the local newspapers, host a press conference, and so on. You may wish to have students create an event centered on a screening of *Edens Lost & Found*. Students could promote the event, arrange for a reception before and perhaps a Q&A session afterward, and invite local media to attend. Please contact *Edens Lost & Found* for help in arranging an educational screening in your area.

EXTENSIONS

Bring in an Expert

Invite someone who either works in public relations or who acts as the media relations director of a local organization to visit your class and share their techniques for bringing their news to the attention of the media. Your local Chamber of Commerce may be of help in identifying the appropriate person.

Talk to the Press

Invite local newspaper editors to come speak to students about the best way to deal with and contact the press. The masthead printed every day in your local newspaper might contain the e-mail addresses and phone numbers of editors.

Invite Community Members

Invite an organizer of a nonprofit community action group to speak about how they first got involved in their organization and what their responsibilities are.

Reduce Global Warming

Since 2005, Mayor Greg Nickels of Seattle has challenged other American mayors to voluntarily adopt the Kyoto Protocol, the international agreement to reduce greenhouse emissions that cause global warming. When this book went to press, the U.S. government had not signed the protocol, saying it would hurt U.S. businesses, but 141 nations have. As of 2008, 902 mayors representing more than 81 million Americans have accepted the challenge. Share the story of Seattle's voluntary adoption of the Kyoto Protocol and challenge students to develop a plan for persuading their city hall to adopt the protocol as well. See the link below and consider viewing the Seattle film in the *Edens Lost & Found* DVD series.

Where's the Controversy?

Train your students to hone in on the controversies behind important laws. Laws such as the Clean Air Act or Clean Water Act are always controversial because each time regulations are reviewed, some politicians want to strengthen them, while others wish to weaken them. The usual objection is that stronger regulations will hurt American businesses forced to implement them, ultimately hurting American jobs. Analyses of the pros and cons behind changes to these laws may provide fodder for interesting classroom debates.

RESOURCES

- **The EPA's guide to the Clean Water Act**
  www.epa.gov/r5water/cwa.htm

- **The EPA's Plain English Guide to the Clean Air Act**
  A website designed to provide information in simple language about the Clean Air Act.
  www.epa.gov/oar/oaqps/peg_caa/pegcaain.html

- **The United Nations and Agenda 21**
  The United Nations' site describing Agenda 21, a comprehensive plan for sustainable development in nations worldwide
  www.un.org/esa/sustdev/documents/agenda21/index.htm

- **Kyoto Protocol and the United Nations**
  The United Nations' background on the Kyoto Protocol.
  unfccc.int/resource/docs/convkp/kpeng.html

- **So Long to CO_2, Seattle!**
  Seattle's website on its mayor's Kyoto Protocol initiative.
  www.seattle.gov/mayor/climate

Reading Handout Unit 17: A LIFETIME OF COMMUNITY ACTION

All Americans have the right to express themselves and affect their local governments through their words and deeds. In this reading, you'll learn about a well-known Chicago resident named Marian Byrnes, who has fought to save wild places in the Chicago area for much of her life. Hers is an excellent example of the power of community action. You can learn more about Ms. Byrnes by watching the Chicago film in the *Edens Lost & Found* DVD set. After you read the story, answer the questions.

If you're going to take a walk with Marian Byrnes, you need to bring along three things: a walking stick, some good shoes, and your conscience. Nothing less will do.

One morning recently, the 79-year-old woman sets out on a walk to a show a visitor some of the wild places in her neighborhood. She lives around the shores of Lake Calumet, on Chicago's southeast side. It's an interesting place because some parts of the neighborhood are famous for the steel factories that once employed many people in the community. The area has long been a site of dumping, legal or otherwise. Other sections of the neighborhood were never developed and are still the original prairie lands that were here before Chicago was born. Ms. Byrnes lived right next door to one of these plots of prairie land most of her life.

The ground under her feet shifts from asphalt to pavement to grass in a matter of minutes, and soon she has left her Jeffrey Manor neighborhood behind. Birds sing in the trees, and the grasses rise as high as your hip in spots. Ms. Byrnes has insisted for decades that she is not a naturalist, and yet she calls out the names of plants as she passes, saluting them as if they are old friends: wild parsnips, wild roses, wild strawberries, fleabane, wild oats, vervain, squirrel grass, and catnip. "Well, since I started work to save the prairie, I learned as much as I could from experts and wildflower identification books," she confesses.

Ms. Byrnes stands straight and has a good spring in her step. Her hair is white, and her voice, when she speaks, is soft and raspy. She grew up on a farm in Indiana and longed for open spaces, even after she came to Chicago to teach school and raise a family.

"I moved here because of the prairie," she says. "I was a widow with three boys to raise. And when we came house hunting, and we came to the house that I live in now, they said, 'This is the house we want!' They hadn't even been inside. They were running toward the prairie. Kids used to come from all over the South Shore to play army on the prairie. The realtor told me that [the prairie] had been a government land grant to the railroad on the condition that it never be sold—that it was supposed to be a buffer between the railroad and any kind of development. But obviously that changed."

In 1979 she came home from teaching school to find a note in her mailbox saying that the Chicago Transit Authority planned to build a bus garage on the north half of her family's beloved Van Vlissingen prairie. "Naturally I was very upset about that, and I went to the public meeting and found a number of my neighbors there who were also upset about it. So we formed an organization, the Committee to Protect the Prairie, to keep it open. We somehow managed to succeed in doing that for 20 years, until the city decided it was also a good idea and took over the project."

In 20 years, Ms. Byrnes and her growing number of friends have thoughtfully prodded the city to do the right thing a number of times in the Calumet region. Ms. Byrnes' Committee has since morphed into the Southeast Environmental Task Force and is part of a larger coalition called the Calumet Stewardship Initiative, which draws its strength from 30 area organizations. They defeated a city proposal to build a dump at Big Marsh, a major birding site. They opposed construction of a garbage incinerator on the site of an old steel mill. And they defused the mayor's plan to build the city's third major airport in the region. At the time, the city thought an airport was the best solution to decades of waste dumped in the region: cover it up with asphalt. Countless other projects, or threats of projects, have sent residents back into the meeting halls.

Eventually area residents forced the city to stop thinking of the Calumet as a dumping ground. The turning point came in 1998, when the National Park Service announced the Calumet was suitable for designation as a National Heritage Area. Two years later, the city and state announced it would allocate funds to save the area. At first they split the acreage down the middle: 3,000 acres of the best land to be preserved for nature and 3,000 acres of the already damaged or denuded fields to be set aside for appropriate industrial use. (The preserve area has since grown to 4,800 acres.) Much of the discussion these days is how to find the right industries to take over the old factories again. The ideal company is one that would add value to the region, not harm it in any way. Part of what makes Ms. Byrnes such an effective crusader is her acceptance of this sensitive balance. "I think we all now agree and accept that industry has a right to be in the region. That's what made the Calumet. But it has to be done intelligently."

Ms. Byrnes is pleased that younger citizens who started on local grassroots organizations have since moved up the ranks in city government themselves, paving the way for a smoother passage of environmentally friendly legislation. And an even younger generation is learning to love the area that is being saved.

"School classes come here for field trips, and the kids have a wonderful time because they can gather any specimens they want to. There's no restriction on gathering," Ms. Byrnes says. "Kids find things here that I've never seen. Last time I brought a group out here, a little boy found a Western frog, a little tiny brown frog, and caught it. We examined it for a while. So it's a really important educational resource."

Recently project leaders have set up ten experimental plots, treating each one differently to find out which works best for prairie restoration. The results will help scientists fine-tune their treatment of the area. At the same time, area residents have started meeting with local factory managers and discovered they too are interested in talking to local residents about the future of the neighborhood.

Ms. Byrnes steps gingerly around the edge of a small pond and spies a small frog and some tadpoles in the water. It's a rare thing to get a private tour with her these days. At an age when most people would consider slowing down, Ms. Byrnes attends at least a meeting a day, shuttling to each on public buses. "It's hard to say what keeps me going," she says. "But there's always something else to do."

But just how has she done it? Ms. Byrnes says she has managed to be an active and powerful voice for so long because she never lost hope. From the very first meeting she attended, she learned something she has never let go of. "I found that there were plenty of people just like me who felt the same way—my own neighbors. If you put us in a room and got us talking, something would happen."

For a while, she thought she would be able to give up her work, now that the city has finally seen the wisdom of saving the Calumet. But she has since changed that notion. The city needs to hear the voices of its people—including Ms. Byrnes's soft but firm one—or it will lose touch again. Much has improved since she became accidentally involved in community action, but she still grieves for what was lost.

She gestures at the land around her. "I'm sorry; I'm probably going to cry. The 'Eden lost' is what used to be here," she says. "Where we are now was once all cattail marsh. I have friends who remember going out with their fathers in the 1930s in a rowboat, where Van Vlissingen Prairie is. They went fishing among the cattails. And now it's all filled in. But the 'Eden found' is the prairie people have worked to restore. The 'Eden found' can be considered superior to the 'Eden lost.' The birds, especially the big birds, have managed to survive. They've become accustomed to this territory, and they persist in returning even though their habitat has shrunk and shrunk. And there definitely are less of them than there were in the original Eden."

Ms. Byrnes continues coming back to the land and the bargaining table, again and again, to raise her voice. Ultimately, she feels, her work will triumph. "I'm hopeful for the future of this area because the city of Chicago and the state [of Illinois] have made this pledge to save thousands of acres of land for permanent preservation, and I think they will hold to that pledge. It makes me feel very happy, wonderful."

And she heads back home through a field of thistle, daisies, and strawberries.

Questions

1. Where in Chicago does Ms. Byrnes live, and why is her neighborhood special?

___

___

2. What environmental problems does the Calumet face?

___

___

3. What happened in 1979 that inspired Ms. Byrnes to take a more active role in her community?

___

___

4. Over the years, what are some of the things the city wanted to do with land in the Calumet area that Ms. Byrnes and her neighbors convinced the city not to do?

____

____

5. How did the big change in the city's attitude come about? What was the turning point for the Calumet?

____

____

6. What do you think of the city and state's solution to protect half of the Calumet's acres?

____

____

7. Many people in the neighborhood need and want jobs. What kinds of business do you think would be perfect for the Calumet area?

____

____

8. How does Ms. Byrnes answer the question about what keeps her going? Do you think she trusts and respects her neighbors? How do you know that?

____

____

Name ______________________ Date ______________

Names in Group______________________________

HANDOUT 17.1: Guidelines to Policy Profiles

POLICY PROFILES

Read the following questions and statements below and use them as a guide to help you and your group complete your policy profile. Write your answers on a separate page.

Policy assigned to your group: ______________________

1. What does the policy say? What is it designed to accomplish?
2. How did the policy come into existence? What is its history?
3. Who does the policy govern (businesses, national/international governments, and the like)?
4. How is the policy enforced? Are there punishments for infractions? If so, what?
5. How well is this policy being followed and enforced? Give evidence to support your position.
6. Do you feel the policy is fair to all? Why or why not?
7. What changes would you make to the policy itself? Is it strict enough? Too strict? Do you think there are potential loopholes in this policy?
8. What changes, if any, would you make to the enforcement? Why?
9. Be sure to include visuals to support your presentation: charts, graphs, maps, video, photos displays, collages—be creative!
10. List the role each team member played in the preparation of this report and presentation. Divide the responsibilities fairly and work together.

Extra Credit:
Create a timeline that shows milestones from the policy's history (when it was first proposed, where it was presented, and when it was ratified by various countries, and so on).

HANDOUT 17.2: Community Action and the Media

WHAT IS A COALITION?

A coalition is an alliance or association of individuals or groups that come together to pursue a common cause. Governments and nations can form coalitions. Churches and neighborhood organizations can form coalitions. Political groups can form coalitions.

Coalitions are often brought together by common concerns and goals. When one group joins with another, the resulting coalition can often accomplish much more.

Most often, small action groups and the coalitions they form will be nonprofit. The terms *nonprofit* and *not-for-profit* are used interchangeably. The big difference between, say, a local chess club and a local church, synagogue, or scouting organization is that the last three are chartered organizations. They have a legal status and are recognized on the state or federal level. Environmental groups such as the Sierra Club are chartered groups. They are legal entities, much like corporations, except that they do not exist to earn a profit. They may earn money but only to pay their employees and run their organizations.

When small action groups band together to form coalitions, they share their strengths. One group may be able to provide excellent media contacts. Another may be great at planning rallies and preparing signs for marchers to wave. Another group may excel at meeting with local business leaders or local politicians.

HOW TO ORGANIZE A COMMUNITY MEETING

Organizing meetings can be more difficult than you think. Who speaks when, what is on the agenda, and how to follow up are all-important issues to consider if you want to get the most out of a meeting—and maintain everyone's enthusiasm.

Before the Meeting

- Know what you want the meeting to accomplish. Set distinct goals.
- Know who you want to attend.
- Know how you want to follow up.
- Publicize appropriately, especially if there are special speakers attending.
- Allow plenty of time to find a meeting place.
- Make sure the meeting time works with the majority of key people.

During the Meeting

- Have a sign-in sheet where attendees can leave their contact information.
- Have an agenda and do your best to stick to it.
- Plan who will speak when and how much time each person will be allotted.
- Assign someone to keep the "minutes" of the meeting.
- Let everyone know how and when there will be follow-up.
- Always end on a positive note, with a plan to get together again.

After the Meeting

- Send the minutes to those who attended and those you think would be interested in knowing more about the topics discussed or any future plans and actions that are coming up.

Welcome feedback—good and bad—about how things can be improved.

Keep the momentum going!

HOW TO WRITE A PRESS RELEASE

A well-written press release can be the key to garnering publicity—and possibly support—for your cause or event.

What to Include

- Contact person's name and phone number and/or e-mail
- The date and location
- Information about the who, what, where, when, why of your event
- Information about the organizing group or groups
- Informational quotes from key people

What *Not* to Include

- Anything that sounds like begging: "Please come! If you don't, no one will know how hard we've worked . . . "
- Also, do not condescend to the reader: "If you don't come, it just proves your lack of concern for real news stories . . . " You get the picture.

In short: Stick to the facts, keep it professional and adhere to the format.

Whom to Send It To

Send your press release to anyone at local newspapers, magazines, TV and radio stations, and government agencies whom you feel needs to get your message or who can give you the kind of publicity and support you need.

Contact the organization. If it's a newspaper or magazine, ask for the editorial department. Explain you have a press release you would like to send them and ask if they prefer e-mail or fax. It is always best if you can address a press release to a particular person. Do some research. Look at the paper and determine who covers the issues your press release discusses. Do the same with TV and radio.

After you've sent your release, do some follow-up leading up to the event itself. First, be sure to phone the office and make sure the appropriate person received your material. Then be sure to stay in touch with them leading up to your event. Media professionals keep very hectic schedules, so it's important you and your organization stay in the forefront of their minds.

The Format

The format for a press release is fairly standard. Basically, you want to sum up the key points of your issue or event on one page. Provide the appropriate contact information and a bit of background so that a prospective media professional can get started researching and developing their coverage. Templates are available on the web; so are services that will send your press release out for you (for a fee, of course). But all you really need is e-mail access or a fax machine and you're in business. By simply following the format and inserting your own information, you can have a professional press release ready to go. Here's a sample:

Contact: Iman Organizer
555-555-5555
contactme@xxx.net

<u>FOR IMMEDIATE RELEASE</u>
June 1, 20__

OUR GREEN TOWN COALITION LAUNCHES INDEPENDENT STUDY OF PROPOSED LANDFILL PROJECT

Local coalition to mobilize investigation of environmental impact

Our Green Town, USA—In June 20__ the Our Green Town Coalition announced a year-long process to analyze unresolved issues surrounding the proposed landfill at Yuck Mountain in Your Fair County. This effort will draw on the technical expertise of more than 10 specialists in ecology, hydrogeology, structural geology, groundwater hydrology, and environmental engineering. The information gathered will be used to critique the basis that the Department of Trash is using to justify the proposed landfill.

"By harnessing the resources of a group of scientists like this, I think we will come up with a more profound statement on how we view the technical basis for the Yuck Mountain Landfill project," said Our Green Town founder Izzy Green. "It is vital that we assure the citizens of our county that if the landfill is built, it will be successful and our citizens will not need to be concerned for their health and safety."

For nearly 20 years, a number of consulting firms, national laboratories, universities, and other experts have been developing and analyzing information to address the landfill's potential environmental impact on nearby communities.

"At this point in time, the scientists will look at all the information and address it from the technical perspectives that *we've* developed and our own understanding of the situation in Your Fair County," Green explained. "Ultimately, this may result in our raising these issues during their license application."

Your Fair County is one of the largest counties in the United States at 15,000 square miles, and it is home to some 200,000 people. Your Fair County has passed several resolutions approving the landfill, contingent on the employment of construction and operation techniques that assure the health and safety of Your Fair County's citizens.

Our Green Town Coalition is a community-based consortium focused on ensuring sustainable practices in the business and private sector. Established in 2005, it seeks to bring individuals, government, schools, and businesses together to work toward a sustainable future.

# # #

HOW TO RUN A PRESS CONFERENCE

When there is an announcement to be made, an issue to be discussed, an event to be highlighted or just some great news to share, a press conference (preceded by a press release, of course) is a great way to do it.

A press conference should not run on too long—maybe a half hour at the most—because the conference is a way of introducing the press to the topic and, ideally, giving them just enough information to get them excited about covering it further.

Decide who will run the press conference. This person should act as a master of ceremonies, welcoming everyone to the press conference, introducing key speakers, and then answering any questions. A schedule should be set well ahead of your conference.

Get confirmation of attendees. This can also help bring more people to the conference. If you find out Mr. Scoop is coming from the local paper, be sure to let people from the TV or radio station know that.

In fact, be sure you know the names of all press members attending. This is key to developing ongoing relationships.

Before and after the press conference, make members of the press aware of any important "angles" they might not know about that would make covering your event more enticing.

Always be professional and polite when dealing with reporters and aware of any deadlines they may have. Make it as easy as possible for them to give your story the coverage it deserves.

Always follow up with key people as a way of cultivating your relationship with the press and making sure they haven't forgotten about your organization.

HOW TO WRITE AN OP-ED OR LETTER TO THE EDITOR

One of the best ways to get involved is to interact with your newspaper. This seems foreign to some people, but the structure of a newspaper does provide for this. First, it is important to become familiar with the editorial or opinion section of the newspaper. It has several main parts.

Editorials

This is where the opinions of the newspaper are expressed. There is no byline or name attached to these opinion pieces, as they are the opinion of the editorial board as a whole. This is where you might find an endorsement for a political candidate, for example.

Letters to the Editor

These are short responses to the editorials or other content in the newspaper by any reader.

Op-Eds

Editorials or op-eds are done by newspaper staff columnists and "guests," who could be anyone who can write a solid, engaging piece of 500 words or so that expresses a particular stance or position.

Every newspaper has information on how to submit letters to the editor and op-eds. To make your submission the strongest, be sure to back up your opinion and position with research and statistics if possible. Think of it as writing the pro or con portion of a debate.

GETTING THE WORD OUT

You can get the public's attention in many ways, and not all of them follow any particular set of rules or guidelines. In fact, the more creative you are, the more likely you are to get attention. Here are just a few ideas:

- Hold a rally and invite motivational speakers.
- Have a bake sale featuring goods from local farmers and bakers.
- Dress up as trees and stand on the corner of a busy intersection to raise awareness of a tree-planting initiative you're sponsoring.
- Give free talks at your local library.
- Start a blog or a podcast where you discuss topics important to your organization.

In all of these cases, being creative is key and staying positive even in the face of adversity will increase your rate of success.

UNIT 18

SUSTAINABLE COMMERCE

TOPIC BACKGROUND

Throughout this curriculum, your students have been introduced to numerous businesspeople who are doing things the "green way" and succeeding. Nearly every issue we explored in this curriculum could potentially be turned into an economic opportunity, if creative entrepreneurs take it to the next level.

In case you have not yet used these units with your students, we mention a few here that delve into sustainable commerce:

- In Unit 4, we saw how green building is transforming American home building.
- In Unit 5, we learned about entrepreneurs who are turning vegetable oil into biodiesel and corn into ethanol to power American vehicles.
- In Unit 8, we read about Chicago residents who are restoring the Calumet region and are hoping to turn their neighborhoods around through ecotourism. The same neighborhood hopes someday to turn some of its old steel factories into museums for paying tourists.
- In Unit 12, we read about how farmers such as Mary Corboy and chefs are growing food in cities. In the Philadelphia film, your students met Judy Wicks, a local restaurateur who committed to supporting local agriculture—and hopes her customers will support her in turn. Our nation will always need people to raise and prepare food.
- In Units 13 and 15, we saw how there is money to be made in redesigning parks, suburbs, and cities with creative landscaping to eliminate the problems of overheated neighborhoods, flooding, and toxic runoff. In the story of Andy Lipkis, we saw how cisterns, trees, and smart planting can turn a city around.
- In Unit 17, we saw how Marian Byrnes and her neighbors are improving community relations by meeting regularly with factory managers and working to find businesses to take over old abandoned factories.

Your students will recall other examples if they viewed the films as well. There's the story of the sustainable fishing off the coast of Seattle. And in the companion DVD to this unit, students are introduced to the world of green bankers, who help find money to give new green businesses a head start.

Green businesses such as the Seattle biodiesel firm use some form of sustainable technology that is energy efficient and less taxing on the planet. But technology doesn't have to play a role in a business for sustainable commerce to prosper. Small, local businesses reduce our dependence on foreign oil because goods and services don't need to be shipped from far-flung locales. The concept of buying locally produced goods not only reduces our carbon footprint but supports local economies as well. When you patronize small, independent businesses, you help your community reduce its dependence on giant chain businesses. When new industries stick with American cities (rather than moving elsewhere) they inject life into cities, which is good for the company, its employees, and the economy of that city. When shopkeepers open a coffee shop or business in a depressed community, they stand a chance of reviving that neighborhood.

Purpose

In this unit, students develop an understanding of the ideals of sustainable commerce. They observe some local businesses and determine ways that those businesses can be greener. Lastly, they determine ways that they can support green businesses.

GROUP ICEBREAKERS

1. Ask students: "Have any of you thought about how much money could be made by working to make our world more sustainable? What 'green'

businesses have you heard of?" (Answers will vary.)

2. Ask: "Think of some of the stories you read or the speakers you heard in our *Edens Lost & Found* classes. What are some of the businesses these people run?" (Listen to some responses and then share some of the examples listed above to jog their memories. Do not be too specific, as you will want them to recall much of this on their own in one of the following unit activities. If they do not recall a specific unit or you have not yet used it, reassign it or have them read it now.)
3. Then ask: "Do you think you could invent a 'green' product or come up with some ideas for a green business?" (Have them suggest some things and write them on the board.)
4. Ask: "Do you think being green or sustainable would be good or bad for a business?" Have them list some ways it would be good. Then list some ways it would be bad. (Accept all answers and try to get a discussion going.)
5. Encourage them to read the business pages or watch some business news programs on TV, keeping an eye out for sustainable businesses, ideas for green businesses, or interesting stories about companies that have changed their policies in order to become more sustainable. (For example, some major fast-food restaurants and discount stores have started selling organic foods and sustainably caught fish.) Have them share the news stories in class.

ACTIVITY 18.1

Interview a Green Business Owner

Materials

pen or pencil
library or Internet research materials
Handout 18.1

Primary Subject Areas and Skills

social studies, science, writing, critical thinking, public speaking

Purpose

Students interview a local business owner to learn what steps that person has taken to make his or her business more sustainable. They also learn how local citizens can help that business achieve its goals.

To perform this activity, your students must have access to a businessperson who is either (a) running a "green" business (a business engaged in green technology such as installing solar panels or providing goods and services in a sustainable way such as growing and selling organic products) or (b) running a conventional business that is taking steps to be more sustainable or green—such as mall store that is implementing recycling practices, encouraging carpooling among its employees, switching its lights to compact fluorescent lightbulbs, or using sustainably manufactured materials.

Ideally, students should research and identify the businessperson they will interview on their own; but, in the interest of expediency and safety, you may wish to research, identify, and prescreen potential businesses on your own. (If this is difficult, enlist the assistance of a parent or another teacher.) Then you could provide your students with a "safe" list of businesses to choose from. It may be necessary for you to choose one particular day that students will shadow their mentor. Arrange with your principal for the necessary passes and excuses from school.

You may also choose to have students conduct a phone or e-mail interview so they do not need to leave school grounds. *Note:* If you choose this route, an adult must supervise all phone calls or approve all e-mail communications between students and the businessperson. Parents of all participating students should sign a permission form.

1. Have pairs of students choose a business on their own or from a preapproved list of local businesses, and contact this professional by phone, e-mail, or conventional mail.
2. Distribute Handout 18.1, which gives students instructions for the activity and a list of questions they are expected to ask the businessperson. Answer any questions they may have about how to proceed.
3. Compile a list of the professionals whom students have chosen to profile and confirm the appointments. If students are interviewing the individual in person, have them select a suitable chaperone to accompany them on that day. Parents of each child must be informed of, and approve of, the chaperone.
4. Discuss conduct, manners, and dress code for the interview.
5. Have students share the results of their interviews with the class.

ACTIVITY 18.2

The Green Business Plan

Materials

pen or pencil
library or Internet research materials
Handout 18.2

Primary Subject Areas and Skills

social studies, science, writing, critical thinking, art, public speaking

Purpose

Students write a business plan for creating their own green business.

Taking sustainability issues out of the realm of the nonprofit and into the business world of for-profit is a key element to establishing many of the practices and philosophies described in this curriculum. From environmental law to green building, a growing number of jobs and businesses have come out of the sustainability movement.

Before a banker will loan money to a new business, he or she needs to read a business plan, which is a document that describes the business and predicts how much it will earn.

In this activity, students will design and flesh out the practical business applications associated with green economics.

1. Ask students if they think committing to sustainable practices can be profitable or whether it costs more money. Why do they think the way they do?
2. Review some of the green jobs and businesses that were contacted in Activity 18.1 or discussed in the course of this curriculum. (We mention several in the Topic Background at the beginning of this unit.)
3. Divide students into groups of three or four or have the class work on this project as a whole. Have them spend some time brainstorming the type of business they wish to start. How will it be green or sustainable?
4. Explain that they will next work together to write a business plan.
5. Distribute the business plan guide to students. The plans should be well researched and should be written in presentation form. Additional information is included in the resources section, with links to the U.S. Small Business Administration and www.bplans.com. Both provide sample plans and extensive information on how to write a business plan of one's own.
6. Once plans have been completed, students should present their plans to their classmates.
7. You may wish to present two or three plans per day and have the students who are not presenting vote on the plan they think is best. Constructive criticism and a question-and-answer period should follow each presentation.
8. Ask students how they think their business would work in their community specifically. Is there a need? Would there be support? Why or why not?

Extension

Have students refine and expand their business plans for presentation to the school. How open is your town or region to green businesses? Arrange for students to present their plans to your local Chamber of Commerce or other business group.

EXTENSIONS

Search the Chamber

Invite a member of your local Chamber of Commerce or appropriate organization to come speak to your class on present and possible future "green" jobs in your area.

Social Responsible Investing

Have students investigate the world of "socially responsible investing," in which stock market investors buy stocks only in companies that are following desirable business practices. For example, an investor could choose not to invest in companies that manufacture tobacco products and invest instead in companies that promote renewable energy products.

RESOURCES

How to Write Business Plans

This helpful website includes sample business plans you can view and download, as well as other helpful hints on writing business plans: www.bplans.com.

Small Business Primer

The U.S. Small Business Administration provides this primer that describes how to create an effective business plan:

www.sba.gov/starting_business/planning/basic.html.

Check your local office supply stores

Many of them sell business plan templates at very low prices. You could purchase one and share it with all the students in the class.

The Earth Policy Institute

This organization is "dedicated to providing a vision of an environmentally sustainable economy—an eco-economy." Titles such as *Eco-Economy: Building an Economy for the Earth* are available for purchase on the site:

www.earth-policy.org/index.htm

Reading Handout Unit 18:
MEET A GREEN BANKER

How important are banks that invest in green businesses? In this reading, you'll meet a banker who lends money to large and small sustainable businesses, as well as to ordinary homeowners with green goals in mind. His bank has made a difference all over the United States. You can learn more about Peter Liu by watching the companion DVD for Unit 18. After you read the story, answer the questions.

Not long ago, a man was interested in opening a bank account at a new bank he'd heard of in San Francisco. New Resource Bank opened in September 2006 to rave reviews. The prospective customer sent an e-mail to the bank, asking a couple of questions. Normally, such questions are answered by customer service representatives who are low on the totem pole in terms of the bank's hierarchy. But when the customer received a note back, he was astonished to find that the person who had responded to his simple question was none other than Peter Liu, the bank's vice chairman and the man who started the bank in the first place.

For Mr. Liu, whose dream was to open a bank that would cater to the needs of green consumers and green businesses, it was just another day at the office. He routinely reads e-mails from customers and enjoys responding whenever he has a chance.

Mr. Liu's unusual bank is on the cutting edge of the green economy, helping green and sustainable businesses get a foot in the door.

Everyone knows that banks can make you money. If you deposit $100 in a savings account paying 4 percent interest at the beginning of the year, at the end of the year you will have $104 in your account. The extra $4 is money that the bank paid you for allowing them to handle your money. The more you save, the more you earn. If you have $1,000 in your bank account at the beginning of the year, at the end you'll have $1,040. In a year, $10,000 will grow to $10,400. Without lifting a finger, you will earn that money simply for allowing the bank to hold onto your money.

But what does the bank do with your money while it is sitting in your savings account? Simple: Banks make money by lending your money to people who need it. These people use the money to start businesses, hire employees, buy goods and materials they will need to get their new ventures started. Then, over time, they pay that loan back to the bank, adding a little bit of interest to each payment. Interest is the fee the bank charges for loaning that money to someone in the first place. Eventually, a portion of the money a bank earns on interest will make its way into your bank account as well—a well-deserved "thank you" from the bank for loaning them the money.

Mr. Liu's bank is unusual or different because it focuses almost exclusively on loaning money to businesses that other banks don't find desirable. Historically, small companies that take the green route find it difficult to borrow money because they have been wrongly perceived as unprofitable. In the past, that may have been true, but it is certainly no longer true, as more and more Americans are thinking of patronizing these types of businesses first.

New Resource Bank finds it profitable to loan money to lots of green organizations and businesses, from small companies that make organic cheese to companies that install solar panels on people's roofs. The bank partnered with a California solar power company to offer home owners loans so they can install solar panels on their roofs without having to come up with the money on their own. The bank loans the home owner the money, and he or she pays it back over time. In the meantime, the home is generating free energy, courtesy of the sun.

Mr. Liu, who was born in Taiwan, first worked in the United States as an engineer on oil refinery projects. The job paid well, but he was unhappy because he knew that these kinds of projects were damaging the environment and making people in Texas and Louisiana sick as well. So he gave it up and started working at banks. Eventually, he wooed investors who could help him make his dream of starting his own green business bank a reality.

"We believe that green has evolved from a social movement to a major market opportunity," he says. "We are providing solutions for green businesses."

Say an entrepreneur wants to sell organic produce to restaurants but needs money to buy trucks and other equipment to help transport her vegetables to market. She fills out an application at New Resource Bank, and if she meets the bank's criteria, she'll be allowed to borrow money from them. She'll sign some documents agreeing to pay the loan back over time at a certain interest rate. In no time, she'll be on her way, earning money for herself and the bank in turn.

"Our bank matters because our clients matter," Mr. Liu says. "We don't like to take too much credit for what we do. Our clients are out there providing the marketplace with products that are cleaner, greener, and more efficient than what's already out there. They tell us, 'Your banking is allowing us to make a difference.' So that's what we do: We are helping our clients matter."

New Resource Bank sees more growth in this market than ever before. "People are caring more abut the food that they put in their bodies," Mr. Liu says. "They care about how things are produced. They care about climate change. So there's nothing visionary about what we're doing. It's very practical, and very commonsensical. We need to build our economy without destroying our surroundings and poisoning ourselves."

As long as there are people who want cars that don't pollute, food that's organic, clothes that are healthy, and energy-efficient homes, a market will exist to make those things and sell them. But to do all those things, smart entrepreneurs need someone to lend them the money to make these dreams possible.

New Resource Bank is funding the new green economy.

Questions

1. How do bank customers make money from a bank?

____

____

2. How do banks make money with their customers' money?

____

____

3. In banking terms, what is "interest"?

____

____

4. How does a bank loan work?

____

____

5. The authors suggest that Mr. Liu is an important part of the new world of green, sustainable commerce. Do you agree? Why or why not?

____

____

6. Mr. Liu doesn't make any of the products he is talking about, such as organic produce, solar panels, and so on. So how can he be an important part of sustainable commerce?

____

____

7. What kinds of businesses does Mr. Liu lend money to?

____

____

8. Why do you think other banks have not wanted to lend money to small, green businesses?

____

____

9. Can you think of three groups of people Mr. Liu is helping?

____

____

10. Do you think sustainable commerce is a positive thing? Why or why not?

____

____

Name ______________________________ Date ____________________

HANDOUT 18.1: Guidelines: Interview a Green Business Owner

1. Use phone books or contact the local Chamber of Commerce to help you locate appropriate professionals.

2. Contact professionals via phone, e-mail, or conventional mail explaining that, for a school project, you would like spend a day with them at work and interview them about their career.

3. Report to your teacher whom you are going to interview and when.

4. All paper correspondence should be written as a form letter, with the appropriate forms of address and salutations.

5. On the day you interview your subject, be sure to dress appropriately. Follow the same rules you must follow at school regarding dress. Leave MP3 players and other such devices at home. Use your best manners. Bring a notebook to keep notes.

6. Before interviewing your subject, create a list of questions that you want to ask him or her. In addition to your own questions, you must ask the following:

    - What is your name?
    - What is your business?
    - We are learning about sustainable business practices in school. What sustainable or green business practices does your company follow?
    - Can you tell me more about them?
    - What is easy about that the green practices you are following?
    - What is challenging about them?
    - Are you planning anything in the future that will help your company be more sustainable?
    - Can you tell me more about these plans?
    - How can local people like me support your efforts to be green and sustainable?

7. After the interview, present your findings to the class.

Name ______________________ Date ______________

HANDOUT 18.2: The Green Business Plan

Use the guidelines below to help you create a business plan for the business of your choice.

- ❒ Choose your business. Give it a name and location.
- ❒ Refer to websites and other resources for sample business plans.
- ❒ Include graphics: charts or graphs, perhaps a flowchart or projected growth plan.
- ❒ Divide up your responsibilities fairly. Include with your report a work breakdown for all of your group members.
- ❒ Write your business plan.
- ❒ Be as creative as possible. Have fun!

The business plan must answer:

- ❒ What services or products will you provide?
- ❒ How will the services be delivered?
- ❒ What need does this fill in your community?
- ❒ Who will buy your service or product?
- ❒ Why will they purchase from you?
- ❒ How will you advertise your service or product?
- ❒ What will your start-up costs be?
- ❒ Where will you get the financial resources to start your business?
- ❒ How many employees will you start out with? What will be their functions? What will they cost to employ?
- ❒ How will your number of employees grow over the next year? Over the next five years?
- ❒ What's the opportunity?
- ❒ What is your expected revenue the first year? In five years?
- ❒ Who are your competitors?
- ❒ What are the weaknesses in your plan? What are the plan's strengths? What are the threats to your plan?

Your report must include:

1. Table of contents
2. Executive summary (one page)
3. A description of the structure of your company
4. A description of your product or service
5. An analysis of the market in your area for this product or service
6. A marketing and publicity plan
7. Operations plan (how will your day-to-day work flow?)
8. Financial loan
9. Management plan
10. Visuals: charts, graphs, statistics related to your business

UNIT 19

GREEN-COLLAR CAREERS

TOPIC BACKGROUND

Your students are the future of sustainable commerce and business opportunities. Much like the generation of young people who learned the importance of recycling, and who as adults cannot imagine tossing a plastic bottle in the trash, the education your students receive today will shape the way they behave in the workplace. We need them to graduate, sharpen their talents, and take part in making new green businesses thrive. We need them to care enough about Earth's future that they will be moved to squeeze wasteful practices out of the workplace. There are smart energy-saving tips and technologies that need to be invented, developed, and brought to the American public and the world. We want your young people to launch initiatives that will benefit all humans and their planet.

Workers of all skill levels and training are needed in sustainable businesses. In the realm of urban park or urban forestry, for example, we need people with college degrees (landscape architects, designers, arborists, and the like) as well as people with simply a high school degree and a desire to work (nursery workers, maintenance people, groundskeepers, and the like). The opportunities are endless. Throughout this curriculum, we have tried to highlight as many career development possibilities as possible, with suggestions that you invite some people from the community to visit the classroom.

The world of sustainable commerce is now—and will be—rich and rewarding, and all adults should be encouraging young people to explore these opportunities. Careers are typically described as being "blue-collar" or "white collar," but new jobs in the world of ecobusiness are whimsically dubbed "green collar" jobs."

Purpose

In this unit, students will develop an appreciation for green collar jobs, observe some grown-ups in the workplace, and consider whether green collars jobs are for them. In the spirit of service learning, we ask your students to dream up and execute ways of educating students younger than themselves about environmental issues.

GROUP ICEBREAKERS

1. Ask: "Does anyone here think he or she would like to work in a job that supports green practices?" (Help the students to see that even if their dream job is not overtly "green," such as a lawyer, car salesperson, or accountant, they may well have green-business clients or serve green causes indirectly. For example, they may work for a bookstore that adopts green practices to save energy.)
2. What does it mean to have a blue- or white-collar job? (Blue-collar jobs are traditionally those that don't require a great deal of schooling, such as factory or manufacturing jobs. White-collar jobs—which refer to office and management-type positions—usually require an advanced degree.)
3. What do they think a green-collar job is? (A green-collar job is any one that is focused on green technology and sustainable goods and services such as the ones studied in the last unit.) Does it sound like a job they would want to have? (Answers will vary. Encourage students to imagine if their current dream jobs could also be a green-collar job? Most usually can.)
4. Can they think of ways to find out about green-collar jobs before they have to make a decision about choosing a particularly field of study? (The best way is to meet and talk to grown-ups already doing those jobs. They can also research and read about green-collar jobs in business magazines, newspapers, and online.)

ACTIVITY 19.1
The Green Workplace

Materials

pen or pencil; Handout 19.1

Primary Subject Areas and Skills

science, social studies, writing, critical thinking

Purpose

Through conversation and note keeping, students brainstorm types of green-collar jobs that grow naturally from various green or sustainable businesses.

A popular misconception is that pursuing environmental causes—alternative power, recycling, and so on—doesn't help the economy. But this is far from the truth, if one simply thinks about the kinds of jobs and economy that could be created by intelligently exploring environmentally sound business ventures.

In this activity, students think about what kinds of jobs might be created by various environmentally conscious businesses. The activity works best with active conversation and discussion in support of the written brainstorming itself.

On Handout 19.1, students should list a topic area, such as biodiesel fuel, and a job that might stem from that topic area, in this case bioengineer or gas station owner. Students should list as many jobs as they can think of, especially jobs that require varying levels of education and those that serve different sections of the community.

1. Ask students if they can think of any jobs in their community that have a sustainable focus. Write answers on the board.
2. Ask students what kind of education they think this job requires and what sector of the community it serves.
3. Distribute Handout 19.1 to students. Explain that they will be thinking of jobs that might arise from some of the industries related to sustainable practices. We have given an example on the handout. (If students need prompting, suggest they review the unit titles in this book, or the various businesses mentioned in the Topic Background for Unit 18, on page 271.)
4. After students have completed the activity, share answers in a class discussion and brainstorm more industries and job markets together, feeding off the ideas that have been generated by the students themselves. How many of these jobs require a high school diploma only? A college education? An advanced degree? Which jobs are good for an individual business owner? How do these industries create jobs and support the economy?

Extensions

1. Use the answers gathered to create a report for your local chamber of commerce, a local economic development agency, or a college or university to highlight this growing job sector.
2. Ask students if they think there will be more jobs with a sustainable or green focus by the time they graduate from college. What trends do they anticipate? Why do they think—or why *don't* they think—jobs in this sector might grow? What might prevent such growth?
3. Ask students to write an essay about a job they might want that has a sustainable focus. They are free to create a job that does not exist yet.

ACTIVITY 19.2
Shadow a Green Mentor

Materials

pen or pencil
library or Internet research materials
Handout 19.2

Primary Subject Areas and Skills

social studies, science, writing, critical thinking, hands-on

Purpose

Students learn about the value of a good mentor, how to select one, and how to shadow one for a close look at that person's day-to-day responsibilities.

The role of quality mentors in the life of a student should not be undervalued. While high school students may spend a good deal of time thinking about the future and what they want to pursue, they are most often doing this in the abstract. Often there is little time to explore any of their work interests. Shadowing, which involves a student following alongside a professional in a field of interest, gives the student a firsthand look at what day-to-day work might be like if she or he pursues that career.

It may be necessary for you to choose one particular day that students will shadow their mentor. Arrange with your principal for the necessary passes and excuses from school.

If the appropriate individuals are not in your town or region, have students choose another field or conduct a series of phone or e-mail interviews. *Note:* Shadowing should be done in pairs or with a school-approved chaperone of some kind. Parents of all participating students should sign a permission form.

1. Ask students to describe a sustainable career that interests them.
2. Request permission from your school and signed permission from parents before allowing students to participate in this activity.
3. Have pairs of students identify a professional in their field of interest and contact this professional by phone, e-mail, or mail.
4. Distribute Handout 19.2, which gives students guidelines to follow. Answer any questions they may have about how to proceed with their profile or shadow day.
5. Keep a list of the professionals that students have chosen to profile, as well as confirming the appointments. Have students select a suitable chaperone to accompany them on that day. Parents of each child must be informed of, and approve of, the chaperone.
6. Discuss conduct, manners, and dress code for the shadow day.
7. If you opt to have students profile their subjects rather than shadow them, instruct them to do two interviews. The first interview will be a get-acquainted followed by asking a list of questions. A follow-up interview allows the student time to go over questions that need further clarification or to ask questions forgotten in the first round.
8. Have students share the reports of their conversations and experiences with the class.

ACTIVITY 19.3
Sharing What You've Learned

Materials

pen or pencil
library or Internet research materials
Handout 19.3

Primary Subject Areas and Skills

science, social studies, math, writing, art, critical thinking, public speaking

Purpose

To help excite an even younger group of students with the promise of green-collar careers, students will create, design, and implement an environmental education unit on green jobs for younger students in their school and/or town.

Service learning is a theme that has been present throughout this curriculum. As the name implies, it involves learning through service, sharing, or teaching. In this activity, students will have a chance to take service learning to a new level.

One job students may not have considered is one that you, their teacher, have tacitly been modeling for them: environmental education teacher. As more and more citizens become interested in environmental issues, the role of educators will become more important. Teaching people about environmental issues will be a fast-growing and rewarding green job.

Your students will have a chance to see what this job is like by doing this activity.

In preparation for this activity, you may wish to refer to the Chicago DVD in the *Edens Lost & Found* DVD series, the Chicago chapter of the companion book, or the unit 19 DVD. These all contain information about teacher Deb Perryman and her students in Illinois. (There is also information about Ms. Perryman and her approach to service learning in the introductory material for this curriculum. See pages xxix–xxxi.) She and her students shared their newly acquired knowledge with younger students in their community.

1. Review the idea of service learning with students. Remind them of some of the service learning activities that they have participated in up to this point.
2. Review any materials related to service learning contained in this curriculum, listed above. After your students have watched Ms. Perryman's students work with younger kids in the Chicago DVD, ask your students: How is what they're doing service learning? Who benefits and how? (By teaching younger students, Ms. Perryman's high schoolers are reinforcing what they themselves have learned in class, and they are educating others.)
3. Divide students into groups of three or four. Explain that they will be working together to create an environmental education activity for sixth graders (or another age group of your choice) based on what they've learned about green-collar jobs or about any other topic.

4. Distribute Handout 19.3 to students. Explain that they will choose a topic and, based on that topic, write a two-page article with questions, create two handouts, and design an outdoor activity. Answer any questions students have about the assignment.
5. Remind students that the topic they choose should ideally be relevant to your community or region. Keep a list of which groups are covering what topics. If possible, each group should focus on a different topic. If this is not possible, guide groups into focusing on different aspects of the topics as they create their units.
6. The plans should be written up clearly, in presentation form. Students should take turns "teaching" their units to the other students in the class and receiving feedback from them.
7. Have students improve or revise their unit, if necessary, based on their classmates' feedback.
8. After students have revised their units, arrange for each group to teach their unit to the designated group of students that you have chosen. Perhaps a field trip could be arranged in conjunction with the younger students your class will be instructing. During this time, each group of "teacher" students could work with their younger pupils in the morning and participate in their outdoor activity in the afternoon.

Extension

Have students publish their units in a bound format and make it available in your larger school district. Your goal should be to have other schools in your district make use of the materials, so be sure to publish them in a clear, legible format.

EXTENSIONS

Have Jobs Come to You

Host a Green Career Day at your school. Invite those professionals who were profiled by your students and any others who might have interesting input. Ask them to talk about what they do and how their job might evolve in the future.

Teacher Conferences

If your students' unit is truly excellent, you might consider having them present it with you at your next local or regional teacher's conference. Have them prepare handouts ahead of time to share with teacher-attendees.

RESOURCES

Environmental Career Opportunities

This site publicizes available positions. This is a good place for students to see the many jobs available that are concerned with the environment. Students can browse to what skills and education are required, as well: www.ecojobs.com/index.php

Environmental Jobs and Careers

As well as more extensive jobs listings, this site also offers a directory of government agencies, environmental companies, law firms, nonprofits, and state jobs involved in the environment: www.ejobs.org/

The North American Association for Environmental Education

For 35 years, the NAAEE has been the nation's premier clearinghouse for environmental education resources. Each year it sponsors a national environmental education conference. This is a great place for students to see environmental education tools and see what might be appropriate for different grade levels, and it has wonderful links: www.naaee.org/

Reading Handout Unit 19:
ALL ABOUT GREEN-COLLAR JOBS

Do you want to earn money and save the planet too? You might be interested in a so-called green-collar job. In this reading, we find how these types of jobs have the power to transform the workplace, employ vast swaths of people, and usher in a better, healthier planet for us all.

Peter Liu, the banker you met in the last unit, once worked as an engineer in the oil industry. But he left that career because he felt strongly that large-scale petroleum businesses were doing more harm than good. Now he's happily employed as a banker funding hundreds of new ecobusinesses and the green-collar jobs that they will undoubtedly generate.

Citizens have historically described themselves as having blue-collar or white-collar jobs. A blue-collar job usually takes place on the assembly line in factories or in other manufacturing settings. White-collar jobs are office jobs, and most of their practitioners have college degrees. Today we stand at the dawn of a new era of green-collar jobs. A green-collar job is a job that is good for Earth. It's the type of job you've seen or read about throughout the *Edens Lost & Found* series. A green-collar worker is an engineer who is helping to create new fuels. It's a mechanic who is working on new hybrid, flex-fuel, or hydrogen-fuel cars. It's a farmer who grows only organic produce, a grocer who sells it at market, and a chef who prepares that food in an organics-only restaurant. Green-collar workers are standing on America's rooftops, installing solar panels; they're building and designing green homes; and they're creating appliances and technologies that are more and more energy efficient.

Some of these jobs—farmers, grocers, chefs, architects, designers, and so on—have existed for a long time. But because the expertise required to make something more efficient, clean, and green represents a new way of doing things, these careers are being transformed forever. Young people emerging from school may actually have better training, be more passionate about, and be quicker to adapt to the needs of the new eco-economy. That means that you and your fellow students are getting in on the ground floor of the next big economic boom.

The environmental and human rights activist Van Jones thinks green-collar jobs may also save the youth of America's cities. He points out that poor neighborhoods are often the first to be exposed to the worst of the nation's bad environmental legacy. Cancer and asthma, he says, are the result of illegal dumping in poor neighborhoods. But now, if the young people of these neighborhoods can be taught new trades, they may find a green pathway out of poverty.

"If you can teach them to put up solar panels, they're on their way to becoming electricians or electrical engineers," says Mr. Jones, whose organization, Green For All, has launched a national campaign to publicize this issue. "If you can teach a young person how to weatherize a building so it won't leak so much energy and we can bring those greenhouse gases down, that young person is on the way to becoming a glazer [a person who installs and creates windows]. These are the jobs of the future."

The best part about all of these jobs is that they have to be done here, in the United States. The jobs cannot be exported to foreign nations because the work needs to be performed on land, structures, homes, and office places where we all live.

Everybody knows that oil prices are on the rise. The time is coming, says Mr. Jones, when every building is every city in Americas will need to be insulated to reduce its energy consumption and make its operation more affordable for landlords, home owners, and tenants everywhere. Who will do that work? Who will build the green roofs of tomorrow?

The answer is YOU.

Questions

1. What is a blue-collar job? What is a white-collar job? How do they differ?

___

___

2. How do the authors define a green-collar job?

___

___

3. What are some of the jobs the authors describe as being green-collar jobs?

___

___

4. How can a person's job be good for Earth?

___

___

5. Why do the authors think people your age have an edge over older people when it comes to green-collar jobs?

___

___

6. Why does Mr. Jones think green-collar jobs can help citizens in the poorest neighborhoods?

___

___

7. How have poor neighborhoods been poisoned by bad environmental decisions?

____

____

8. Why does Mr. Jones think green-collar jobs cannot be exported overseas?

____

____

9. What is the relationship between the high cost of oil and the green-collar economy?

____

____

10. Are you interested in a green-collar job? If so, what type of job?

____

____

Name ______________________________ Date ____________________

HANDOUT 19.1: The Green Workplace

Fill in the chart below. List any sustainable topics and areas of focus you can think of and then list jobs you can imagine might stem from that industry. List as many jobs as you can think of. An example is done for you.

Subject Area or Area of Study	Jobs
Biodiesel Fuel	Bioengineering research, gas station owner, fuel distributor

HANDOUT 19.2: Shadow a Green Mentor

1. Use phone books or contact the local chamber of commerce to help you locate appropriate professionals.

2. Contact professionals via phone, e-mail, or mail explaining that, for a school project, you would like spend a day with them at work and interview them about their career.

3. Report to your teacher whom you are going to shadow and when.

4. All correspondence should be written as a form letter, with the appropriate forms of address and salutations.

5. On the day you shadow your subject, be sure to dress appropriately. Follow the same rules you must follow at school regarding dress. Leave MP3 players and other such devices at home. Use your best manners.

6. Before interviewing your subject, create a list of important questions that you want to ask him or her.

7. Bring a notebook on your trip and take detailed notes about what a day is like in the work life of your subject.

8. Write a full report for the class about what you observed and experienced. Be sure to include your impressions of what you learned about this job: What did you like? What did you dislike? Could you see yourself pursuing this career? Why or why not? Did this experience make you more or less determined to pursue this career?

HANDOUT 19.3: Sharing What You've Learned: Guidelines

Use the guidelines below to help you create your environmental educational unit.

1. Choose a topic that interests all of you. You may wish to look at topics that directly affect your immediate area or something that has been recently written in your local paper. Topic idea: How can you interest younger kids in green jobs?

2. Research your topic well. Do not limit yourself to the Internet. Call and speak to someone who knows about the topic—a member of an environmental organization or a journalist who covers the topic.

3. Prepare a two-page reading on your topic. Remember that you are writing for students younger than you. Write for your audience.

4. Include questions based on the reading, some to be answered and others for discussion.

5. Create two handouts to go with your unit. Review some of the handouts you've completed throughout this curriculum and look at other resources in your library.

6. Plan an outdoor activity that can be done with the students. Again, you might want to review activities that have been done in this course, such as planning a nature trail. It is best if you plan something in which everyone can participate and that benefits your community.

APPENDIX A

ANSWER KEY

UNIT 1: UNDERSTANDING SUSTAINABILITY

READING HANDOUT: The Path to Sustainability

1. Examples may include: Everything that is discarded is reused and channeled into new growth. Nothing is wasted, whether the material is logs, water, or leaves. As animals die and decompose, they break down and return to the soil.
2. An urban ecosystem is one that develops from human-made structures in cities, such as buildings, roads, playgrounds, and so on. The system uses human-made materials in additional to naturally occurring materials.
3. A forest can more easily break down and absorb waste as nutrients. A city cannot do this as efficiently because much of its natural environment has been covered by materials that are not porous and do not break down and allow natural substances to penetrate them. For example: A city's asphalt does not absorb rain as efficiently as the forest floor and therefore contributes to runoff.
4. Many of the materials used to construct buildings and roads absorb and radiate heat rather than dissipate it.
5. Answers will vary. Cities can certainly be more sustainable. The more trees that are planted, the more energy efficient appliances that are in use, the more mass transportation employed—and all of these raise the level of sustainability of a city.
6. If people plant more trees, build more green and open spaces, and commit to collecting, recycling, and mulching their organic materials, cities can more closely resemble the sustainable system that is a forest.
7. Answers will vary. Students' answers should include the idea of growing without jeopardizing the needs of future inhabitants and organisms.

ACTIVITY 1.1: The Sustainability Game

Discussion question responses included in text.

ACTIVITY 1.2: Making Connections in the Urban Ecosystem

Question responses included in text.

ACTIVITY 1.3: Sustainable or Not Sustainable?

Handout 1.3

1. a. You are recycling organic material in your garden.
   b. You are using gasoline to power your mower.
2. a. The low-flow fixtures reduce water intake.
   b. Water is wasted by taking such a long shower and not turning off the faucet while you brush your teeth.
3. a. Maintaining your automobile helps increase its fuel efficiency, using fewer fossil fuels.
   b. When the oil is washed down the sewer, it enters and harms the water system, where it can also damage animal and plant life.
4. a. The growing of organic food is less harmful to the soil and to the bodies of animals and humans.
   b. Shipping the food from far away uses fossil fuels in transportation. The frequent trips to the store also use more gas than if fewer trips were taken.
5. a. Your donation to protect your natural resources will help preserve valuable ecosystems.
   b. SUVs are highly fuel inefficient.

Extension

Answers will vary. Students should exhibit an understanding of the complex nature of sustainable living in modern society. Some may find it easier to act individually because it is difficult to monitor other people's behaviors. Others may find strength in numbers and appreciate the ability to work together toward a goal. Students should demonstrate an understanding that, no matter what, humans have an impact on the world in which they live.

ACTIVITY 1.4: Global Perspectives

Reproducible 1.4

1. North America and Western Europe (EU-25) have the largest footprint.
2. Africa has the smallest footprint.
3. The Asia-Pacific region—which includes China—has the largest population. But this region has a footprint that is smaller compared to the regions in questions 1 and 2. Specifically, the Asia-Pacific region has a footprint that is almost 6.5 times smaller than North America's.
4. Central and Eastern Europe (other Europe) are closest to North America in population (about 23 million more people). Yet North America's footprint is about 3 times greater.
5. Earth would quickly run out of resources if residents of the Asia-Pacific region adopted a lifestyle that was closer to that of citizens in North America. You might tell students that this issue is currently dominating world affairs, as China and other Asia-Pacific nations begin to improve their lifestyle and acquire goods in a way not seen at any other time in history.
6. Answers will vary. Some students will not be surprised, while others will be quite shocked by the disparity between our region and the rest of the world.

Extension

Question responses included in text.

ACTIVITY 1.5: At the Podium

Arguments will vary, but see to it that students stick to the topic at hand and that judges complete their evaluations sheets properly and without bias. Decisions should be made based on arguments, not on personal feelings. You may wish to judge each student on his or her own preparation and presentation.

UNIT 2: BUILDING COMMUNITY

READING HANDOUT: Real People, Real Lessons

1. She found it hard to believe that, after all this time, the city would finally promote change. What changed her mind was finally seeing results, seeing a financial commitment from the city.
2. If law-abiding citizens do not demonstrate an interest in their surroundings, their lack of attention will result in poor upkeep and will attract less desirable elements to the area.
3. A park or area that is well attended sends a signal that there are people working on a regular basis to maintain their surroundings. They are present. They care. They are watching.
4. Now that the area has changed and become cleaner and more attractive, it is more desirable to many more home buyers.
5. Community comes from working together for a common goal. They were a community before, but they didn't know it. Cleaning up the park brought them together.
6. Philadelphia Green gets the satisfaction of knowing that they have contributed to the improvement of their city and that they have helped make it more livable for their citizens. Philanthropic organizations see value in the improvement of their communities, not in making money.
7. Philadelphia Green wants to make sure that the majority of the people are committed to making change happen; otherwise their money would be poorly spent. Money alone can't effect change—the community must be committed.
8. It allows people to come together informally and get to know them. It gets people out of their homes and offices and brings them together in a shared environment.

ACTIVITY 2.1: Thinking about Communities

Check to see that students have filled out their charts based on any communities they could reasonably belong to. Encourage students to examine all areas of their life—home, school, athletics, and so on—as they put their chart together.

ACTIVITY 2.2: Now and Venn

Answers will vary. The example provided on the handout can guide you and students. Be sure that students' description of communities in which they participate is reasonable and all inclusive.

Extension

Answers will vary. Any kind of graphical representation is acceptable providing it is properly labeled and details the amount of time spent with various communities.

ACTIVITY 2.3: The Power of Intention

1. Answers will vary. List all responses on the board. Encourage students to think about the

word *intentional* and what they mean when they "intend" to do something. (Example: It is planned. It is not a community that occurs from circumstance.)

2. Answers will vary. Students should express themselves as clearly as possible when responding to the different types of communities listed.
3. Students should provide basic information about each of the organizations, including the philosophy behind the community's development.
4. Students should have a prepared presentation.
5. a. Example: You can live with like-minded individuals who share your concerns and priorities.
   b. Example: You may feel restricted or subject to a lack of privacy.
   c. They are alike in that individuals have decided to live together based on shared concerns for a particular issue. They differ according to the reason for the community coming together. Some may be energy conservation based. Others may exist for retirement or spiritual reasons.
   d. Answers will vary. Students should comment on their feelings about having shared community responsibilities with people to whom they are not related or whom they do not know that well.
   e. Answers will vary. Students should be specific and should offer detailed, supported reasons why they could or couldn't live in an intentional community.

Extension

Answers will vary. Essays should be as descriptive as possible with an emphasis on the day-to-day realities of living in a shared, communal environment, including any pluses or minuses associated with this lifestyle.

ACTIVITY 2.4: Making Sense of the Census

Part I (Handout 2.4)

1. Answers will vary. Samples might include foreign born, married, Latino, black or African American, civilian veterans, and the like.
2. Answers will vary. Some examples may include European American citizens, citizens with an MA or PhD, divorced citizens, homeless citizens, and the like. Students should explain why they feel the group should be included.
3. Answers will vary. Factors that may affect the accuracy include how honestly people answer, their willingness to participate, whether or not they have homes.
4. To determine financial disbursements to a city for public services, to examine migration patterns from one city to another, to determine the number of government representatives a particular city or state should have.

Part II

2. Answers will vary for all responses. Make sure that students supply evidence of the research they have done and that their answers are supported by their research.
   a. Examples may include ethnic groups, age groups, economic groups.
   b. Examples may include private clubs or organizations, religious groups, businesses, and schools.
   c. Students should be as honest as possible about how they feel. Some may be offended; others may not feel strongly; others may feel it's important to have populations counted.
   d. Students may be able to make assumptions about the economic makeup of a city and if that reflects on the availability of jobs in the area. They may also be able to understand the immigration history of a region
   e. The main difference is that members of a geographic community may not have any shared values or interests.
3. Graphs should be labeled properly; make sure the graph corresponds to their research.
4. Encourage students to think of all the different types of people in their town, not just the ones they come in contact with on a daily basis. This would include shut-ins, homeless people, and the like, who may not be reached by a census survey. Encourage students to think of creative ways to reach these populations and devise ways by which the information could be used to enhance the quality of life for all.

ACTIVITY 2.5: Front Page

Answers to discussion questions are included in the text. Profiles should be well researched and also as well written. You may wish to grade students on their interview preparation and class presentation as well as the written profile itself.

ACTIVITY 2.6: At the Podium

Arguments will vary, but see to it that students stick to the topic at hand and that judges complete their evaluations sheets properly and without bias. Decisions should be made based on arguments, not on personal feelings. You may wish to judge each student on his or her own preparation and presentation.

UNIT 3: WASTE MANAGEMENT AND RECYCLING

READING HANDOUT: Young People in Urban Nature

1. Answers may vary slightly. He tries to eliminate drug dealing, help unemployed people get jobs, purify drinking water, prevent toxic dumping, and build a nature center. He is motivated to improve his community, and there are so many different ways to do that, he never stops.
2. Contractors and builders must pay to dispose of trash and pay more to dump toxic trash. To avoid this cost, disreputable contractors will drive to poor neighborhoods and illegally dump it in the middle of the night, assuming that no one will care or be able to do anything about it.
3. They raised money and bought water-purifying filters for everybody's water pitchers and faucets.
4. Howard had friends in the construction business donate the use of trucks and earth-moving equipment. They picked up the trash and sent it to a proper facility for disposal. He got donations for new topsoil, which they used to "seal" the site.
5. They built a nature center.

ACTIVITY 3.1: Don't "Waste" This Quiz

Handout 3.1

1. b
2. True
3. d
4. d
5. d (Some can take much longer, though.)
6. d.
7. These items should be checked: wood, acid batteries, inkjet cartridges, tires, steel cans, motor oil, plastic bags.
8. True

ACTIVITY 3.2: "Waste-ing" Away

1. Answers will vary. Encourage creativity. Old CDs may be used to make a sculpture or tree ornaments. Plastic bottles can be used for planters or to house detergents or old paint. Students should think outside the box. Accept all reasonable answers.
2. For the trash journal, make sure students show they are thinking carefully about how and when they create waste. They should count *everything*—every gum wrapper, food container, piece of paper. Nothing is too small because when billions of people are throwing the same things away, that can add up pretty quickly. Remind them that organic waste should also be counted (apple cores and the like).

ACTIVITY 3.3: Talking Trash

Discussion Questions

Answers are included in the text.

Handouts and/or reproducibles should be thoroughly discussed in class.

ACTIVITY 3.4: Trash Audit

Check to make sure students have gathered, sorted, and weighed their trash carefully. Check to make sure the charts are properly labeled and clearly depict the results of their audit.

For the evaluation phase, answers and responses will vary; possible responses are included in the text.

ACTIVITY 3.5: Trash Tops the Charts

For discussion questions: Answers will vary, but possible responses are included in the text.

Handout 3.5

1. 29.9 percent
2. 12.1 percent, or 28,556,000 tons
3. 16,077,028 tons
4. 8 percent, or 18,880,000 tons
5. The percentage of total paper waste and yard trimmings is greater by 16.3 percent or 38,468,000 tons.
6. 56.1 percent
7. 40 percent

ACTIVITY 3.6: Landfill Dispute

There is no real "winner" in this dispute. Students should be evaluated on the basis of their preparation and their assessment of the pros and cons of their assigned choice, whether they are environmental, financial, or political.

Be sure that students study all aspects of the map carefully as they form their arguments.

Example: Option A is away from people but is in a natural, forested setting and near a water supply. Option B is not damaging the forest or near the water supply, but it is close to a school.

ACTIVITY 3.7: At the Podium

Arguments will vary, but see to it that students stick to the topic at hand and that judges complete their evaluations sheets properly and without bias. Decisions should be made based on arguments, not on personal feelings. You may wish to judge each student on his or her own preparation and presentation.

UNIT 4: GREEN BUILDING

READING HANDOUT: A City of Green Builders

1. Urban infill housing is the practice of building houses in the city on vacant or abandoned lots between existing homes.
2. Building in the countryside means decreasing the amount of undeveloped, natural land that we have. Green builders choose to build in the cities so as to not further deplete untouched land.
3. When trees and vegetation are removed, she takes steps to prevent soil erosion. If there is an existing building or structure on the site, they will pick through the debris to see what can be salvaged or recycled.
4. Durable goods are large appliances that last for a long period of time, including washing machines, dishwashers, and the like.
5. About $157,500 (5 percent more).

ACTIVITY 4.1: Anatomy of a Green Home

Discussion Questions

Answers are included in the text.

ACTIVITY 4.2: Under the Green Roof

Handout 4.2

1. If the soil doesn't drain well, weight and stress are added to the roof.
2. Winds are higher on a roof, and there is greater exposure to heat and sunlight. This tends to dry plants out more quickly, so they need to be irrigated more often.
3. So that they can sustain the more extreme conditions found on a roof.
4. Native plants are better suited to their specific climate and therefore tend to withstand weather variations better.
5. It could be collected in a cistern and carried down to be used at ground level or reused to irrigate the roof garden when there hasn't been any rain.
6. Besides insulating the building, it can also produce food, flowers, herbs, and other plant products. (Accept all reasonable answers.)

Pros and Cons of a Green Roof

Accept all reasonable answers.

Pros may include excellent insulation, lowering heating and cooling costs, lowering the overall urban heat index of a city, helping purify air and water.

Cons may include very high start-up costs and requires maintenance and eventual replacement when roof wears out.

ACTIVITY 4.3: A Green Light

Handout 4.3

Type of Bulb	100-watt Incandescent	23-watt Compact Fluorescent
Cost per bulb (approximate)	$0.75	$11.00
Life of the bulb	750 hours	10,000 hours
Number of hours burned per day	4 hours	4 hours
Amount of time (in days, months or years) one bulb will last	187.5 days	6.8 years
Number of bulbs needed for 3 years (estimate)	About 6	1
Total cost of bulbs	$4.50	$11.00
Lumens produced	1,690	1,500
Total cost of electricity for 3 years (at 8 cents/kilowatt-hour)	$35.04	$8.06
Total cost over 3 years	$39.54	$19.06
Total savings over three years with the compact fluorescent:		$20.48

Source: U.S. Department of Energy, Energy Information Administration

1. 16.9 lumens per watt
2. 65.2 lumens per watt
3. More than 13 times as long
4. About 51 percent of total electricity costs could be saved over three years.
5. Increase because the cost of the bulb is spread out over a longer period of time.
6. Answers may vary, but in general people resist because CFL bulbs cost more. Students may or may not agree with this.

Extra Credit

First get the total number of watts used based on the total number of hours, then convert to kilowatt hours.

Making Connections

What is the relationship between energy production and pollution? Most electricity is made by burning fossil fuels, which release CO_2 and other pollutants into the atmosphere. If you reduce the amount of energy needed to light your home, you reduce the amount of air pollution released.

ACTIVITY 4.4: Front Page

Essays should clearly communicate the students' opinions and reflect an understanding of the general op-ed form.

ACTIVITY 4.5: At the Podium

Arguments will vary, but see to it that students stick to the topic at hand and that judges complete their evaluation sheets properly and without bias. Decisions should be made based on arguments, not on personal feelings. You may wish to judge each student on her or his own preparation and presentation.

UNIT 5: ENERGY

READING HANDOUT: French Fries in the Tank?

1. Biodiesel is made from plant-based oils, such as vegetable oils, soybean oils, and the like. Gasoline is made from petroleum.
2. Engines running on biodiesel run 78 percent cleaner than petroleum-filled engines and produce no carbon dioxide. The vehicle gets similar or better mileage and has a longer engine life because plant oils do a better job of lubricating and scrubbing the interiors of engines.
3. Plants are a renewable resource, whereas petroleum is a nonrenewable resource. Plants can be grown virtually anywhere, but petroleum exists only in certain parts of the world.
4. Capital is money or other financial support, and it helps new businesses get off the ground.
5. The one thing that needs to change is the source of the energy. The four things that do not have to change are the technology, the infrastructure for delivering fuel, our vehicles, and our behavior.
6. Answers will vary but might include high cost, environmental concerns, and international relations.
7. Answers will vary. It would make the community more independent. A farming community, for example, could grow its own fuel and support and grow its own economy rather than relying on outside suppliers and businesses.
8. Answers will vary. Students may mention ethanol, electric cars, hybrid cars, and solar cars.

ACTIVITY 5.1: Sun, Sky, and Earth: Where We Get Our Energy

Discussion Questions

Answers are included in the text.

Group Activity

For the class reports and presentations, evaluate students based on how well they addressed all of the bullet points listed. Students should be able to show their research.

ACTIVITY 5.2: I Love My Car—I Just Hate Paying for Gas!

Handout 5.2

1. Answers will vary.
2. Answers will vary. Students should address the man's concern for lack of availability of alternative fuels and the "power" he associates with gasoline-fueled vehicles.
3. Answers will vary. Students should support their choice with facts from the handout.
4. Answers will vary. Students should support their choice with facts from the handout.
5. Answers will vary. Students should support their choice with facts from the handout and take into consideration the flexibility in choice that arises when you don't need to drive your car long distances.
6. The biodiesel car.

ACTIVITY 5.3: Charting Energy

Handout 5.3A

1. China; about 1.3 billion
2. No
3. Germany; about 500 billion kWh
4. United States, Russia, Japan, and Germany
5. India, Indonesia, Pakistan, Bangladesh, and Nigeria
6. About 700,000,000 more live in India than the United States.
7. The United States uses almost 3 trillion more kWh (3,000 × 1 billion) than India.
8. The amount of kWhs used by the United States (about 3.25 trillion) is more than the amount used by India, China, and Russia together (about 1.65 trillion).
9. Answers will vary but may include: Population does not determine energy consumption; the United States uses far more energy than most other countries.

Handout 5.3B

1. About 2.9 gallons per day per capita
2. About 1.4 gallons per day per capita
3. About 2.4 gallons more per day per capita are consumed by the United States and Canada than by the entire world on average.
4. The United States and Canada consume about 1.25 gallons per day per capita more than other industrialized countries and nonindustrialized countries combined.
5. Answers will vary. Students should be clear and support their positions.

ACTIVITY 5.4: Home Energy Survey

Answers will vary greatly depending on the individual homes, the habits of the students' families, and the students' attention to details while conducting the survey. Students should do their best to complete the survey as fully as possible. In discussing results, have students pay particular attention to any common problems that plagued a large number of families.

UNIT 6: AIR QUALITY

READING HANDOUT: Tree Boy

1. While he was at summer camp, a naturalist told him about the plight of the trees.
2. Answers will vary but may include: Smog chokes the trees, and it comes from power plants and cars. Pollutants include sulfur dioxide and ozone.
3. Answers will vary. Most students will say "no," California isn't the only state with problems. You may wish to check with your local EPA office about air quality statistics for your area.
4. Answers will vary but should include reducing smog and other pollutants.
5. Answers will vary but may include *determined, stubborn, committed. Perseverance* is one word that well describes how he was able to accomplish his goals.
6. Answers will vary but may include: Andy's plan didn't sound like a good investment; he was a young boy, and they didn't want to take a risk on him. Once he became popular and the community supported him, the banks would have appeared foolish had they not done the same. Their behavior shows that fundraisers will have better luck with large businesses if they have first built strong community support.
7. Answers will vary.

ACTIVITY 6.1: The Air Up There

Handout 6.1

1. Troposphere
2. Stratosphere
3. About 31 miles
4. About 0° Celsius, 32° Fahrenheit
5. The stratopause
6. Pressure decreases

ACTIVITY 6.2: Smog, Ozone, and the Greenhouse Effect

Discussion questions

Answers are included in the text.

ACTIVITY 6.3: How Much CO_2 Do You Contribute?

All answers will vary depending on students' bills. Make sure students have performed the correct calculations by checking their totals against their bills.

ACTIVITY 6.4: Understanding the Greenhouse Effect

If students are working individually or in pairs, monitor them to be sure they are following the instructions carefully.

Answers to classroom discussion questions are included in the text.

UNIT 7: WATER QUALITY

READING HANDOUT: Restoring the Chicago River

1. South. Answers will vary. Though it may not contaminate the direct source of Chicago's drinking water, it now flows into the Mississippi, which takes the toxic waste past many more communities on its way south.
2. It was an eyesore, heavily polluted, unpleasant to look at and to smell. They chose to ignore it.
3. Chlorine. It was used in the sewage treatment plants to treat runoff. It is used in the home as a bleach and cleaner and to treat swimming pools.
4. Native plants are those that naturally thrive in a given area. They are an indicator of the overall environmental health of a region; these plants support other species in the ecosystem that feed off of them, use them for shelter, and so on.
5. Answers may vary; but, based on the story, these plants presented a problem because they were choking out the native plants that used to be in the area.
6. Answers will vary. It's important for people to see the river as a part of their daily lives and something that is important to them. If they use the river for fun and it is something they value, they will be more likely to take care of it.
7. Marquette and Joliet were explorers who paddled the river in the 1600s, about 400 years ago. They were looking for trade routes and to establish trading of items like beaver pelts with Native Americans and other inhabitants in the area.

ACTIVITY 7.1: Understanding Earth's Water Supply

Handout 7.1

1. 3 percent
2. 30.1 percent
3. .3 percent
4. groundwater
5. .3 percent
6. 89 percent
7. groundwater
8. Answers will vary. Most students are surprised to learn that the pollution that can seep into the ground will most adversely affect their drinking water. If groundwater is the most affected, it should drive home for students the importance of protecting groundwater from pollution by runoff and other dangers.

Water Cycle Discussion

Answers are included in the text.

Watershed Discussion

Answers are included in the text.

ACTIVITY 7.2: Town Meeting: Watershed at Risk

For the town meeting, be sure both sides have done their research. Students should be able to show evidence of the facts that their argument is based on. You may wish to encourage students to include charts and graphs or lists of pros and cons.

It is easy to be on the "right" side in this argument (against the development), so the activity can be more challenging for the students who are assigned to support the development. If they get stuck, encourage them to put themselves into the shoes of the developers and think like many cities do: The development will bring in needed tax dollars that can support schools and roads, the development could adhere to green building design, the development will bring jobs to the area and lower unemployment rates.

Both sides should be thorough. At the end of the exercise, all students should come away with an understanding of the complexities presented by both sides of the issue.

Answers to extension discussion questions are included in the text.

ACTIVITY 7.3: The Rundown on Runoff

For some of the products that students identify, the potential harmful effects on living things are not listed. You may wish to have students research them for class discussion. There are a variety of tox-free substitutes for many home products. You may wish to create a chart in class that lists the most common household products found by all your students with a tox-free substitute that corresponds to each. (Try to encourage students to find product substitutes—such as the vinegar and water glass cleaner—not merely environmentally friendly brands.)

Handout 7.3

Answers will vary. Students should be as thorough as possible and think of as many different ways possible that the products could get into the groundwater (not just by washing them down the drain).

ACTIVITY 7.4: Understanding Water Quality

Handout 7.4

1. Dichloromethane
   a. 5 parts per billion
   b. Pharmaceutical and chemical factories and insecticides
2. 15 parts per billion
   a. One of the 44 tested
3. The cloudiness of the water
4. 12.50 parts per billion
   a. Erosion, runoff from orchards, glass and electronics production wastes
5. 1.3 parts per *million*
6. fertilizers, septic tanks, sewage, and erosion

ACTIVITY 7.5: Swales and Cisterns

Handout 7.5

SWALES

1. Gravel, crushed rock, and sand; they are better because they allow water to drain more readily.
2. It is sloped so that the water can be funneled toward the swale.
3. It will pool up in the swale because everything is angled to collect the water there. It will eventually be absorbed.
4. Ones that thrive in wet conditions.
5. The bottom of the "V" next to the tree because it is the lowest point in the area and forms a natural basin.
6. Probably not, because the curb is blocking it.
7. I would remove the curb and gently slope the road and sidewalk to feed into the "V."
8. Answers may vary. The plants, shrubs, and so on need to be pruned and weeded. Water will also wash any garbage (gum wrappers and the like) into the swale, which needs to be cleaned out from time to time.

CISTERNS

1. It enters through a drain and passes through a filter before entering the cistern.
2. It's pumped out.
3. To better funnel the water into the cistern
4. A home owner might use it for watering the lawns and gardens, for washing the car, for washing pets. A school could use it to water the school grounds and playing fields. A business can use it to wash its exterior windows and to water its landscape plants. A park or golf course can use it to water its extensive lawns or perhaps fill a pond.
5. A cistern is on your property and uses completely free water. Once the cistern is installed, the water is free, and all you need to do is pay for filters and maintain your system. The water also does not require the chemical treatment if it is for outdoor (nonpotable) use only, which reduces pollutants.
6. The swale only requires gravel, rock, sand, a shovel, and some plants, which you may already have. A cistern requires more equipment and mechanical maintenance over time, which will add to expenses.

UNIT 8: SOIL QUALITY

READING HANDOUT: Restoring Chicago's Calumet Region

1. The Calumet is known for being a factory area, and tourists don't usually visit those kinds of areas. It was once the site of Chicago's steel industry.
2. The steel factories used to dump slag outdoors, which hardened and covered the ground and vegetation.
3. Slag is a by-product of steel manufacturing. It looks and feels like a thick slab of concrete that coats an area when it hardens. It would stop most things from growing in an area where it was dumped.
4. They are coating the area with several inches of new topsoil and then planting alfalfa, which has thick roots and is expected to eventually break the slag down over time.
5. Ecotourism, canoeing, hiking, and the like
6. Answers will vary. Encourage students to be honest about how feasible they think his plans are.
7. Answers will vary depending on students' passions and interests.
8. Answers will vary. It may not sound very pretty now, but it has great potential to be restored to a beautiful wetland area. The slag would have to be removed; the buildings may need to be restored or renovated; other pollution and trash would have to be cleaned up; wildlife and native plants would have to be encouraged and reinstated.
9. Answers will vary. Some students may find it very interesting. If they're city kids, it might sound like fun to paddle a canoe no matter what. If you live

in an area with access to nice rivers and streams, it might sound less appealing, unless students are curious about the environmental impact.

10. Answers will vary; however, students may not want to wait for the alfalfa to do its job, opting for a physical removal of the slag, which would be more expensive. If students wanted a parking lot for their building, for example, they would not necessarily have to remove that area of slag.

ACTIVITY 8.1: Hydroponics to the Rescue

Handout 8.1

1. What are the chemical formulas for the following suggested plant nutrients?

Calcium nitrate	$Ca(NO_3)_2$
Potassium nitrate	KNO_3
Magnesium sulfate	$MgSO_4$
Potassium phosphate	K_3PO_4

2. Answers may vary, and students should note any and all changes. Students may notice that the water is depleted more quickly in the sun. The plant may also exhibit a phototropic ("light-loving") response and lean toward the light source. This may also be seen if the plant is out of, but near, a direct light source.
3. Answers will vary.
4. CO_2 plays an important role in photosynthesis, which creates energy for the plant (and releases oxygen for us). Increasing it can boost growth.
5. Answers may vary slightly. Here are some examples:

   Nitrogen is part of chlorophyll and helps with rapid growth and seed and fruit production.

   Phosphorus is an important part of photosynthesis and helps with root formation and blooming.

   Potassium helps build proteins and helps fight disease.

   Calcium helps make up cell walls, helps with transport, and maintains plant strength.

   Magnesium is part of chlorophyll needed for photosynthesis and helps activate enzymes.

   Sulfur helps in the production of chlorophyll and protein, helps foster good roots and seeds, and helps plants resist cold.

ACTIVITY 8.2: Alfalfa and the Importance of Cover Crops

Handout 8.2

1. Answers may vary, but most students will say "no."
2. They have grown, and the roots have penetrated small holes in the slag. They are getting their nutrients from the new soils on top of the slag and the old soil below the slag, as well as from the air.
3. Rainwater that has entered the cracks in the slag can freeze and thaw and freeze again, helping to break up the slag. The alfalfa roots have helped by widening the cracks in the slag. When the alfalfa dies back in winter, it also fertilizes the soil.
4. The thickness of the root helped it push through the slag; the moisture-loving roots took up a lot of the water in the marshy area and the water trapped beneath the slag; and the natural fertilizer helped repair old soil and also prepare it for the following year's plants.
5. Yes, they could have used bulldozers and fertilizer. This way, they accomplish the breaking up of the slag and the fertilizing of the area by sowing the seeds of just one type of plant. This costs less, requires less hands-on time, and doesn't damage any surrounding plants (which a bulldozer may do). Also, bulldozers use gas, another waste of fossil fuels.

ACTIVITY 8.3: Soil, Air, and Water

Reproducible 8.3

Students should study the reproducible before working on their op-ed pieces. It is important to explain the difference between direct and indirect contamination.

Examples from the reproducible:

1. The rainwater comes down the mountain and passes through cultivated farmland, where it can pick up pesticides, fertilizers, and animal waste and wash them into local waterways.
2. The sewage pipe directly dumps untreated waste into the water.
3. The drum is leaking oil and waste into the soil where it can seep into the groundwater and then enter the water supply.

Op-ed pieces should reflect an understanding of the form and arguments—whether students are writing "pro" or "con" pieces—should be based in some fact, in addition to the students' opinions.

UNIT 9: PARKS AND OPEN SPACES

READING HANDOUT: The Great Public Space

1. It was a railroad yard. Most people did not particularly like the old railroad yard. Doing something about it seemed too expensive. It required a lot of imagination and planning to do something different with the site.
2. He built a parking garage and used the money from the garage to pay for the construction of the park. Also, much of the art and other special features (pavilions, stage, gardens) were purchased with donations from major corporations and private philanthropists.
3. The CO_2 reduction achieved by a single tree has a dollar value of $920 per ton per year, yet it costs only a few dollars to buy and plant.
4. The plants in the park capture water for the city and allow it to seep into the ground. The trees remove CO_2 from the atmosphere and help reduce the urban heat island effect.
5. Answers will vary. Students should look for large municipal projects (transit systems and the like).
6. Answers will vary. No matter their opinion, students should back up their thoughts with evidence from the article.
7. Answers will vary. The metal boxes are cars, trains, and buses. Students should be able to comment specifically on the daily lives of individuals they know and discuss whether or not they believe them to have adequate time to interact with their community.
8. Answers may vary. Parks provide a space where people who may not normally interact have an opportunity to meet.

ACTIVITY 9.1: Thinking about Parks

Answers will vary. Check to see that students have used reasonable criteria to create their chart of the three chosen parks. If you have them report on a world-famous park, be sure they have adequately researched the history of the park and encourage students to get beyond the look and feel of the park and to think about the purpose the park served for those who used it.

ACTIVITY 9.2: Open Space Around the World

Answers to discussion questions are included in the text.

Make sure reports include a historical perspective on the role of these open spaces in the larger city over time.

ACTIVITY 9.3: Mapping the Millennium

Handout 9.3

1. Answers will vary.
2. 4
3. Answers will vary.
4. Answers will vary but may include: Many famous parks have areas for outdoor performances, formal gardens, fountains, and ice rinks. Not all parks have art galleries, theaters, or roof gardens.
5. The parking garage is the main one. But the concessions, bike stations, skating rink fees, and concert tickets are other examples of revenue-generating features.
6. Answers will vary.
7. The trees provide a valuable service by cleaning air (taking up CO_2) and capturing water. The park also allows a place for people to gather, meet, exercise, and socialize. Some answers may vary.

ACTIVITY 9.4: Design a Park or Piazza

Be sure that park plans address all of the individual features and concerns listed in the activity overview.

ACTIVITY 9.5: One Space, Two Plans

As with the debate activities, be sure students have done their research. Financial considerations should be included, where possible, and arguments must be based on some factual realities as well as personal opinion or emotion. Encourage students to role-play if they would like. (For example: A student for development could pretend to be a council member concerned with unemployment in the "town" in which he or she lives.)

UNIT 10: TRANSPORTATION

READING HANDOUT: Living with One Less Car

1. CO_2 emissions, rubber dust from tires, copper dust from brake pads
2. It is a program in which families agree to use one less car for a period of time to investigate whether or not they would like to do so permanently.
3. They reduced the miles they drove and increased mass transit use, biking, and walking. They also saved money. They definitely increased their use of alternative transportation. The story cites that participants increased their mass transit use by 125 percent, upped their walking mileage by 38

percent, and boosted their biking mileage by 30 percent.

4. Answers will vary. Families in which both parents work on different schedules and in different parts of town, is one example. The availability of mass transit will also affect this.
5. There is less need for cars; they are healthier; and they get to interact more with their neighbors. Towns can increase the availability of sidewalks and other pedestrian-friendly routes.
6. They have to create space for, build, and maintain roads, garages, and parking spaces. You also have to pay employees to maintain all of these features.
7. People should car pool, use mass transit, walk, or bike.
8. Answers will vary. Some people find traffic stressful. Traffic also requires concentration. You can feel very susceptible to things that are out of your control—traffic, accidents, construction delays. On mass transportation, someone else is driving. You can read, nap, listen to music, and so on.

ACTIVITY 10.1: Passenger Costs

Answers to discussion questions are in the text.

Handout 10.1

1. Commuter rail
2. Cars
3. 206,690 BTUs
4. 911
5. 38,251
   a. 2,741
   b. The second number drops markedly because you factor in the number of people who are being transported for each BTU being expended.
6. The more people there are in the car, the fewer BTUs per person are consumed.
8. One bus carrying 40 people uses 41,338 BTUs per vehicle mile. Forty individual cars use 253,920 BTUs. The cars use 212,582 more BTUs than the bus.
9. Transporting more people in one vehicle (bus, train, or the like) decreases the number of BTUs per person that are consumed. This means that CO_2 emissions and other pollutants resulting from fossil fuel combustion will also drop.

ACTIVITY 10.2: Pros, Cons, and in Between

Answers will vary. Costs will vary depending on what sources the students locate.

Here are some considerations for completing the pros and cons:

- Senior citizens are going to be affected by anything that raises their taxes and will not want to have to pay more on their fixed income. However, many don't drive and may get around more easily with better public transportation.
- College students will usually vote for anything that makes their lives easier—mass transit if they don't have their own car—and many aren't paying takes yet. However, the higher taxes on food and drink might bother them, too. As a group, they may be more concerned with the environmental impact of any of these choices.
- Business owners along the new line will likely be excited that more customers can be brought to their doors. But they won't like the fact that during construction—which could last years—their business may be adversely affected. They may also lose parking.
- A politician seeking reelection will most likely swing with whatever she or he feels the majority of voters want. Large projects costing money—and raising taxes—tend to be unpopular with voters.
- A resident living along the new transportation line may be happy to have added transportation options. However, he or she may not want the streets dug up and may not want the increased traffic in the neighborhood and the noise it may bring (if a train or light rail).

ACTIVITY 10.3: One Less Car Challenge

Answers will vary greatly, as this is a long-term, multifaceted project. Develop a method for assessing all the completed sheets (diary, car cost sheet, depreciation calculation) and making sure that they have been filled out properly. Be sure to check students' math.

ACTIVITY 10.4: Writing about It and Making Connections

Answers may vary greatly. Students' responses should be well thought out and crafted. This is a writing and creative thinking exercise, and students should put equal effort into both.

UNIT 11: BIODIVERSITY

READING HANDOUT: Restoring the Prairie

1. A prairie is an ecosystem in which native grasses and plants are more abundant than trees.
2. The prairie will help the city feel cool on summer days, have cleaner air, and be less vulnerable to dangerous floods.
3. The prairie acts like a buffer around the city. Prairies absorb water. The grasses absorb more CO_2 than many trees because they grow so rapidly and need a constant source of energy.
4. Their roots are very dense and will absorb tremendous amounts of water and therefore reduce the risk of runoff.
5. A fire on a prairie doesn't kill the plant; the plant grows back. But the struggling young trees may die off.
6. Having no trees on the landscape helped them in two ways: First, it made it easier to hunt prey; second, the grasses fed the animals they were trying to hunt.
7. The Europeans made houses out of wood, which burns, so therefore the Europeans believed they should suppress all fires because they were endangering their houses. So, when the prairie began to burn, the Europeans would put the fires out quickly, which resulted in the growing of more trees and loss of prairie.
8. To help return the area to native grasses and plants, to reduce the chokehold that exotic, invasive plants, trees, and shrubs have on the native grasses and prairie lands.
9. Answers will vary but may include: The prairie is a snapshot of our environmental history; the prairie protects the soil and protects against flooding.
10. Answers will vary.

ACTIVITY 11.1: Classmate Diversity Scavenger Hunt

As students fill out their scavenger hunt work sheets, make sure they talk to as many different classmates as possible—not just their friends. It's a good opportunity for students to get to know people outside their normal circle. Students should have a wide variety of classmates on their sheets, not the same person filling five different categories.

ACTIVITY 11.2: Getting to Know Your Own Backyard: An Urban Survey

Answers will vary greatly depending on your area. Be sure you have adequately discussed your region's ecosystems before heading out and that students have an idea of the resources available to them for researching the plants and animals in your area. They should identify something by sight and then be able to find it in a book or online. If students choose to, they may wish to photograph what they see.

Relevant extension answers are included in the text.

ACTIVITY 11.3: Overfished and Contaminated: Crisis in Our Seas

Handout 11.3

Students' Venn diagram should have the following:

- In the "safe to eat" circle: Anchovies, Butterfish, Calamari (squid), Crab (king), Crawfish/crayfish, Flounder, Haddock, Hake, Herring, Lobster (rock), Perch (ocean), Scallops, Shad, Sole, Trout (freshwater), Whitefish
- In the "plentiful" circle: Char, Arctic (farmed), Crab (Dungeness; imitation; snow; stone), Halibut (Pacific), Mussels (farmed), Striped Bass (wild and farmed), Trout (rainbow, farmed), Tuna (albacore, bigeye, skipjack, yellowfin caught on lines)
- In the intersection of the two circles: Catfish (farmed), Caviar (farmed), Clams (farmed), Lobster (spiny), Oysters (farmed), Pollock (wild-caught), Salmon (wild-caught), Sardines, Shrimp, pink (from Oregon), Sturgeon (farmed), Tilapia (farmed)

Answers to relevant extension questions are in the text.

ACTIVITY11. 4: Ecosystem Swap Meet

Be sure to check everyone's cards to be sure they contain the correct information. Also, be sure that students have created complete ecosystems, including plants, some insects and birds, and so on (not just mammals). Then be sure that the creature cards were matched to the correct ecosystem.

UNIT 12: URBAN AGRICULTURE AND COMMUNITY GARDENS

READING HANDOUT: Philadelphia Eats What Philadelphia Grows

1. A brownfield is an area that is considered so contaminated that it cannot be easily used. It may be the former site of a factory or dump, for example.

2. The main characteristic is that they are grown without soil.
3. Answers will vary. Ms. Corboy wanted to use the land but wanted to avoid the possibility of contaminated soil.
4. It supports the small family farms that have been going out of business, helping them make a living, and saving farmland from developers.
5. At first, she wasn't as engaged with her neighborhood. Then, when she began selling to others in her neighborhood, she developed a closer relationship with members of her community.
6. Answers will vary, but most will say "yes." If they do, people will understand the quality of the food they are buying and how supporting the local farmers in their area is also supporting their entire economy.
7. She is performing two major roles: She is growing and selling food to people in her community, and she is also putting a previously useless piece of land to good use.
8. Answers will vary, but most will say "encouraged." They use land that has been neglected, and they bring quality food to city neighborhoods, instead of having them shipped from other out-of-town farms or businesses (which uses fossil fuels). Encouraging local farms also supports the local economy and builds community.

ACTIVITY 12.1: Growing All Around

Make sure students have access to adequate resources to answer the questions. Don't neglect phone books and the library.

Handout 12.1

1. Answers will vary.
2. No; organic produce is grown all over the world and shipped to supermarkets everywhere.
3. No; not all local farmers may have gotten their organic certification from the government (it is often expensive to do so).
4. Answers will vary.
5. Answers will vary.
6. Answers will vary.
7. Answers will vary.
8. Answers will vary.
9. Answers will vary.
10. Answers will vary.

ACTIVITY 12.2: Zoning in on Hardiness

Answers to relevant discussion questions are included in the text.

Handout 12.2

1. Answers will vary.
2. Zone 11
    a. Above 40°F
    b. Answers will vary. One example is Honolulu, Hawaii.
3. Zone 1
    a. –50°F
    b. Answers will vary. One example is Fairbanks, Alaska.
5. Answers will vary but should be colder zones (1 to 5). Students may be familiar with cold-weather crops, such as spinach and kale. They may also answer "trees" to this question. See if they can name any specific trees (birch, for example).
6. Answers will vary. One example is Des Moines, Iowa.
7. Answers will vary. Possible answers include: Some places are at higher altitudes than others, making them colder. Some places are warmed by being closer to the ocean. Some places are colder because they are more susceptible to weather patterns, such as cold fronts that come down from Canada into the Midwest.
8. 4.44° to –1.1° Celsius.

ACTIVITY 12.3: Food on the Move

Detective and Local Food activities should be graded based on level of detail and research as well as writing and presentation.

Answers to relevant extension questions are included in the text.

Handout 12.3

1. Grapes travel the farthest, 2,143 miles. One state supplies them.
2. 37 percent
3. Tomatoes are supplied by 18 states.
4. Grapes and green peas are supplied by only one state.
5. Answers will vary. In general, the fewer the states that supply it, the more it costs, unless you live in that state.
6. Answers will vary. In general, the farther it travels, the more expensive it is. However, some larger companies can afford to sell produce more cheaply because of the volume they deal in,

regardless of the distance they have to ship their goods.
7. One trip for spinach is 2,086 miles. That will require 298 gallons of gas, which will result in 5,662 pounds of CO_2 emitted into the atmosphere for just one trip.
8. Answers will vary. Make sure students show their work.

ACTIVITY 12.4: Your Own Garden

Answers to relevant discussion questions are included in the text.

UNIT 13: URBAN FORESTRY

READING HANDOUT: The Magic of Trees

1. *Urban forestry* refers to the management and use of trees in an urban environment.
2. Their shade can cool animals, plants, humans, and buildings, and it reduces the smog that often worsens when temperatures rise by absorbing CO_2.
3. Their root systems can hold many gallons of water, helping to prevent runoff.
4. Their absorption of CO_2 helps fight global warming and climate change.
5. Green infrastructure is a city's investment in trees, and it can be "built," in a sense, by planting more of them.
6. Crime went down, housing prices went up, and more people wanted to live in those neighborhoods that had planted trees.
7. Answers will vary. It's prettier. The trees shade you while going from store to store on hot days.
8. Answers will vary.
9. Do you think it is fair to fine people who harm older trees on work sites? If so, how much should they be fined? If not, what should be their penalty? Answers will vary. Students should support their answers with clear reasons.

ACTIVITY 13.1: Did You Know . . . ? Fun Facts about Urban Forestry

Handout 13.1

1. b
2. Answers will vary.
   a. They absorb CO_2 through their leaves for photosynthesis.
   b. Trees shade and cool buildings, which means that air conditioners are run less often, which corresponds to less electricity consumption and therefore fewer CO_2 emissions.
3. True
4. c
5. True
6. b
7. Answers may vary.
   a. The roots absorb many gallons of water, helping prevent erosion and runoff.
   b. The roots also hold onto the soil, making it more difficult for it to wash away.
8. c

ACTIVITY 13.2: Getting to Know Your Trees

Handout 13.2

Answers will vary.

Check that students have performed their calculations accurately. Encourage students to find as many different kinds of trees as possible.

Question 3: The calculation is derived from the formula for circumference: $c = \pi d$ (or $c/\pi = d$)

For the writing assignment, encourage students to gather as much information as possible about the history of the area during the life of the tree.

ACTIVITY 13.3: Trees and You: A Numbers Game

Handout 13.3

1. Runoff volume over time. More trees decrease the total runoff volume.
2. 8.7 percent. Landfills are a source of methane. (Answers will vary.)
3. CO_2 is the main greenhouse gas contributing to climate change; 84.6 percent of greenhouse gases are CO_2 (83 percent + 1.6 percent). Answers will vary. Breathing produces CO_2. Electricity production produces CO_2. Cars produce CO_2.
4. *Energy-related* refers to processes that involve the production of energy and conversion of fossil fuels into other forms of energy.
5. a. 100
   b. They will go down as trees shade in summer and protect from wind in winter.
   c. Answers will vary but may include: Property values will rise; people will drive more slowly through the streets; the neighborhood will become more desirable; the neighborhood will be cooler in summer, encouraging people to spend more time outside.

6. a. $30
   b. $90
   c. $900
7. a. $20
   b. $60
   c. $600.
8. Answers will vary. Students should mention the reduction of CO_2 that results not only from the absorption of CO_2 by the trees but by the reduced need for electricity (which reduces the combustion of fossil fuels in order to generate electricity). This reduction in CO_2 emissions helps reduce the buildup of bad ozone and therefore reduces the global warming effect. Air quality is improved, which is also beneficial to animals, plants, and humans. Soil quality is improved and is held in place better.

ACTIVITY 13.4: At the Podium

Arguments will vary, but see to it that students stick to the topic at hand and that judges complete their evaluations sheets properly and without bias. Decisions should be made based on arguments, not on personal feelings. You may wish to judge each student on her or his own preparation and presentation.

UNIT 14: URBAN PLANNING

READING HANDOUT: William Penn's "Green Countrie Towne"

1. Because of his religious beliefs
2. Philadelphia was the first city in America to guarantee all citizens equal rights under the law, regardless of race, gender, or religion.
3. Free press, free enterprise, trial by jury, education for both men and women, and religious tolerance
4. The idea that all men are created equal
5. Because wooden homes were more susceptible to fire
6. So that they could have a little garden around their houses
7. So that there would always be some wild spaces or park areas in their settlements
8. Fairmount Park was deliberately built to act as a buffer and purify rainwater that fell in and around Philadelphia. It was the first park of this kind to be built this way.
9. Answers will vary. Environmentally, Philadelphia is ahead of other cities. Because the city incorporated a huge park and many little parks into its design when it was still young, citizens there now enjoy green spaces and wilderness right in the heart of town. And the big park system continues to serve as a giant filter that purifies the city's water. Modern-day planners find this legacy inspiring and do not have to struggle to try to find space in the city for parks—they're already there.
10. Answers will vary.

ACTIVITY 14.1: Our Town: A Closer Look

Handout 14.1

Answers will vary. Students should be as descriptive as possible when examining their chosen surroundings.

ACTIVITY 14.2: Penn's Plan: A Look at Old Philly

Answers to extension questions are included in the text.

Handout 14.2

1. Four
2. City Hall
3. The two rivers (the Schuylkill and the Delaware), which are located to the east and west of the city. The rivers provided transportation and good access for trade ships.
4. Answers will vary. It enhances traffic flow.
5. No. Answers will vary as to why. One reason is that some buildings required larger parcels of land. The larger parcels may have been for municipal buildings or farms, while small blocks would have been for row houses or small businesses.
6. Answers will vary. Many major cities still follow the grid pattern, which is easy for residents and visitors to navigate.
7. Answers will vary. The biggest difference is that in Penn's day they didn't have large trucks and other vehicles—there were horses and carts. Today, we have more vehicles that need to fit on the streets.

ACTIVITY 14.3: You're the Planner

Make sure students adhere to the checklist and minimum requirements for their plan. Plans should be well thought out with organization, research, and presentation being the cornerstones of your evaluation.

ACTIVITY 14.4: Frederick Law Olmsted's Vision

Students should be evaluated on both their writing

style and their ability to blend their memory and opinions with some research and fact.

UNIT 15: POPULATION GROWTH AND INTEGRATED RESOURCE MANAGEMENT

READING HANDOUT: Water In, Water Out

1. They weren't saving any water; they were simply buying it from other places.
2. 85 percent
3. Everything that grows and dies in the forest goes back into the ecosystem to nourish it.
4. Cisterns are large underground tanks that capture water and hold it until it is needed. Swales are special landscaping devices that direct water to flow into a kind of ditch so that it can be properly drained.
5. Because the money saved on water alone would pay back the cost of the cistern
6. Answers will vary, but may include: They installed swales at the front of the yard; they installed a cistern in the backyard; they heavily mulched the beds; and the like.
7. It means that all the water that fell on the house and property stayed on the grounds and wasn't washed away. And some of it could be reused.
8. Answers will vary. In general, *integrated* means various parts are united into a single, larger unit.
9. Answers will vary. An integrated approach means to link all the parts of a particular problem together so that one approach solves many problems. By using cisterns, swales, and trees, Mr. Lipkis solves several problems at once: prevents runoff, conserves water, conserves energy, and lays the groundwork for future employment, as these structures need maintenance.
10. Because the city was spending millions every year to buy water, they had less money to spend on social services such as employment programs.
11. Answers will vary. If, instead of disposing of green waste, they used it as mulch, they could help absorb and keep some of their much-needed water.
12. It decomposes, and its nutrients feed successive generation of the forest.
13. Because it could be used for composting, mulching, and retaining water. It also wastes the fossil fuels necessary to have these large trucks drive to these locations and pick up this waste and transport it to the dump. And it wastes space in the landfills.
14. Two different agencies in Los Angeles were performing two separate functions regarding the water supply: One was responsible for removing water so that it didn't flood the city, the other was responsible for buying water so that the people would have water to drink.

Activity 15.1: Pop Goes the Population

Answers to discussion questions and reproducibles are included in the text.

Handout 15.1

1. 1,015,529,498 more people live in China.
2. 237,835,062 more people live in the United States.
3. Africa
4. One (the United States)
5. Four
6. India

Activity 15.2: Growing, Growing . . . Gone?

Answers to discussion and extension questions are included in text.

Handout 15.2

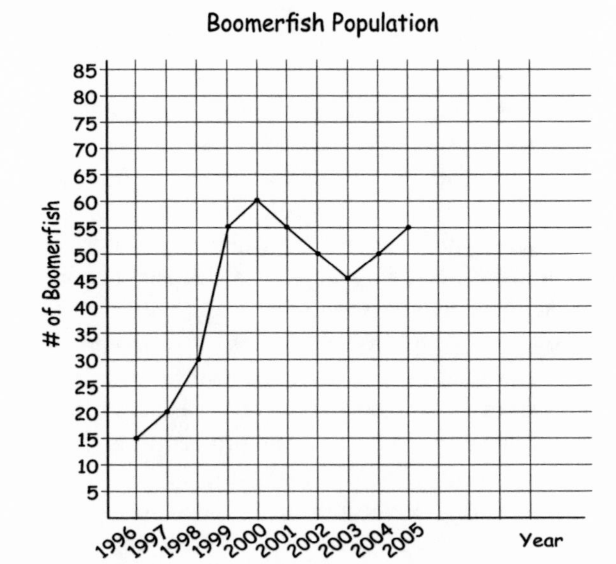

1. See completed graph above.
2. It increased by 40 fish.
3. Between 1998 and 1999
4. It drops, then goes up slightly. But the last five years are all lower than the peak (60) in 2000.
5. Answers will vary, but lack of resources and crowded habitat should be mentioned.

ACTIVITY 15.3: Greening Your School

Answers to discussion questions are included in the text.

ACTIVITY 15.4: The Population Debate

Answers will vary throughout, depending on the countries that the students were assigned and their interpretation of that country's policies. Check that students have adequately researched their countries, and they must support all of their positions with facts gleaned from their research and not be influenced by their individual feelings or opinions. Students should also be evaluated on their presentation and speaking skills.

UNIT 16: ENVIRONMENTAL JUSTICE

READING HANDOUT: Toward a Wireless Chicago

1. A person can use the Internet to research health information and educational opportunities and to interact with local government.
2. WiFi, an abbreviation for wireless fidelity, allows one high-speed Internet connection to be shared by multiple wireless users. The technology works the way walkie-talkies do: High-frequency radio signals radiate from a central transmitter in concentric circles and are picked up by other computers equipped with a computer card that receives and broadcasts back its own signal.
3. With dial-up, a phone number calls into a central server. With high-speed, a cable or high-speed phone line maintains a constant connection with a server and operates at a higher transfer rate. Some answers will vary, but high-speed is faster.
4. Answers will vary. Most will say "yes" because it allows people to have access to computers that they wouldn't be able to buy themselves.
5. Regardless of your race, gender, or income level, everyone deserves the same protection and freedom from environmental hazards and access to healthy environments. They are stretching it because they include having high-speed Internet (and therefore more ability to live, work, and learn) as being necessary to a healthy environment.
6. Access to important health and educational information and information about local community issues and their government
7. Answers will vary. Students should support their answers with some fact as well as their opinions.

ACTIVITY 16.1: What Is Environmental Justice?

Handout 16.1

Answers will vary depending on personal opinion and your location. Where applicable, students should back up their answers with research.

ACTIVITY 16.2: Environmental Justice on the Map

Handout 16.2

1. 2
2. 2
3. 40 to 100 percent
4. 6
5. Answers will vary.

ACTIVITY 16.3: Environmental Justice Close to Home

Handout 16.3

1 and 2. Answers will vary. Make sure students use facts to support their assertions.

3. Answers will vary.
4. Make sure locations on maps correspond to the data that students found.
5. Answers will vary.
6. Answers will vary. Direct students to agencies that may be able to help them with their research (not just the Internet).

UNIT 17: PUBLIC POLICY AND COMMUNITY ACTION

READING HANDOUT: A Lifetime of Community Action

1. The Jeffrey Manor neighborhood near Lake Calumet, on Chicago's southeast side. It has one of the last remaining prairie ecosystems within Chicago city limits.
2. Because it was an industrial dump site, some of the area is contaminated.
3. The city announced that it was going to build a bus depot on land in her neighborhood. She decided to organize to fight it.
4. In addition to the bus garage, the city wanted to build a dump, a garbage incinerator, and an airport. All plans were defeated.
5. First, the citizens organized, forming the Southeast Environmental Task Force and the Calumet Stewardship Initiative. A big turning point came in 1998, when the National Park

Service announced the Calumet was suitable for designation as a National Heritage Area.

6. Answers will vary. Some students will agree, others may not. Some may think that they should protect more.
7. Answers will vary. Encourage students to think outside the box and focus on jobs that would not harm the area.
8. She says, "There's always something else to do." Yes, she seems to trust her neighbors. She expresses a feeling of camaraderie and shared purpose and the belief that when they come together, good things happen.

ACTIVITY 17.1: Understanding Public Policy

Web links for additional information about the policies listed are available at the end of the unit. Answers included in the reports will vary according to the policy being presented. Students should answer all questions on the handout in their report and be graded according to presentation, as well as research and writing. Encourage students to use a variety of sources (library, web, local environmental agencies, interviews with environmental activists or politicians) in the preparation of their reports.

ACTIVITY 17.2: Getting Organized

Grade students on the quality of their press release and/or their letter to an editor.

ACTIVITY 17.3: Putting It All Together

Sign off on students' event ideas before they begin preparing their campaigns. Try to have variety in the classroom (not five groups all promoting Earth Day). Remind students that they can organize around an event of their own creation, not just a holiday. Make sure students have prepared all components of their campaign (listed in #4). Being able to put these campaigns into action—hosting the events, sending out the press releases, following up with press contacts and supporters—is the best way to drive this unit home.

UNIT 18: SUSTAINABLE COMMERCE

READING HANDOUT: Meet a Green Banker

1. Bank customers make money from a bank in the form of interest paid to them on the money they deposit in their checking and savings accounts.
2. Banks make money by loaning their depositors' money to individuals and businesses that need it. Borrowers pay them back the loan with interest.
3. Interest is a fee paid by a borrower to a lender as payment for a loan. It's also the money a bank pays a depositor on money in that depositor's account.
4. A prospective borrower fills out a loan application; then a bank investigates the person's creditworthiness and disburses the funds. After a short period, the borrowers start paying the loan back with interest.
5. Answers will vary. Most students will say Mr. Liu plays a critical role because, without him, most small green businesses would not be able to find funding. Be open to divergent views on this matter.
6. By loaning money, Mr. Liu is making these businesses possible. He is allowing them to exist and get off the ground floor. Without his help, they wouldn't get off the ground.
7. The story mentions a solar power company, an organic produce company, and a cheese company.
8. Small businesses always find it difficult to get loans. Green businesses, however, have historically been perceived as being a greater risk because banks thought most consumers weren't interested in such products. But this has changed dramatically in the last two decades.
9. Mr. Liu helps borrowers, his bank, and his depositors.
10. Answers will vary. Most students will say sustainable commerce is a positive thing because it helps Earth and is healthier for the planet and for most communities. Be open, however, to different points of view.

ACTIVITY 18.1: Interview a Green Business Owner

Make sure students have asked all the questions indicated of their interviewee and that those answers have been filled out on the handout or an attached document. They should be graded on presentation of their findings as well as preparatory research and writing.

ACTIVITY 18.2: The Green Business Plan

Make sure students have completed all applicable areas of their business plans and have included all parts of the report (numbers 1 through 10 included on handout). They should be graded on presentation as well as research and writing. The question-and-answer period is very helpful and allows for good discussion of the feasibility and potential reception of the business ideas being presented.

UNIT 19: GREEN-COLLAR CAREERS

READING HANDOUT: All about Green-Collar Jobs

1. A blue-collar job is one that usually takes place on the assembly line in factories or in other manufacturing settings. White-collar jobs are office jobs, and most of their practitioners have college degrees.
2. A green-collar job is a job that is good for Earth.
3. Answers will vary. According to the article, green-collar workers are engineers, mechanics, organic farmers, organic grocers, chefs, solar panel installers, green home builders and designers, and many more. Accept all professions students can think of, provided they can justify their choices.
4. When a person performs a job that helps clean the environment, supports healthy lifestyles and choices, and engages in sound, sustainable decisions, it can benefit Earth as a whole.
5. The authors seem to think that young people emerging from school may actually have better training, be more passionate about, and be quicker to adapt to the needs of the new eco-economy.
6. Mr. Jones thinks green-collar jobs will expand the job market and create more jobs for those in poor neighborhoods.
7. As we have seen in earlier units, toxic materials are most often illegally dumped in poor neighborhoods.
8. Mr. Jones thinks green-collar jobs cannot be exported overseas because most of the work rehabbing America's infrastructure must be done locally.
9. High oil costs have resulted in higher utility bills, necessitating the renovation of millions of homes, businesses, and other structures to be more energy efficient. Millions of green-collar jobs will grow out of this need.
10. Answers will vary. Help students to see that even a job that does not immediately seem to tie into the green economy can in fact benefit Earth.

ACTIVITY 19.1: The Green Workplace

Answers will vary. Students should remember all of the units that were covered in the curriculum and all of the potential jobs that would be created if all of these best practices were put into place.

See that Handout 19.1 has been filled out properly and fully. Some job examples that draw from information in the curriculum include excavating and designing swales, installing cisterns, engineering for both, maintenance for both, consultants for both, biodiesel mechanics and suppliers, green roof construction, green architects, planting parks, horticulturists, city planners, teachers, recycling programs and recycled products managers and entrepreneurs.

ACTIVITY 19.2: Shadow a Green Mentor

Answers will vary. Reports provided by students should be detailed and include as much information and insight as possible regarding their experience shadowing their subject.

ACTIVITY 19.3: Sharing What You've Learned

Answers to relevant discussion questions are included in the text. Make sure students follow all of the guidelines listed on the handout in creating their environmental education unit. It is very important that students have the opportunity to "teach" their units, even if only to their own class. Being able to teach it to local sixth graders would be ideal.

APPENDIX B

SCOPE and SEQUENCE

The National Science Standards for Grades 9–12 (NSTA) 312

The National Social Studies Standards for Grades 9–12 (NCSS) 312–315

Guide to Science and Social Studies Standards for All Activities 316–337

The National Science Standards for Grades 9–12 (NSTA)

Standard A: Science as Inquiry
- A1. Abilities necessary to do scientific inquiry
- A2. Understand about scientific inquiry
- A3. Design and conduct scientific investigations
- A4. Use technology and mathematics to improve investigations and communications
- A5. Formulate and revise scientific explanations and models, using logic and evidence
- A6. Recognize and analyze alternative explanations and models
- A7. Communicate and defend a scientific argument

Standard B: Physical Science
- B1. Chemical reactions
- B2. Structure and properties of matter
- B3. Motion and forces
- B4. Conservation of energy
- B5. Interactions of energy and matter

Standard C: Life Science
- C1. The cell
- C2. The molecular basis of heredity
- C3. Biological evolution
- C4. The interdependence of organisms
- C5. Matter, energy, and organization in living systems
- C6. The behavior of organisms

Standard D: Earth and Space Science
- D1. Energy in the Earth system
- D2. Geochemical cycles
- D3. The origin and evolution of the Earth system

Standard E: Science and Technology
- E1. Identify a problem or design an opportunity
- E2. Propose designs and choose between alternative solutions
- E3. Implement a proposed solution
- E4. Evaluate the solution and its consequences
- E5. Communicate the problem, process, and solution

Standard F: Science in Personal and Social Perspectives
- F1. Personal and community health
- F2. Population growth
- F3. Natural resources
- F4. Environmental quality
- F5. Natural and human-induced hazards
- F6. Science and technology in local, national, and global challenges

Standard G: History and Nature of Science
- G1. Science as a human endeavor
- G2. Nature of scientific knowledge
- G3. Historical perspectives

The National Social Studies Standards for Grades 9–12 (NCSS)

Theme 1: Culture
- A. Compare, analyze, and explain the similarities and differences in the ways groups, societies, and cultures meet human needs and concerns.
- B. Explain why individuals and groups respond differently to their physical and social environments and/or changes to them on the basis of shared assumptions, values, and beliefs.
- C. Compare and analyze societal patterns for preserving and transmitting culture while adapting to environmental or social changes.
- D. Construct reasoned judgments about specific cultural responses to persistent human issues.

Theme 2: Time, Continuity, and Change
- A. Identify and describe selected historical periods and change within and across cultures, such as the rise of civilizations, the development of transportation systems, the growth and breakdown of colonial systems, and others.
- B. Systematically employ processes of critical historical inquiry to reconstruct and reinterpret the past, such as using a variety of sources and checking their credibility, validating and weighing evidence for claims, and searching causality.
- C. Investigate, interpret, and analyze multiple historical and contemporary viewpoints within and across cultures related to important events, recurring dilemmas, and persistent issues, while employing empathy, skepticism, and critical judgment.
- D. Apply ideas, theories, and modes of historical inquiry to analyze historical and contemporary developments and to inform and evaluate actions concerning public policy issues.

Theme 3: People, Places and Environments
- A. Elaborate and refine mental maps of locales, regions, and the world that demonstrate under-

standing of relative location, direction, size, and shape.
B. Create, interpret, use, and synthesize information from various representations of Earth, such as maps, globes, and photographs.
C. Use appropriate resources and geographical tools such as aerial photos, satellite images, geographical information systems (GIS), map projections, and cartography to generate, manipulate, and interpret information such as atlases, data bases, grid systems, charts, graphs, and maps.
D. Estimate; calculate distance, scale, area, and density; and distinguish spatial distribution patterns including population density and distribution.
E. Locate, describe, differentiate, and explain the relationships among various regional and global patterns of geographical phenomena such as landforms, soils, climate vegetation, natural resources, and population.
F. Describe physical system changes such as seasons, climate, weather, and the water cycle and identify geographical patterns associated with them. Use this knowledge to explain patterns in geographical phenomena.
G. Describe how people create places that reflect culture, human needs, government policy, current values, and ideals as they design and build specialized buildings, neighborhoods, industrial parks, shopping centers, and urban centers.
H. Examine, interpret, and analyze physical and cultural patterns and interactions, such as land use, settlement patterns, cultural transmissions of customs and ideas, and ecosystem changes.
I. Describe and assess ways that historical events have been influenced by, and have influenced, physical and human geographic factors in local, regional, and global settings.
J. Analyze and evaluate social and economic factors of environmental changes and crisis resulting from phenomena such as floods, storms, and droughts.
K. Propose, compare, and evaluate alternative policies for the use of land and other resources in communities, regions, nations, and the world.

Theme 4: Individual Development and Identity

A. Articulate personal connections to time, place, and social/cultural systems.
B. Identify, describe, and express appreciation for the influences of various historical and contemporary cultures on an individual's daily life.
C. Examine the interactions of ethnic, national, or cultural influences in specific situations or events.
D. Analyze the role of perceptions, attitudes, values, and beliefs in the development of personal identity.
E. Compare and evaluate the impact of stereotyping, conformity, acts of altruism, and other behaviors on individuals and groups.
F. Work independently and within groups and institutions to accomplish goals.
G. Examine factors that contribute to and damage one's mental health and analyze issues related to mental health and behavioral disorders in contemporary science.

Theme 5: Individuals, Groups and Institutions

A. Demonstrate an understanding of concepts such as role status and social class in describing the interactions and connections of individuals and social groups.
B. Analyze group and institutional influences on people, events, and elements of culture in both historical and contemporary settings.
C. Identify and analyze examples of tensions between expression of individuality and efforts used to promote social conformity by groups and institutions.
D. Describe and examine belief systems basic to specific traditions and laws in contemporary and historical movements.
E. Evaluate the role of institutions in furthering both continuity and change.
F. Analyze the extent to which groups and institutions meet individual needs and promote common good in contemporary and historical settings.
G. Explain and apply ideas and modes of inquiry drawn from behavioral science and social theory in the examination of persistent issues and social problems.

Theme 6: Power, Authority, and Governance

A. Examine persistent issues involving the rights, roles, and status of the individual in relation to the general welfare.
B. Explain the purpose of government and analyze how its power is acquired, used, and justified.

C. Analyze and explain ideas and mechanisms to meet needs and wants of citizens, regulate territory, manage conflicts, establish order and security, and balance competing conceptions of a just society.
D. Compare and analyze the ways nations and organizations respond to conflict between forces of unity and forces of diversity.
E. Analyze and evaluate conditions, actions, and motivations that contribute to conflict and cooperation within and among nations.
F. Explain and apply ideas, theories, and modes of inquiry drawn from political science to the examination of persistent issues and social problems.
G. Evaluate the extent to which governments achieve their stated ideals and policies at home and abroad.
H. Prepare a public policy paper and defend it before an appropriate forum in school or in the community.

Theme 7: Production, Distribution, and Consumption

A. Explain how the scarcity of productive resources (human, capital, technological, and natural) requires the development of economic systems to make decisions about how goods and services are to be produced and distributed.
B. Explain the role that supply and demand, prices, incentives, and profits play in determining what is produced and distributed in a competitive market system.
C. Consider the costs and benefits to society of allocating goods and services through private and public sectors.
D. Describe the relationship among various economic institutions that comprise economic systems such as households, business firms, banks, government agencies, labor unions, and corporations.
E. Analyze the role of specialization and exchange in economic processes.
F. Compare how values and beliefs influence economic decisions in different societies.
G. Apply economic concepts and reasoning when evaluating historical and contemporary social developments and issues.
H. Distinguish between the domestic and global economic systems and explain how the two interact.
I. Apply knowledge of production, distribution, and consumption in the analysis of public issues such as health care and energy consumption. Devise an economic plan for accomplishing a socially desirable outcome related to that issue.

Theme 8: Science, Technology, and Society

A. Identify and describe both current and historical examples of the interactions and interdependence of science, technology, and society in a variety of cultural settings.
B. Make judgments about how science and technology have transformed the physical world and human society and our understanding of time, space, place, and human–environment interactions.
C. Analyze how science and technology influence the core values, beliefs, and attitudes of society and shape scientific and technological change.
D. Evaluate various policies that have been proposed as ways of dealing with social changes resulting from new technologies, such as genetically engineered plants and animals.
E. Recognize and interpret varied perspectives about human societies and the physical world using scientific knowledge, ethical standards, and technologies from diverse world cultures.
F. Formulate strategies and develop policies for influencing public discussions associated with technology-societal issues, such as the greenhouse effect.

Theme 9: Global Connections

A. Explain how language, art, music, belief systems, and other cultural elements can facilitate global understanding or cause misunderstanding.
B. Explain conditions and motivations that contribute to conflict, cooperation, and interdependence among groups, societies, and nations.
C. Analyze and evaluate the effects of changing technologies on the global community.
D. Analyze the causes, consequences, and possible solutions to persistent, contemporary, and emerging global issues, such as health, security, resource allocation, economic development, and environmental quality.
E. Analyze the relationship and tensions between sovereignty and global interests in such matters as territory, economic development, nuclear and other weapons, use of natural resources, and human rights concerns.

F. Describe and evaluate the role of international and multinational organizations in the global arena.
G. Illustrate how individual behaviors and decisions connect with global systems.

Theme 10: Civic Ideals and Practices

A. Explain the origins and interpret the continuing influence of key ideals of the democratic republican form of government, such as individual human dignity, liberty, justice, equality, and the rule of law.
B. Identify, analyze, interpret, and evaluate sources and examples of citizens' rights and responsibilities.
C. Locate, access, analyze, organize, synthesize, evaluate, and apply information about selected public issues, identifying, describing, and evaluating multiple points of view.
D. Practice forms of civic discussion and participation consistent with the ideals of citizen in a democratic republic.
E. Analyze and evaluate the influence of various forms of citizen action on public policies.
F. Analyze a variety of public policies and issues from the perspective of formal and informal political actors.
G. Evaluate the effectiveness of public opinion in the influencing and shaping public policy development and decision making.
H. Evaluate the degree to which public policies and citizen behaviors reflect or foster the stated ideals of a democratic republican form of government.
I. Construct a policy statement and an action plan to achieve one or more goals related to an issue of public concern.
J. Participate in activities to strengthen the "common good," based on careful evaluation of possible options for citizen action.

Unit Number and Title	Activity	Science Standards (for each activity)	Social Studies Standards (for entire unit)	Reproducibles (suitable for overhead)
Unit 1: Understanding Sustainability	Reading: "The Path to Sustainability"	A: 2,6; C:3-6; D: 2; F: 1-6	1: C, D; 2: A-D; 3: A-K; 4: A, F; 5: A-G; 6: A-H; 7: A-I; 8: A-F; 9: A-G	
	1.1: The Sustainability Game	A: 4,6; C:4,6; E:5; F1-6		
	1.2: Making Connections in the Urban Ecosystem	A: 2,6,7; C; 4,6; D: 2; E: 1-3,5; F:1,3-6		1.2A: "A City Suffers"; 1.2B: "A City Comes Back"
	1.3: Sustainable or Not Sustainable?	A: 6,7; C: 4,6; E: 1, 5; F: 1,4,5		
	1.4: Global Perspectives	A: 4,6; C: 4-6; E: 1,2,5; F: 1-3,5,6; G: 1-3		1.4: "Global Perspectives"
	1.5: At the Podium	A: 7; C: 4-6; E: 1-5, F: 1-6		
Unit 2: Building Community	Reading: "Real People, Real Lessons"	A: 2; C: 4, 6; E: 1,5; F: 1,3,5,6	1: A-D; 2: D; 3: A,B,G; 4: A-G; 5: A-G; 6: A,H; 10: B, C, J	
	2.1: Thinking about Communities	C: 4; F:1		Teacher-created chart
	2.2: Now and Venn	A: 2,4; C:4; F:1		2.2 "Venn Diagram Sample"
	2.3: The Power of Intention	C:4; E:5; F:1		
	2.4: Making Sense of the Census	A: 4-7; C: 4,6; F: 1,2,6; G: 1-3		

Handouts (for student work)	Expected Outcomes	Extensions	Assessment
"The Path to Sustainability" with questions	To learn how sustainability occurs in nature; to apply it to human settlements		Questions assess writing and reading comprehension
	To learn how Earth's resources must be shared wisely by all	To extend what students have learned to world situation	Game, discussion questions and follow-up assess strategic and critical thinking
	To understand that problems in ecosystems are often interconnected; to become familiar with potential solutions	Research city agencies responsible for mitigating environmental problems; consider possible solutions for students' cities	Discussion questions and follow-up assess students' abilities to analyze visual data, research issues and problem-solve
1.3: "Sustainability in the Gray Area"	To decide which everyday behaviors are sustainable or unsustainable	Thinking about the feasibility of living sustainably as an individual or a group	Handout requires students to think analytically and express their opinions.
	To carefully read a bar graph; to analyze data showing nations' ecological footprints	Students calculate their "ecological footprint" using Internet resources; think critically about their use of resources	Reproducible assesses chart-reading skills; discussion and follow-up assess topic comprehension
1.5 Judges' Evaluation Sheet	To prepare for and participate in two debates about sustainability		Research preparedness, debate presentation, speaking skills, Judges' Evaluation Sheet
		Unit-wide extensions	
"Real People, Real Lessons" with questions	To understand how people can come together as a community to solve problems; to understand philanthropy		Questions assess writing and reading comprehension
	To understand the various kinds of communities that exist; to understand any common purposes community members share		Classroom discussion assesses students' comprehension of term *community*, critical thinking about their own roles
	To learn how community is a part of daily life; to understand Venn diagrams	To explore other graphical representation of community involvement	Diagram assesses students' graphing abilities and understanding of community
	To learn about intentional communities	Writing exercise further explores students' understanding of concept	Research preparedness and writing abilities, critical thinking and reflection
2.4: "U.S. Census (2004)"	To interpret data about community populations; to understand the difference between types of communities; to extend understanding to their community		Chart-reading and comprehension, critical thinking on topic, research, writing and presentation skills.

Unit Number and Title	Activity	Science Standards (for each activity)	Social Studies Standards (for entire unit)	Reproducibles (suitable for overhead)
	2.5: Front Page	A:1-3,7; E:5; F:1,6; G:1,3		
	2.6: At the Podium	A:7; C:4; E:1,2,5; F:1; G:1,3		
Unit 3: Waste Management and Recycling	Reading: "Young People in Urban Nature"	C:4,5; E:1,5; F:1,4,5; G:1	1: C,D; 2: D; 3: A-K; 4: E,F; 5: A-G; 6:A-H; 7: A-I; 8:A-F; 9: A-G; 10: A-J;	
	3.1: Don't "Waste" This Quiz	B:1,2,5; C:5; D:2; E:5; F:3-6; G:3		
	3.2: "Waste"-ing Away	A:7; B:1,2; C:5; D:2; E: 1,2,5; F:3-5		
	3.3: Talking Trash	A:1,2,6,7; B:1,2; D:1,2;E:1-5; F:3-6; G:1,3		3.3A: "Cross-Section of an Active Landfill"; 3.3B "Is All Trash the Same?"; 3.3C "The Breakdown Rates of Common Household Items"
	3.4: Trash Audit	A:1-7; B:1,2; C:5; D:1,2; E: 1-5; F:1,3,4,5; G:1-3		
	3.5: Trash Tops the Charts	A:6,7; B:2; D:2; E:1,2; F:3-6; G:1,3		
	3.6: Landfill Dispute	A:1,2,5,7; B:1,2,5; D:2; E:1,2,4,5; F:1,4,5; G:1,3		3.6: "Landfill Map"
	3.7: At the Podium	A:7; C:4; E:1,2,5; F:1,4,5; G:1,3		

Handouts (for student work)	Expected Outcomes	Extensions	Assessment
	To interview a community member; to write a profile	To share findings by publishing results in paper or blog	Interviewing and writing skills, critical thinking
2.6: Judges' Evaluation Sheet	To prepare for and participate in a debate about community involvement		Research preparedness, debate presentation, speaking skills, Judges' Evaluation Sheet
		Unit-wide extensions	
"Young People in Urban Nature" with questions	To see how one person can make a difference in a community; to learn about water and soil contamination		Questions assess writing and reading comprehension
3.1: "Don't 'Waste' This Quiz"	To see how much students know about recycling issues		Self-test assesses knowledge of fun recycling facts
	To study everyday items and assess the likelihood of their being reusable or recyclable		Personal or class tables, personal trash journal assess understanding of recycling and indicates level of creativity
	To learn how a landfill works, what types of trash are suitable for disposal in one, and how humans dispose of trash		Discussion questions and follow-up assess reading comprehension, comprehension of key concepts, and critical thinking
	To conduct an audit of materials discarded by students and the class; to use data to brainstorm ways of changing future behavior	To devise a program to provide for recycling and/or composting bins on school premises	Discussion questions, follow-up, class tables and charts, and personal trash journals assess research skills
3.5: "Trash Tops the Charts"	To learn how much waste is produced by Americans, and what percentage is recycled		Handout and discussion assess knowledge of key concepts, understanding of percentages, pie chart, and bar graph
3.6: Judges' Evaluation Sheet	To evaluate the objections that arise when a (fictional) landfill needs a home, and no ideal locations exist		Research preparedness, public presentation and critical and logical thinking
3.7: Judges' Evaluation Sheet	To prepare for and participate in a debate about imposing limits on waste generation		Research preparedness, debate presentation, speaking skills, Judges' Evaluation Sheet
		Unit-wide extensions	

Unit Number and Title	Activity	Science Standards (for each activity)	Social Studies Standards (for entire unit)	Reproducibles (suitable for overhead)
Unit 4: Green Building	Reading: "A City of Green Builders"	A:7; D:1; E:5; F:1,3, 5,6; G:1	1: C,D; 3: A-K; 4: E,F; 5:A-G; 6:A-H; 7:A-I; 8: A-F; 9:A-G; 10: A-J	
	4.1: Anatomy of a Green Home	A: 4,6,7; C:5; D:1,2; E: 1, 2,5; F:1,3-6; G:1		4.1: "Elements of a Green Home"
	4.2 Under the Green Roof	A:2,6,7; B:2,3; C:5; D:1,2; E:1,2,4,5; F:1,3-6; G:1-3		4.2: "What's Under the Green Roof?"
	4.3 A Green Light	A: 6,7; B:4; D:1; E:4,5; F:3,5,6; G:1,3		
	4.4: Front Page	A:1,2,7; E:1,5; F:1,3,5,6; G:1,3		
	4.5: At the Podium	A:7; D:1; E:1,2,5; F:3-6; G:1,3		
Unit 5: Energy	Reading: "French Fries in the Tank?"	A:2,6,7; B:1; D:1; E:5; F:1,3-6; G:1,3	1: A-D; 2:D;3:D,K; 4:A-D,F; 5:B-G; 6:C,D; 7:A-I; 8:A-F; 9:B-E,G; 10:B,C,E,F,G, J	
	5.1: Sun, Sky and Earth: Where We Get Our Energy	A:2,6,7; B:1,2,4,5; D:1-3; E: 1,2,4,5; F:3-6; G:1-3		5.1A: "Fossil Fuels"; 5.1B: "Sun and Wind: Renewable Energy Sources"
	5.2: I Love My Car—I Just Hate Paying for Gas!	A:2,7; B:1; D:1-3; E:1,4,5; F:3-6; G:1-3		

Handouts (for student work)	Expected Outcomes	Extensions	Assessment
"A City of Green Builders" with questions	To gain an understanding of green building		Questions assess writing and reading comprehension
4.1: "Green Home Smarts"	To learn the attributes of a green-built home	To prepare a reports on one green building feature	Discussion assesses students' comprehension of key concepts; extensions assess research and writing abilities
4.2: "Under the Green Roof"	To understand how a green roof is constructed; to learn how such roofs help cities combat heat in summer and conserve energy costs	To investigate sustainable commerce applications of green roofs	Completion of questions and worksheet on handout; critical and creative thinking about feasibility of green roofs; well-thought-out arguments for and/or against concepts
4.3: "CFL Quick Facts"	To understand the cost/benefits of using compact fluorescent light bulbs	Showing math behind cost/benefits; connecting idea of air pollution and energy production	Successfully completed handout; understanding of key vocabulary; understanding of relationship between energy consumption and pollution
	To analyze articles from the opinion section of the newspaper; to write an opinion essay calling for more green building in their town or city	To publish pieces in school or local paper	Demonstrate understanding of the op-ed writing form; understanding of green building concepts; demonstrate support of argument with clear writing
4.5: Judges' Evaluation Sheet	To prepare for and participate in a debate about mandatory CFL use		Research preparedness, debate presentation, speaking skills, Judges' Evaluation Sheet
		Unit-wide extensions	
"French Fries in the Tank?", with questions	To understand biodiesel fuel and its potential impact on the economy		Questions assess writing and reading comprehension
	To become familiar with traditional and alternative energy sources		Research projects address all bulleted points; demonstrated understanding of various kinds of fuel; critical thinking about role of alternative energies in our society
5.2: "I Love My Car—I Just Hate Paying For Gas!"	To learn types of energies humans use to power motor vehicles; to create researched readings	To investigate fuel facts not mentioned in the activity	Completed questions demonstrate understanding of topic; critical thinking and clarity of expression

Unit Number and Title	Activity	Science Standards (for each activity)	Social Studies Standards (for entire unit)	Reproducibles (suitable for overhead)
	5.3: Charting Energy	A:2,4,7; D:1-3; E:1,4,5; F:1-6; G:1-3		
	5.4: Home Energy Survey	A:1-3,6,7; D:1-3; E:1-5; F:1,3-6; G:1,3		
Unit 6: Air Quality	Reading: "Tree Boy"	A:2,7; B:1; C:1,3,4; E:4,5; F:1,3-6; G:1-3	1:A-D; 2:C,D; 3:A-F, H,I,K; 5:E,G; 6:A,C,E,F; 7: A-I; 8:A-F; 9:C,D,E,G; 10:B,C,E,G, J	
	6.1: The Air Up There	A:2,4; B:1; D:1-3; E:1,5		
	6.2: Smog, Ozone, and the Greenhouse Effect	A:1,2,4,7; B:1-3,5; D:2,3; E:1,4,5; F:1,3-6; G:1-3		6.2: "Smog Diagram"
	6.3 How Much CO_2 Do You Contribute?	A:1,2,4,7; B:1-3,5; D:2,3; E:1,4,5; F:1,3-6; G:1-3		6.3: "Calculating Your CO_2"
	6.4 Understanding the Greenhouse Effect	A: 1-3,5-7; B:1,2; D:1-3; E:1,4,5; F:1, 4-6; G:1-3		
Unit 7: Water Quality	Reading: "Restoring the Chicago River"	A:2,7; C:3,4,6; E:1,4,5; F:1,3-6; G:1-3	1: A-D; 2:C,D; 3:A-K; 4:E,F; 5: E,F,G; 6:A,C,F,H; 7:A,D,F,G,I; 8: A-F; 9:B-E; 10:B-G,I, J	
	7.1: Understanding Earth's Water Supply	A:1,2,4; B:2,3; C:3,4,6; D:2; E:1,4,5; F:1,3-6; G:1-3		7.1: "The Water Cycle"

Handouts (for student work)	Expected Outcomes	Extensions	Assessment
5.3A: "How Much Energy Do We Really Use?" 5.3B: "Global Consumption of Oil Per Capita: 2001"	To understand which nations use the most energy; to understand use of the term "per capita"	Critical thinking and pro/con essay on energy use	Successful completion of all handout questions; demonstrate understanding of key concepts through essay and class discussion
Handout 5.4: "Home Energy Survey"	To conduct an energy survey; to better understand energy efficiency at home	Anticipating outcomes of widespread implementation of energy surveys and follow-ups	Successful completion of home survey; clear presentation of findings; understanding of potential remedies
		Unit-wide extensions	
"Tree Boy," with questions	To understand how trees affect air quality		Questions assess writing and reading comprehension
6.1: "The Air Up There"	To understand the layers of our atmosphere	Additional research into layers of the atmosphere; larger-scale reproduction of Earth's atmosphere	Successful completion of handout
	To understand smog, ozone, and the greenhouse effect; to understand the ways to reduce their detrimental effects on air quality	Additional research into chlorofluorocarbons	Discussion questions demonstrate understanding of topics; demonstrated understanding of smog, ozone and greenhouse gases individually as well as together
	To estimate how much CO_2 their families release during a particular time period	Devise a strategy for reducing families' CO_2 emissions; possible presentation of plan to media and civic leaders	Successful completion of calculations based on bills gathered; clear reasoning for CO_2 reduction strategy
	To duplicate the greenhouse effect in a classroom experiment		Demonstrate understanding of the greenhouse effect and how it is exhibited in this experiment
		Unit-wide extensions	
"Restoring the Chicago River," with questions	To learn how a group of dedicated citizens restored a major waterway; to understand the role of rivers in everyday life		Questions assess writing and reading comprehension
7.1: "The Distribution of Earth's Water"	To understand Earth's water cycle; to learn about watersheds		Successful completion of handout; discussion questions demonstrate understanding of water cycle and watershed pollution

Unit Number and Title	Activity	Science Standards (for each activity)	Social Studies Standards (for entire unit)	Reproducibles (suitable for overhead)
	7.2: Town Meeting: Watershed at Risk	A:1,2,5-7; B:3; D:2; E:1-5; F:1,3-6; G:1-3		
	7.3: The Rundown on Runoff	A:1-3,6,7; B:1,3; C:5,6; D:2; E:1,2,5; F:1,3-6; G:1-3		
	7.4: Understanding Water Quality	A:2,4,7; D:1-3; E:1,4,5; F:1-6; G:1-3		
	7.5: Swales and Cisterns	A:1,2,6,7; B:1-3; C:5; D:2; E:1,2,4,5; F:1,3-6; G:1-3		
Unit 8: Soil Quality	Reading: "Restoring Chicago's Calumet Region"	A:2,7; B:3; C:4; D:2; E:1,2,4,5; F:1,3-6; G:1-3	1:A-D; 2:A-D; 3:C, E-K; 4:A,F; 6:A,C,F; 7:A,D,F,G,I; 8:A-F;9:B,D; 10:B,C,E,F,G	
	8.1: Hydroponics to the Rescue	A:1-7; B:3,5;C:1,6; D:1,2; E:1-5; F:1,3,5,6; G:1-3		
	8.2: Alfalfa and the Importance of Cover Crops	A:1,2,4-6; B:3,5;C:1,6; D:1,2;E:1,4,5; F:1,3-6; G:1-3		
	8.3: Soil, Air, and Water	A:2,6,7; B:1,3; C:1,4-6; D:2; E:1,2,4,5; F:1-6; G:1-3		8.3: "How Pollution Travels"
Unit 9: Parks and Open Spaces	Reading: "The Great Public Space"	A:1,7;B:3; C:1,3-6; D:2;E:1,4,5; F:1,3-6; G:1-3	1:C,D;2:A-D;3:A-K; 4:E,F;5:A-G;6:A-H; 7:A-I;8:A-F; 9:A-G; 10:A-J	
	9.1: Thinking about Parks	A:1,2; B:3; C:1,4-6; D:2;E:1,4,5; F:1,6; G:1-3		

Handouts (for student work)	Expected Outcomes	Extensions	Assessment
	To conduct research in preparation for a debate on risks to a watershed posed by development	To investigate freshwater in students' area, noting any water-quality issues; making connections between daily habits and pollution	Research preparedness and citing of sources; presentation and/or role-playing skills; clarity of argument
7.3: "Take-Home Audit: Runoff Pollutants"	To understand runoff; to perform a take-home audit to identify runoff problems at home		Successful completion of audit; detailed research on chemicals and tox-free substances; clear reports on findings
7.4: "Water Quality Analysis"	To become familiar with water quality reports; to analyze components of water pollution	Obtain water-quality ratings and reports from students' region	Successful completion of handout; demonstrated understanding of water pollutants
7.5: "Swales and Cisterns"	To learn how swales and cisterns work		Successful completion of handout; demonstrated understanding of swales and cisterns
		Unit-wide extensions	
"Restoring Chicago's Calumet Region," with questions	To understand the challenges of restoring a contaminated site		Questions assess writing and reading comprehension
8.1: "Questions for Hydroponics Experiment"	To conduct a hydroponics experiment; to see how plants and vegetables can grow without soil		Successful completion of experiment and handout questions; demonstrated understanding of hydroponics
8.2: "Alfalfa and the Importance of Cover Crops"	To learn how cover crops restore wetlands damaged by steel factory waste		Successful completion of handout
	To understand how soil contaminants can also pollute water and air; to write pro/con essays on the topic		Completion of essay; demonstrated understanding of interrelation of soil, air and water contamination; demonstrated understanding of opinion essay form
		Unit-wide extensions	
"The Great Public Space," with questions	To understand how public space improves the health of cities and builds community		Questions assess writing and reading comprehension
	To engage in a conversation about, and increase understanding of, the value of public parks	To write a report on a world-famous park, its history and change in use over time	Successful completion of chart of park attributes; clear expression of thoughts and opinions in discussion

Unit Number and Title	Activity	Science Standards (for each activity)	Social Studies Standards (for entire unit)	Reproducibles (suitable for overhead)
	9.2: Open Space Around the World	A:2; B:3; C:4,6;E:1,5; F:1,6; G:1-3		9.2 "Diagram of a Piazza"
	9.3: Mapping the Millennium	A:2,4,7; B:3; C:3-6; D:2;E:1,4,5; F:1,3-6; G:1-3		
	9.4: Design a Park or Piazza	A:1-7; C:4,6; E:1-5; F:1,3-6; G:1-3		
	9.5: One Space, Two Plans	A:1-7; C:4-6; E:1-5; F:1-6; G:1-3		
Unit 10: Transportation	Reading: "Living With One Less Car"	A:2,7;B:1; C:4,6; E:1,4,5; F:1,3-6; G:1-3	1: C,D; 2:C,D 3:A-K; 4:E,F 5:A-G; 6:A-H; 7:A-I; 8:A-F; 9:A-G;10: A-J	
	10.1: Passenger Costs	A:2,4,7; B:1-5; D:1,2; E:1,2,4,5; F:1,3-6; G:1-3		
	10.2: Pros, Cons, and in Between	A: 2,5-7; E:1,2,4,5; F:1-6; G:1-3		
	10.3: One-Less-Car Challenge	A:1-7; B:1,4; E:1-5; F:1,3-6; G:1-3		
	10.4: Writing about It and Making Connections	A:2,6,7; B:1; C:1; D:2;E:1,2,5; F:1-6; G:1-3		

Handouts (for student work)	Expected Outcomes	Extensions	Assessment
	To compare and contrast the design and uses of public spaces in the U.S. and great European cities		Successful completion of group report; adherence to all questions posed; good group participation; demonstrated critical thinking
9.3: "Mapping Millennium Park"	To analyze a map of a park; to determine what features Americans consider essential to new parks		Successful completion of handout; participation in class discussion; ability to expand on role of parks in community
	To design a park or other open space		Successful completion of design, encompassing all attributes listed in activity; research preparedness; good group participation; presentation skills, including visual aids
	To develop two different plans for the same parcel of undeveloped land	To devise one plan based on a compromise	Successful completion of plan; good group participation; demonstrated understanding of issue at hand and idea of "compromise"; presentation skills
		Unit-wide extensions	
"Living With One Less Car," with questions	To understand the potential challenges and benefits faced when living with one less car		Questions assess writing and reading comprehension
10.1: "Energy Efficiency Chart"	To determine how many BTUs are produced or saved in several different driving scenarios	To convert BTUs into other units of measurement	Successful completion of handout; participation in classroom discussion; clear expression of ideas
10.2: "Transportation Chart"	To research and compare the attributes different types of transportation; to write about their findings	To complete more in-depth research into a chosen form of transportation	Successful completion of the handout; demonstrated understanding of mind-set of assigned roles; clearly written essay
10.3A: "Car Cost Worksheet" 10.3B: "Daily Transportation Diary"	To consider and/or practice living with one less car and address results		Successful completion of handouts; accurate compilation of data; demonstrated critical thinking and comprehension through discussion questions
10.4: Writing about It and Making Connections: "Thinking about Transportation"	To summarize, in essays, the connections between transportation and sustainability		Successful completion of essay; demonstrated understanding of chosen topic; clearly supported ideas and opinions
		Unit-wide extensions	

Unit Number and Title	Activity	Science Standards (for each activity)	Social Studies Standards (for entire unit)	Reproducibles (suitable for overhead)
Unit 11: Biodiversity	Reading: "Restoring the Prairie"	A:2,6,7; C:3,4,6; D:3;E:1,4,5; F:1-6;G:1-3	1: C,D; 2:C,D; 3:A-K; 4:A,F; 7:A,B, F,I; 8:A-F;9:D, E,G; 10:C,E, F,G, J	
	11.1: Classmate Diversity Scavenger Hunt	A:1,7; E:1,3,5; F:1; G:1-3		
	11.2: Getting to Know Your Own Backyard: An Urban Survey	A:1,2,7; C:3,4,5; E:1,5; F:1,3-6; G:1,3		
	11.3: Overfished and Contaminated: Crisis in Our Seas	A:1,2,4,6,7; B:1; C:3-6; D:2; E:1,5; F:1,6; G:1-3		
	11.4: Ecosystem Swap Meet	A:1,2,4,7; C:3,4,6; E:1,3,4,5; F:1-6; G:1-3		
Unit 12: Urban Agriculture and Community Gardens	Reading: "Philadelphia Eats What Philadelphia Grows"	A:2,7; B:1,3; C:1,3-6; D:1; E:1,4,5; F:1,3-6; G:1-3	1:A-D; 2:C,D; 3:A-K; 4:F; 5:F,G; 6:F; 7:A-I; 8:A-F; 9:B-G; 10:J	
	12.1: Growing All Around	A:2,7; B:1; C:3,4,6; D:1,3-6; E:5; F:1,6; G:1-3		
	12.2: Zoning in on Hardiness	A:2,4,7; B:3; C:3-6; D:2,3; E:5; F:1,3-6; G:1-3		
	12.3: Food on the Move	A:2,4,6,7; B:4,5; C:4-6; D:1; E:1,2,4,5; F:1,3-6; G:1-3		
	12.4: Your Own Garden	A:1-7; B:3; C:3-6; D:1-3; E:1-5; F:1,3-6; G:1-3		

Handouts (for student work)	Expected Outcomes	Extensions	Assessment
"Restoring the Prairie," with questions	To understand the role of prairies as ecosystems; to learn the benefits of restoring areas damaged by invasive plants		Questions assess writing and reading comprehension
11.1: "Classmate Diversity Scavenger Hunt"	To develop an understanding of biodiversity by learning more about classmates		Successful completion of handout; evidence of speaking to a variety of classmates; demonstrated understanding of humans as similar to animals
11.2: "Getting to Know Your Own Backyard"	To gain an understanding of the diversity of flora and fauna in students' surroundings	To write a well-argued essay about weeds	Successful completion of handout; active participation in class discussion; understanding of biodiversity
11.3: "Overfished and Contaminated"	To create a Venn diagram listing fish that are both plentiful and safe to eat	To research safety of fish locally; to design a wallet-sized fish-buying guide	Successful completion of Venn diagram; demonstrated understanding of difference between availability and health risks of fish
	To create trading cards and posters depicting various ecosystems; to understand how these ecosystems thrive	Creating additional cards for other classes or exhibits	Successful completion of ecosystem design and trading cards; supporting research; good group participation
		Unit-wide extensions	
"Philadelphia Eats What Philadelphia Grows," with questions	To understand how food can be grown and sold in urban environments; the importance of eating locally		Questions assess writing and reading comprehension
12.1: "Growing All Around"	To learn what role local agriculture does—or does not—play in students' daily life		Successful completion of handout; supporting research; participation in classroom discussion
12.2: "Zoning in on Hardiness and Growing Seasons"	To analyze a USDA Hardiness Zone map; to learn about different growing zones		Successful completion of handout; demonstrated understanding of growing seasons
12.3: "Food on the Move"	To learn how far certain foods have traveled to reach a central U.S. market	To write about local food; to research food ingredients at home; to discuss benefits of local produce and mandatory food labeling	Successful completion of handout; participation in group discussion; critical thinking and communication of opinions
	To assess the feasibility of building and tending a school garden.		Well-researched plan and design for garden; good group participation; critical and logical thinking about garden feasibility
		Unit-wide extensions	

Unit Number and Title	Activity	Science Standards (for each activity)	Social Studies Standards (for entire unit)	Reproducibles (suitable for overhead)
Unit 13: Urban Forestry	Reading: "The Magic of Trees"	A:2,7; B:1,3; C:1-6; D:1-3; E:1,4,5; F:1,3-6; G:1-3	1:C,D; 2:D 3:A-K; 4: E,F; 5:A-G; 6:A-H; 7:A-I; 8:A-F; 9:B-G; 10:A-J	
	13.1: Did You Know...? Fun Facts about Urban Forestry	A:2,7; B:1,3; C:1,3-6; D:2; E:1,5; F:1,3-6; G:1-3		
	13.2: Getting to Know Your Trees	A: 1-7; C:3,4,6; D:2; F:1,3-6; G:1-3		
	13.3: Trees and You: A Numbers Game	A:1,2,4-7; B:1; C:3,4-6; D:2; E:1,2,4,5; F:1,3-6; G:1-3		
	13.4: At the Podium	A:1-7; B:1; C:1,3-6; D:2; E:1,2,4,5; F:1-6; G:1-3		
Unit 14: Urban Planning	Reading: "William Penn's 'Green Countrie Towne'"	A:2,7; C:4; E:1,5; F:1,3,4,6; G:1-3	1:A-D; 2:A-D; 3:A-K; 4:A-D,F; 5:A-F; 6:B,C,H; 7:A,D,F,G,I; 8:A-F; 10:B-J	
	14.1: Our Town: A Closer Look	A:1,2,7; E:1-5; F:1,3,4,6; G:1-3		
	14.2: Penn's Plan: A Look at Old Philly	A:2,7; E:1,2,4,5; F:1,3-6; G:1-3		
	14.3: You're the Planner	A:1,2,7; E:1-5; F:1,3-6; G:1-3		

Handouts (for student work)	Expected Outcomes	Extensions	Assessment
"The Magic of Trees," with questions	To understand the importance of trees in the urban ecosystem		Questions assess writing and reading comprehension
13.1 "Eco Facts Urban Forestry Quiz"	To test students' knowledge of trees' role in the urban landscape and their effects beyond environmental realm	To investigate trees in students' area; to examine how forest canopy has changed locally over time	Successful completion of the handout; participation in classroom discussion; demonstrated understanding of key concepts
13.2: "How Old Is That Tree?"	To identify various species of trees locally and to estimate their ages	To report on a tree near students' homes; to write an essay describing the life of that tree	Successful completion of handout; demonstrated understanding of major species of trees in students' area
13.3: "Trees and You: A Numbers Game"	To interpret a graph and chart that quantify the beneficial role that trees can play in a community		Successful completion of handout; good classroom discussion participation; demonstrated understanding of relationship between trees, runoff and greenhouse gases
13.4: Judges' Evaluation Sheet	To prepare for and participate in a debate about establishing minimum tree-planting guidelines in new communities		Research preparedness, debate presentation, speaking skills, Judges' Evaluation Sheet
		Unit-wide extensions	
"William Penn's 'Green Countrie Towne'," with questions	To learn about William Penn's master plan for Philadelphia		Questions assess writing and reading comprehension
14.1: "Your Town: A Closer Look"	To perform a local walking tour; to collect data on town attributes; to brainstorm improvements	To engage the local public works department; to create a proposal for revitalizing town	Successful completion of handout; clear, well-conceived ideas for improvement
14.2: "Penn's Plan: Old Philly"	To study a city plan created by William Penn for the city of Philadelphia; to understand urban organization	To place Penn's plan in a modern-day context; to think/write about how city planning has changed and why/how it can be sustainable or unsustainable	Successful completion of the handout; good classroom discussion participation; demonstrated understanding of city planning concepts
14.3: "Checklist"	To map out a town, focusing on issues and characteristics that are important to students	To create a diorama or other visual display of plans; to present plans to school and/or local officials	Successful completion of city plan, addressing all items on checklist; clear and creative thinking about city attributes and features

Unit Number and Title	Activity	Science Standards (for each activity)	Social Studies Standards (for entire unit)	Reproducibles (suitable for overhead)
	14.4: Frederick Law Olmsted's Vision	A:1,2,7; E:1-5; F:1,3-6; G:1-3		
Unit 15: Population Growth and Integrated Resource Management	Reading: "Water In, Water Out"	A:2,6,7; C:4,6; D:2; E:1,4,5; F:1,4-6; G:1-3	1:A-D; 2:A-D; 3:A-K; 4:F; 5:E,F,G; 6:A,C,D,E,F,G,H; 7:A,D,F,G,I; 8:A-F; 9:B-G; 10:A-J	
	15.1: Pop Goes the Population	A:2,4,6,7; C:4,6; E:1,4,5; F:1-6; G:1-3		15.1: "Projected World Population Growth to 2050"
	15.2: Growing, Growing...Gone?	A:2,4,7; C:3,4,6; E:1,4,5; F:1-6; G:1-3		
	15.3: Greening Your School	A:1-3,6,7; C:3,4,6; D:2; E:1-5; F:1,3-6; G:1-3		15.3 A-E "School #1: Before and After," "School #2: Before and After," "The Technology... Cistern," "School #3: Before and After," "Artists'... Redesign"
	15.4: The Population Debate	A:1,2,4-6; C:3,4,6; E:1,2,4,5; F:1-6; G:1-3		

Handouts (for student work)	Expected Outcomes	Extensions	Assessment
	To learn about Frederick Law Olmsted; to write about memorable experiences in nature	To publish resulting essays	Successful completion of essay; clear presentation of ideas; good classroom participation in discussion and presentation
		Unit-wide extensions	
"Water In, Water Out," with questions	To understand the meaning of integrated resource management		Questions assess writing and reading comprehension
15.1: "Pop Goes the Population"	To read charts and learn about the world's population and how it will continue to rise; to understand the relationship between population and resources		Successful completion of handout; participation in classroom discussion; demonstrated understanding of relationship between resource availability and population growth
15.2: "Growing, Growing...Gone"	To graph and calculate proportions that illustrate the carrying capacity of a group of organisms	To understand limiting factors in a population and as related to the human species	Successful completion of handout; demonstrated understanding of carrying capacity; participation in classroom discussion
	To apply principles of integrated resource management to students' school		Successful participation in classroom discussion; well-conceived suggestions for school retrofit; demonstrated understanding of how design can affect environmental factors
15.4A: "Human Population Summit: Know Your Nation" 15.4B: "Human Population Mandates: The Future in Your Hands"	To participate in a mock summit on population issues, role-playing as delegates from various different nations	To write an essay discussing how countries might work together on population issues	Research preparedness; good group participation; good presentation skills, including visual aids; classroom discussion; critical and logical thinking; well-formed arguments
		Unit-wide extensions	

Unit Number and Title	Activity	Science Standards (for each activity)	Social Studies Standards (for entire unit)	Reproducibles (suitable for overhead)
Unit 16: Environmental Justice	Reading: "Toward a Wireless Chicago"	A:2,6,7; E:1,4,5; F:1,6; G:1-3	1:A-D; 2:C,D; 3:A,B,C,E,G,H,I, J,K; 4:E,F,G; 5:A,B,C,E,F,G; 6:A,C,D,F,G,H; 7:F,I; 8:E,F; 10:A-J	
	16.1: What Is Environmental Justice?	A:2,7; C:4,6; E:1,4,5; F:1,4-6; G:1-3		
	16.2: Environmental Justice on the Map	A:1,5,6; E:1,5; F:1,4-6; G:1-3		
	16.3: Environmental Justice Close to Home	A:1,2,6,7; C:4,6; E:1,2,4,5; F:1,3-6; G:1-3		
Unit 17: Public Policy and Community Action	Reading: "A Lifetime of Community Action"	A:1,6,7; C:3,4,6; E:1,4,5; F:1,3-6; G:1-3	1:A-D; 2: C,D; 3:G-K; 4:A-F; 5:A-G; 6:A-H; 7:F,G,H,I; 8:A-F; 9:B-G; 10:A-J	
	17.1: Understanding Public Policy	A:1,2,6,7; C:3,4,6; D:2,3; E:1,2,4,5; F:1-6; G:1-3		
	17.2: Getting Organized: Making the Most of the Media	A:2,6,7; E:1,5; F:1,3,4,6; G:1-3		
	17.3: Putting it All Together	A:2,6,7; E:1,2,5; F:1,4-6; G:1-3		

Handouts (for student work)	Expected Outcomes	Extensions	Assessment
"Toward a Wireless Chicago," with questions	To understand the role that communications plays in our society		Questions assess writing and reading comprehension
16.1: "Understanding Environmental Justice"	To understand and reflect on the definition of environmental justice		Successful completion of handout; well-conceived responses to questions that demonstrate an understanding of the concept; clear communication in classroom discussion and essay, if applicable
16.2: "Environmental Justice on the Map"	To analyze a map depicting neighborhoods that may or may not be affected by toxic waste sites	To examine toxic waste in students' communities	Successful completion of handout; good classroom discussion participation
16.3: "Environmental Justice Close to Home"	To gauge the level of environmental health of their neighborhood or region	To prepare a report for local officials regarding students' findings; to contact press about the same	Successful completion of handout based on solid research; good classroom discussion; demonstrated understanding of risks posed to students' neighborhoods
		Unit-wide extensions	
"A Lifetime of Community Action," with questions	To understand how community involvement can impact lives and change communities		Questions assess writing and reading comprehension
17.1: "Guidelines to Policy Profiles"	To understand, research and report on major environmental policies in the U.S.; to learn the role that the U.S. plays in the global arena of environmental policy-making		Research preparedness; successful completion of report; good group participation; good classroom presentation
17.2 "Community Action and the Media"	To understand the different aspects of organizing and publicizing events; to understand community coalitions	To write a sample press release; to write a letter to the editor	Demonstrated understanding of the role of community groups in a society; understanding of the role of the media and its relationship to community groups
	To plan a publicity campaign to bring attention to an important environmental issue	To put students' plans into action and publicize/host an actual event	Successful plan for a publicity campaign that addresses all points listed in activity
		Unit-wide extensions	

Unit Number and Title	Activity	Science Standards (for each activity)	Social Studies Standards (for entire unit)	Reproducibles (suitable for overhead)
Unit 18: Sustainable Commerce	Reading: "Meet a Green Banker"	A:6,7; C:3; E:1,5; F:1,3,4; G:1,3	1:A-D; 2:B,C,D; 3:A-K; 4:F; 5:E,F,G; 6:A-H; 7:A-I; 8:A-F; 9:A-G; 10:A-J	
	18.1: Interview a Green Business Owner	A:2,6,7; E:1,4; F:1,6; G:1,3		
	18.2: The Green Business Plan	A:1,2,6,7; C:4; E:1,2,4,5; F:1,3-6; G:1-3		
Unit 19: Green-Collar Careers	Reading: "All About Green-Collar Jobs"	A:6,7; C:3; E:1,5; F:1,3,4; G:1,3	1:A-D; 2:B,C,D; 3:A-K; 4:F; 5:E,F,G; 6:A-H; 7:A-I; 8:A-F; 9:A-G; 10:A-J	
	19.1: The Green Workplace	A:2,6,7; E:1,4; F:1,6; G:1,3		
	19.2: Shadow a Green Mentor	A:2,7; E:1,2,5; F:1,4,6; G:1-3		
	19.3: Sharing What You've Learned	A:1-3,6,7; B:3,4,6; E:1-5; F:1-6; G:1-3		

Handouts (for student work)	**Expected Outcomes**	**Extensions**	**Assessment**
"Meet a Green Banker," with questions	To understand the role banks play in funding green or sustainable businesses		Questions assess writing and reading comprehension
18.1: "Interview a Green Business Owner"	To learn what steps a local business is taking to introduce greener or more sustainable practices in the workplace		Successful completion of handout; demonstrated understanding of how sustainable practices can lead to new business sectors that create new jobs
18.2: "The Green Business Plan"	To develop a sound business plan for a hypothetical business your students would like to create	To present plans to school and local leaders	Successful preparation of business plan that addresses all points listed on handout; good group participation; good presentation; demonstrated understanding of a business plan and its purpose
		Unit-wide extensions	
"All about Green-Collar Jobs," with questions	To understand how green technology and sustainable practices is creating demand for new jobs		Questions assess writing and reading comprehension
19.1: "The Green Workplace"	To develop a list of different jobs that could arise from sustainable practices	To create a report for the local Chamber of Commerce	Successful completion of handout; demonstrated understanding of how sustainable practices can lead to new business sectors that create new jobs
19.2: "Shadow a Green Mentor"	To learn the value of a good mentor, select one, shadow him or her, and report on the experience		Successful shadowing of chosen subject; well-written/presented report on the experience
19.3: "Sharing What You've Learned: Guidelines"	To create, design and implement an environmental education unit for younger students in school/town	To publish environmental units; to teach units to other students in area	Successful completion of environmental unit, addressing all points listed on handout; good group participation; demonstrated understanding of teaching concepts
		Unit-wide extensions	

Made in the USA